# Technological and Experimental Advances in Microgrids and Renewable Energy Systems

# Technological and Experimental Advances in Microgrids and Renewable Energy Systems

Editors

**Quynh Thi Tu Tran**
**Saeed Sepasi**

Basel • Beijing • Wuhan • Barcelona • Belgrade • Novi Sad • Cluj • Manchester

*Editors*
Quynh Thi Tu Tran
University of Hawaii at Manoa
Honolulu, HI
USA

Saeed Sepasi
University of Hawaii at Manoa
Honolulu, HI
USA

*Editorial Office*
MDPI
St. Alban-Anlage 66
4052 Basel, Switzerland

This is a reprint of articles from the Special Issue published online in the open access journal *Energies* (ISSN 1996-1073) (available at: https://www.mdpi.com/journal/energies/special_issues/T4931Z0VI6).

For citation purposes, cite each article independently as indicated on the article page online and as indicated below:

Lastname, A.A.; Lastname, B.B. Article Title. *Journal Name* **Year**, *Volume Number*, Page Range.

ISBN 978-3-7258-1003-1 (Hbk)
ISBN 978-3-7258-1004-8 (PDF)
doi.org/10.3390/books978-3-7258-1004-8

# Contents

# About the Editors

**Quynh Thi Tu Tran**

Quynh joined the Hawai'i Natural Energy Institute (HNEI) in June 2019 as a Postdoctoral Fellow. She received her PhD in Electrical Engineering from the University of Palermo, Italy in 2019. Her work focuses on studying the production cost model for power systems, with the integration of renewable energy resources and developing optimization algorithms for distribution power systems. She is also participating in the research and development of advanced power systems solutions, to enable the grid integration of renewable energy while achieving efficient, secure, reliable, and resilient grid operations.

**Saeed Sepasi**

Saeed Sepasi is currently an Assistant Researcher at the Hawaii Natural Energy Institute, University of Hawaii at Manoa, where he specializes in the integration of renewable energy into power grids and smart grid optimization. He earned his PhD in Mechanical Engineering from the University of Hawaii at Manoa, focusing on advanced energy storage solutions and grid-scale energy systems. His current research interests include energy storage optimization, sustainable energy solutions, and the development of control strategies to enhance grid reliability and efficiency.

# Preface

In pursuit of achieving net zero emissions by 2050, the integration of microgrids and renewable energy systems stands as a pivotal focus. This Special Issue explores the most recent technological and experimental advancements to significantly enhance our understanding of microgrid and renewable energy integration, grid resilience, and cybersecurity.

**Quynh Thi Tu Tran and Saeed Sepasi**
*Editors*

*Article*

# Development of a Supervisory System Using Open-Source for a Power Micro-Grid Composed of a Photovoltaic (PV) Plant Connected to a Battery Energy Storage System and Loads [†]

Fernanda Moura Quintão Silva [1,2], Menaouar Berrehil El Kattel [1,3] and Igor Amariz Pires [1,4,*] and Thales Alexandre Carvalho Maia [1,5]

[1] Graduate Program in Electrical Engineering (PPGEE), Universidade Federal de Minas Gerais—UFMG, Belo Horizonte 31270-901, MG, Brazil
[2] Department of Mechanical Engineering, University of Texas at Austin—UT, Austin, TX 78712, USA
[3] Department of Electrical Engineering (PPGEE), Universidade Federal do Ceará—UFC, Fortaleza 60356-001, CE, Brazil
[4] Department of Electronic Engineering (DELT), Universidade Federal de Minas Gerais—UFMG, Belo Horizonte 31270-901, MG, Brazil
[5] Department of Electrical Engineering (DEE), Universidade Federal de Minas Gerais—UFMG, Belo Horizonte 31270-901, MG, Brazil
* Correspondence: iap@ufmg.br
† This paper is an extended version of our paper published in: Silva, F.M.Q.; Cardoso, Filho, B.J.; Pires, I.A.; Maia, T.A.C. Design of a SCADA System Based on Open-Source Tools. In Proceedings of the 2021 14th IEEE International Conference on Industry Applications (INDUSCON), Virtual Conference, 15–18 August 2021.

**Citation:** Silva, F.M.Q.; El Kattel, M.B.; Pires, I.A.; Maia, T.A.C. Development of a Supervisory System Using Open-Source for a Power Micro-Grid Composed of a Photovoltaic (PV) Plant Connected to a Battery Energy Storage System and Loads. *Energies* **2022**, *15*, 8324. https://doi.org/10.3390/en15228324

Academic Editors: Saeed Sepasi and Quynh Thi Tu Tran

Received: 28 September 2022
Accepted: 21 October 2022
Published: 8 November 2022

**Publisher's Note:** MDPI stays neutral with regard to jurisdictional claims in published maps and institutional affiliations.

**Abstract:** The importance of renewable energies and energy storage system forming a micro-grid and integrating it to the electrical grid is widely spread. A supervisory system plays a crucial role in controlling, managing, and planning the micro-grid. This paper demonstrates the development of a new custom supervisory system based on Internet of Things (IoT), creating an information sharing environment. The proposed supervisory system is based on open-source tools for a micro-grid, composed of a photovoltaic power plant and a storage system, employing smart devices and making non-smart devices compatible with IoT systems. The new supervisory improves the available system by incorporating new features and devices and increasing the data polling rate when necessary. A comparison between the current supervisory system and the proposed one is performed, showing that the new system is more flexible, easily modified, cost-effective, and more fault-resilient.

**Keywords:** internet of things (IoT); supervisory system; solar power; Raspberry Pi; node-red; Grafana; energy storage

## 1. Introduction

The main challenges to be surpassed regarding renewable energy sources are availability and cost. Solar and wind resources, for instance, are seasonally and hourly dependent. These can greatly influence the energy generation [1]. The mitigation alternatives to the aforementioned challenges could be supervision, control, and energy storage.

Traditionally, a supervisory and data acquisition system (SCADA) is used to supervise processes by acquiring data from field devices. Users can access the data via screens that provide a visual aspect of the whole system. The supervisory system provides a link among the field devices and the control room [2,3]. This system interacts with field devices via inputs/outputs, using communication equipment to create a link among them, and is generally installed in a central computer.

In order to supervise a micro-grid, the requirements are not the same from the ones used in regular grids, since multiple generation, storage, and load systems are independently connected. There is a lack of research papers related to supervision in micro-grid studies. The improvement and renovation of supervision and monitoring features could lead to a more efficient management of power generation and micro-grid operation. Adding new capabilities to the system, such as Internet of Things, plays a vital part in the operation quality of renewable source power plants [4–7].

The Internet of Things (IoT) is an exchange information environment with connected devices in wired or wireless networks, enabling easy remote access. Constant monitoring of environmental and electrical data is an extremely promising application [8–10]. With the aid of IoT, it is possible to improve processes by acquiring and processing large amount of data [11].

This paper presents a customized system capable of monitoring and controlling a micro-grid. The micro-grid used for this study is composed of a 37 kWp photovoltaic power plant, which is connected to a battery energy storage system (BESS) of 15 kW/45 kWh. The motivation of developing a supervisory system comes from the fact that the current one, developed by a third party company, has lower flexibility to incorporate new devices, higher costs, and lower sample rate of data acquisition. The proposed SCADA is designed to integrate IoT devices, allowing better efficiency and flexibility to not only monitor and perform data analysis, but also for the management of the power plant in conjunction with the energy storage system. Thus, information sharing between different grids could be enabled in order to better manage the energy supply for a specific load. The information sharing would lead to an increase in efficiency and better usage of the energy.

Another advantage of the proposed system is the segregation of SCADA services into individual hardware, which are network connected. The proposed change promotes a fault tolerant capability, since eventual failures in a service do not affect the other ones. In addition, if the hardware fails, only one service is compromised and easily replaced.

This paper aims to demonstrate the development of a custom-made remote access control and monitoring system of all available electrical and environmental variables. This supervisory system might be flexible—changes and expansions should be easily incorporated—adaptable to research, more cost-effective, and fault-resilient [12,13].

The contributions of this paper are

- Development of a low-cost open-source flexible supervisory system for the monitoring of the measured values of a photovoltaic power plant;
- Development of a supervisory system that combines the traditional capabilities with IoT concepts;
- Development of a supervisory system that allows remote communication and remote data sharing;
- Development of a supervisory with fault tolerant capability.

This paper is divided in six sections: Section 2 provides a brief bibliographic review, with an overview of traditional supervisory systems and an overview of Internet of Things concepts, such as its structure and smart devices. Section 3 details the methodology implemented in the development of the supervisory system. Section 4 describes the analyzed micro-grid, which comprises of a photovoltaic power plant and the energy storage system. Section 5 shows a comparison between the current supervisory system and the new one. Section 6 concludes this paper.

## 2. Bibliographic Review

### 2.1. Traditional Supervisory Systems

The supervisory control and data acquisition (SCADA) system is built using hardware devices and software that enable the control and monitoring of the analyzed process manually or automatically. It is traditionally a central controller that consists of network interfaces, input/output, communication equipment, and software. The system gathers data from sensors and field instruments and, via a communication interface, the data are sent to the central controller to showcase an overview of the whole process. Figure 1a illustrates a generic architecture for the SCADA system.

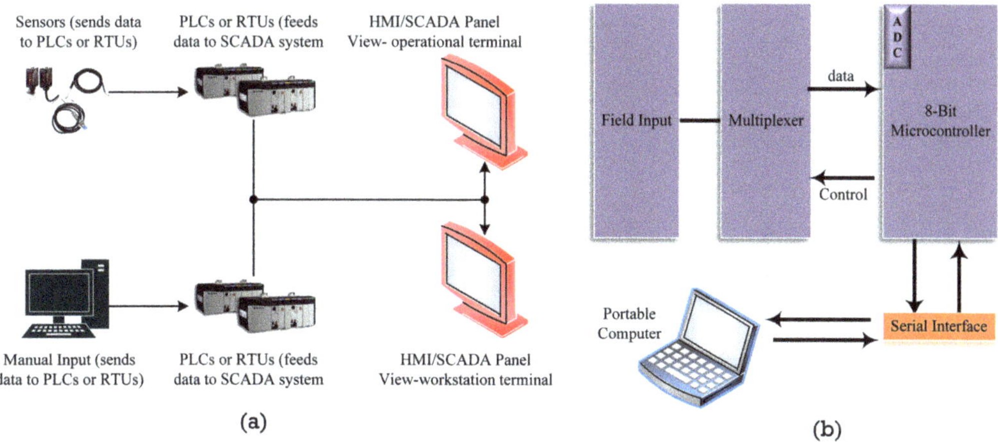


**Figure 1.** SCADA System. (**a**) Representation of the hardware architecture of a SCADA system, including field devices such as PLCs (programmable logic controller) or RTUs (remote terminal units) and the computer with the controller. (**b**) Example of implementation using a Portable Computer and microcontrollers via a serial interface.

The supervisory system acts as a link between the user in the control room and the field equipment, such as intelligent electronic devices, programmable logic controllers (PLC), remote terminal units (RTU), and sensors. These field devices are capable of transmitting data in both directions. They send the field data to the supervisory system and receive commands from the control center [2,3].

Traditionally, data loggers or microcontrollers are used for measurements/data acquisition, which can be transmitted to a user terminal via a serial port (UART/RS232/RS485) [4,5,7,14,15]. A system is developed in order to collect, record, and transmit the measured data to a computer; an example of this implementation is performed in [14], in which the developed structure is summarized in Figure 1b.

The acquisition scheme employs an I/O (input/output) microcontroller to retrieve the measured data. The data are then transmitted to a portable computer via wired connection through a serial port via the communication protocol, and it can be accessed for further analysis. The supervisory and control system are developed on this specific computer using assembly language, and it is only accessible in it. This system allows bidirectional communication, sending field data and receiving commands from the control center [14,15]. There are other literature references that also use this methodology for developing the supervisory system [16–18].

There are limits in the usage of serial communication, such as low sampling rates, and some authors propose the usage of I/O modular devices and data acquisition with high-speed A/D conversion [4]. This project also relies on a local supervisory system with wired-serial connections between the user terminal and the field devices, i.e., the monitoring and access to the system are available only to local users.

More recently, with the technological advancements in wired/wireless networks, the serial communication has been replaced by the establishment of a local network. The use of a local network enables communication between the field and the control room [19,20]. With the aid of this architecture, most operations are automatically executed by the local equipment or by the programmable logic control unit (PLC). This process is illustrated in Figure 2.

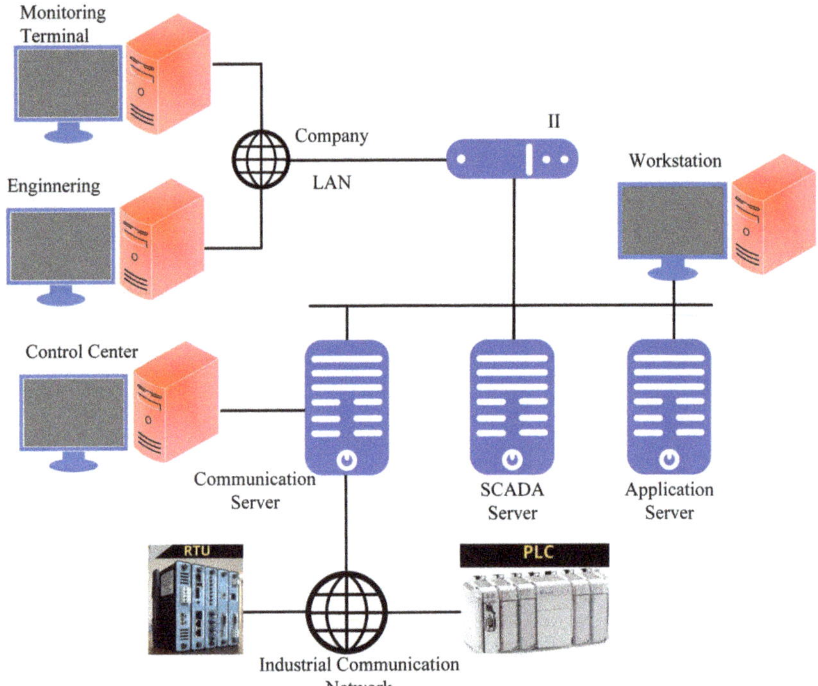


**Figure 2.** Illustration of a SCADA, data-acquisition devices, servers, workstations, and user terminals, in which data are exchanged over a Local Area Network (LAN) between devices.

Data acquisition begins at the RTU or PLC. The measurements and the state of gauges are communicated to the SCADA application. Dedicated devices are responsible for the monitoring and the control. The supervisory system presents the real-time and the historical data graphically in terminals connected to the local network. This structure is a technological advancement when compared to previously shown systems. It establishes a LAN (local area network—wired or wireless) to promote the communication between field devices and the SCADA serves, the communication server, and the user interfaces.

The authors of [21] develop a supervisory system for a photovoltaic system whose structure is similar to the one seen in Figure 2. The field devices communicate with a local gateway through a serial bus. The gateway provides wireless communication services for the devices using wireless sensor network (WSN) technology with a central terminal, where the data are processed. It is interesting to notice in this research work that the devices are not IoT devices—they are merely wireless. However, after the data are processed in the central terminal, it is published to remote users using publishing services [21]. This work shows a step towards the concept of the IoT.

### 2.2. Internet of Things—An Overview

IoT has been identified as one of the emerging technologies in IT [22–24]. It is a global Internet-based information architecture, facilitating the exchange of goods and services in global supply chain networks [25]. IoT allows the interaction among devices inside the global network, enabling devices to communicate with each other and interact with inventory systems, customer support systems, and business applications if required. Due to its communication via Internet, this environment also allows remote control [26,27]. A modified model for the IoT architecture can be seen in Figure 3a.

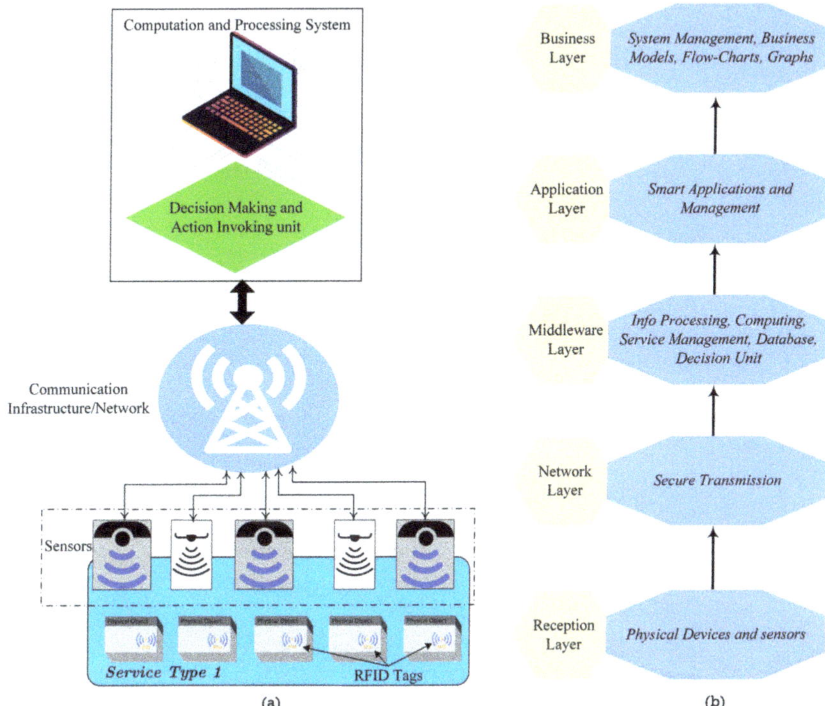



**Figure 3.** Illustration of IoT structure (**a**) showcasing each fundamental block, and (**b**) IoT structural layers, comprised by sensing and data acquisition, up to system management.

The physical objects and sensors block are related to the identification of each individual component when inserted in the context of the network. As it is represented by the arrows, the physical objects can send/receive data to/from the network. The communication infrastructure is related to the communication protocols used to establish the link among devices, devices and services, and devices and users. The decision-making inside the computation and processing unit is related to the ability of extracting information from the devices data in order to activate certain services and to execute local algorithms [28]. The basic structure for IoT is composed of five layers as seen in Figure 3b.

The bottom layer is the perception (or device) layer, and it consists of the physical equipment and the sensing devices. This layer is responsible for the identification of the devices and data acquisition. This information is then passed to the network (or transmission) layer for its secure transmission to the data processing system. The transmission can be wired or wireless. The data are then transmitted to the middleware layer. This layer has links to the database, which is capable of storing the data. The middleware layer processes the data and activates services based on the results of the data processing. The application layer is responsible for global management of the application based on the data processing on the previous layer. Finally, the business layer is responsible for the management of the IoT system, including applications and services [28].

IoT is pushing a whole slew of applications, especially when considering industry and customer-oriented applications, such as automotive and intelligent transportation systems [29,30], remote structural health monitoring [31], smart homes, buildings, and neighborhoods [32–34], smart irrigation in tunnel farming [35], metering monitoring [36], concrete surface, and level of trash monitoring [37]. IoT can also be applied to manage the energy consumption of a building in order to improve costs and efficiency [11].

Smart Devices

The devices and gauges which interact in IoT environment are smart devices. Smart devices are capable of communication and computation. The key features of smart devices are autonomy, connectivity, and context-awareness. Autonomy is the capability of performing tasks without external interference. Connectivity is to establish a link to any type of wired/wireless network to complete a task as, for example, accessing the Internet or sharing information with other devices. Context-awareness means that the device is able to perceive environmental information via sensors or sensing units [38].

Smart devices are able to connect to the Internet via wired/wireless connections, to exchange data with other IoT-gauges. Due to the feature of being smart, it is possible to control and remotely access them; the access can be achieved by users or services. A smart device is a context-aware equipment that can be used as a service provider and can communicate with devices in different networks [38].

Smart devices implement three steps: data acquisition, data process, and communication. In the first step, data acquisition, the device acquires the data measured by its sensing unit. Then, the raw data are processed by the processing unit. In addition, finally, the processed data are transmitted to the network via a communication protocol [39]. Even though this technology has allowed various possibilities to emerge as smart houses and smart grids, there are a number of challenges that must be investigated.

The main challenges in IoT are [28]:

- *Naming and identity management*: Each device connected to the network must has a unique identity over the Internet—an efficient naming and identity system that is capable of managing a large number of devices is required;
- *Interoperability and standardization*: Many manufactures have their own protocols—it is important to standardize them to guarantee interoperability. A network that employs IoT concepts has to deal with heterogeneous elements and different communication protocols [40];
- *Devices safety and security*: It is necessary to prevent security breaches in order to protect the integrity of the devices;
- *Network security*: The data transmission should be secure and capable enough to protect the integrity of the data and to prevent external interference or monitoring.

Even with these challenges, this new technology can provide significant benefits.

## 3. Methodology for the Development of the Supervisory System

To develop a monitoring and supervision system, one has to focus on the computer service that is inherent to the development of a supervisory system. The step of hardware design could be challenging. There are some demands to fulfill this requirement satisfactorily: implementation of backups services and tools, setup of security features, setup of software updates, and usage of uninterruptible power supply (UPS) systems. One important task to consider is also the adequate management of hardware resources, since it is vital to maintain enough available hardware for expansion when dealing with cloud computing.

To overcome the hardware design hurdle, each specialized subsystem of this supervisory is assigned to a container-like system. Specifically, one resource is allocated for data acquisition from the available field hardware, another one is employed for information processing, another to provide a user interface, and so on. With this structure, a single failure in one of the specialized services does not negatively impact the whole system. The main difference of this architecture over container computing is the usage of a dedicated server for each proposed service. Multiple microcomputers (Raspberry Pis) are used instead of developing services in a single multi-core hardware.

The usage of the proposed modular structure brings several benefits. Employing single separate servers for each service prevents a catastrophic failure from happening when one of the servers presents a failure. If one of the server fails, the error could be easily traced to the failed one. This error could be studied and isolated in order to be fixed, without impacting the performance of the other functioning servers. In other words, a singular isolated failure in one service would not translate in a failure of the entire system. Another advantage is the possibility of parallel processing. All services could operate concurrently with no impact on each other. The services are allocated to different servers, meaning that, if there is a failure in a process, it does not hinder others. Increases in the efficiency and in the reliability of the system as a whole could also be counted amongst the benefits of this proposed structure since each server is powered by its own dedicated UPS (uninterruptible power supply).

The Raspberry Pi is a series of compact single-board computers. The choice of using Raspberry Pi in this application is due to the fact that its features align with the requirements for the development of the new supervisory system. They are cost effective and a powerful processing unit in a compact board. The Raspberry Pi has built-ins for the usage of different interfaces (HDMI—High Definition Multimedia Interface, USB—Universal Serial Bus, Ethernet, Wi-Fi, Bluetooth), and it supports Linux and Python, enabling easy development of applications [41–44].

The development of a supervisory system requires the implementation of some specialized services in order to function properly. These services would be required to acquire data, interpret, and translate this data, and display it in a user interface via the monitoring system. The establishment of a communication among field devices, supervisory system, data storage, and backup system is necessary. Three servers are established: (1) communication server, (2) database and application server, and (3) the backup server as seen in Figure 4.

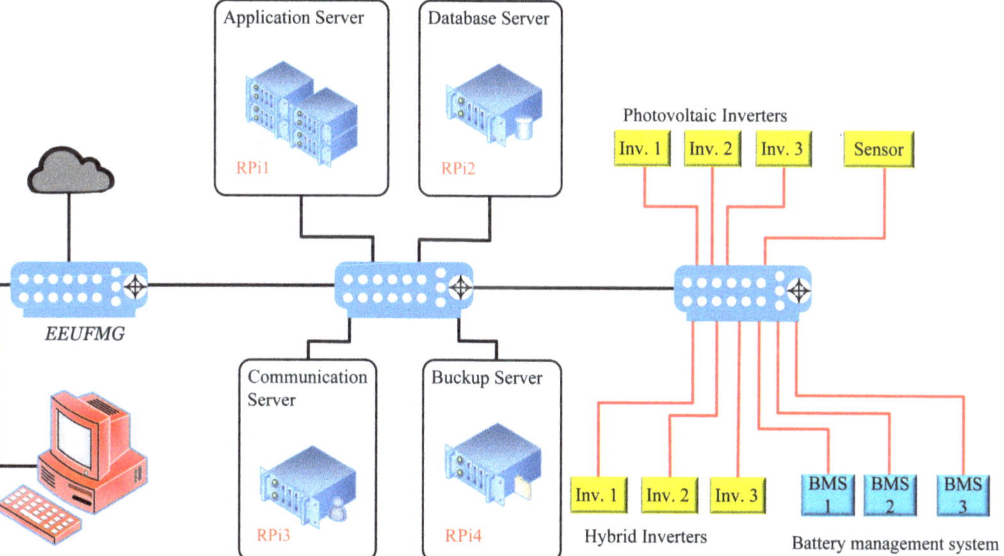


**Figure 4.** Schematic of the proposed server architecture for the supervisory system presenting the implemented system with the RPis (Raspberry Pi), inverters, and batteries.

Another important feature of the supervisory system is the employment of a set of tools capable of providing an easily expandable and flexible monitoring system. Since it has been established that employing IoT-related concepts is beneficial, the chosen tool set must be integrated into its environment.

There are a lot of open-source software options that could be implemented in the development of the system to meet the requirements. The one that is the most popular and the most versatile is Node-Red [45]. Node-Red is an open source tool developed initially by IBM for flow programming using a local host, meaning it is a browser based flow editor [46]. The usage of a flow-based programming (FBP) has many advantages. The main reason is that FBP allows parallelism, providing performance benefits in some situations. In FBP, the dataflow is the main driving force of the program. The logical execution flow is expressed by the block diagram created: when a node receives all necessary inputs, the block produces an output that is transmitted to the next node in the path. FBP treats the applications as black-boxes: the important component for this type of programming is the connections between components that are conducted externally to the processes. Due to this style of programming, FBP lends itself to a plug-and-play approach. Node-Red is also able to connect to hardware devices, such as microcontrollers and the Raspberry Pi amongst many others, and to the cloud environment.

The advantages of using Node-Red when comparing it to other open-source tools come especially from the usage of Node.js, a light and agile open-source tool. Due to its flow-based programming and the ability of building custom functions, Node-Red is able to fulfill the requirement of flexibility [47]. Since the developed supervisory system is built for monitoring electrical and environmental variables, the processing rate of messages inside the code does not need to be extremely fast. For this reason, the speed of one message/second of Node-Red is sufficient and suitable for the application.

In order to build the graphical screens for the development of the supervisory system, it is necessary to use another tool that is more user-friendly for the creation of dashboards. The platform used here is Grafana. Grafana is a multi-platform tool that is able to provide graphics, charts, and real-time measurements based on incoming information from any kind of data source [48]. Grafana is commonly used to display data and alarms for further analysis of any type of application. This platform can display real-time data as well historic measurements and events. To achieve the access to the data, Grafana has embedded a database tool that allows the user to configure a database in which the desired data are stored. The database amongst the pre-configured in Grafana that is more aligned with the goals of this paper and with the type of desired monitoring (measurements of a power plant with smart devices) is InfluxDB. InfluxDB is an open-source time series database, and it uses a programming language similar to SQL. By using the InfluxDB nodes in Node-Red and the pre-configured tool in Grafana for InfluxDB databases, it is possible to create a link between the supervisory system and the data requested by Node-Red from the field devices.

*3.1. Communication Server and Database Server*

The plant where the supervisory system has been developed has smart and non-smart devices. Smart devices are already integrated with the concept of IoT, meaning that they are already able to join an established network and communicate with other smart devices and/or servers directly. The photovoltaic inverters, the hybrid inverter connected to the energy storage system, and the energy meter are smart devices. The non-smart devices cannot communicate in the established network without an interface with an intermediary equipment, such as a data aggregator or a microcontroller; they are not integrated with IoT concepts. The DC-current meter in the photovoltaic inverter's input and the solar meter station are not smart devices. Thus, both of them need an intermediary device to manage the flow of data. In addition, as already stated, the energy storage system communicates with the supervisory system remotely. This means that its communication is conducted via the Internet, via the cloud. The communication architecture can be seen in Figure 5.

Modbus TCP/IP is used for the communication between two of the photovoltaic inverters (Fronius and SMA) and the communication server and between the energy meter and this server. The inverters are capable of measuring electrical data from the DC-bus (the solar panel side) and from the AC-bus (the main grid side) as the energy meter. These data are available via communications protocol for data processing. The Modbus protocol is a highly used flexible open message structure used for the communication between master and slave. The communication is only initiated by the master. The master is responsible for beginning the process by sending a data request to the slave. The connection is conducted point-to-point. The structure of Modbus guarantees that the final recipient receives the message: the master receives an acknowledgment that the request has reached the slave and the slave receives a flag that the master received the requested data. The Modbus protocol is illustrated in Figure 6a.

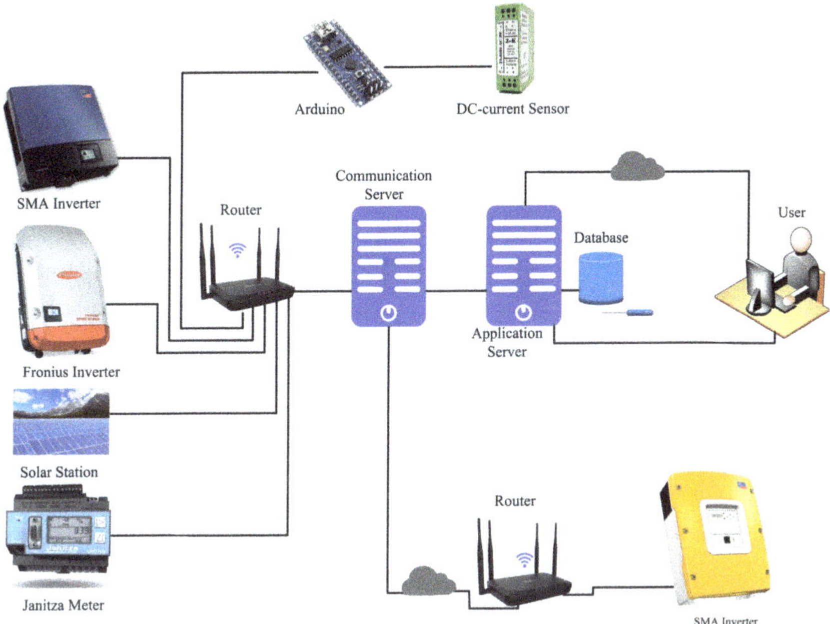


**Figure 5.** Proposed structure for the data acquisition and supervisory system including the data flow path , field devices, router of the control room, application system, and the database.

The third photovoltaic inverter PHB communicates with the server via Serial Modbus RS-485 since it does not have an Ethernet interface and it does not allow for IP configuration. The Serial Modbus follows the same methodology as the Modbus TCP/IP. The physical connection is conducted with a twisted wire pair plus ground (GND).

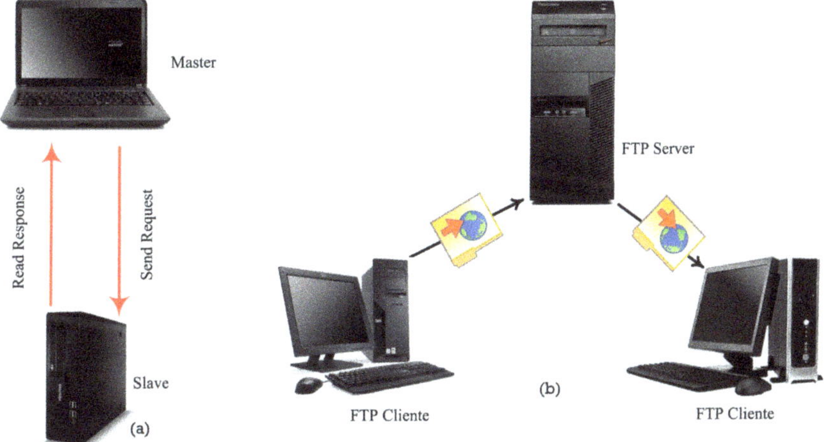


**Figure 6.** Schematic of the process involved in the Modbus protocols, including the (**a**) server and client (Master/Slave) communication, and (**b**) the process to request data using ftp.

Node-Red is equipped with nodes that can be configured to establish a Modbus connection between the communication server (Raspberry installed with Node-Red) and the inverters and the energy meter. That way, the raw data from these devices can be acquired by the request made by Node-Red (acting as a server to modbus clients for certain electrical variables, e.g., AC/DC voltage, current, power, amongst others).

A solar meter station is another gauge incorporated in the supervisory system. It could measure absolute and relative humidity and pressure, ambient temperature, global radiance, and wind direction (a reference cell is monitored separately). This solar station is composed exclusively of non-smart devices, i.e., in order to make their data available, these gauges communicate with an intermediary data-logging device via Modbus RS-485 (Modbus via serial communication). Using the file transfer protocol (FTP), it is possible to transfer the data from these gauges to the communication server. FTP is a secure connection between devices that allow file exchange. FTP also relies on a client–server structure; the client requests access to a certain file and the server grants this access. This protocol is a simple tool that allows high volume of data transfer through a network, and it allows various directories to be transferred at the same instant. The process is shown in Figure 6b.

The idea is to upload the files with the gauges data to the communication server and post it to the application server to incorporate it into the supervisory system. The FTP client needs to act as the FTP link and make requests to the FTP server and receive access to the desired files. In order to supply this link for the current application to retrieve the sensors data from the data-logging device, a Python program is used. This program is uploaded to the communication server as a background service that starts running with the booting of the Raspberry. The upload to the communication server is performed via FTP through the Python service. Figure 7 summarizes this process.

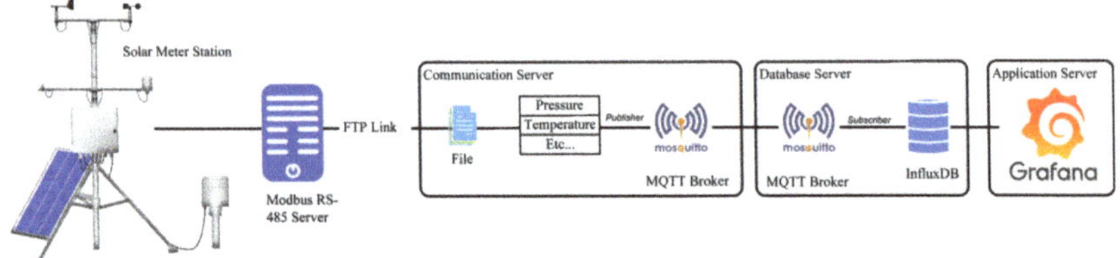


**Figure 7.** Schematic of the communication established between the solar meter station and the application server, showcasing the interaction between FTP, MQTT, and the database.

There are also the DC-current sensors that are not smart devices. In order to include the data from the sensors to the supervisory system, an intermediary device must be added to the system in order to collect the data and provide it to the application system. To make these measurements available for the network, the output of each current sensor is connected as an analog input to an Arduino Nano with an Ethernet Shield (ENC28J60). Each analog input of the Arduino that is connected to the DC-current sensors has its value read in the Arduino. These values need to be converted to engineering units since they are presented in bits as digital outputs. After they are processed, they are transmitted from the communication server to the application server via MQTT. Figure 8 summarizes this scheme.

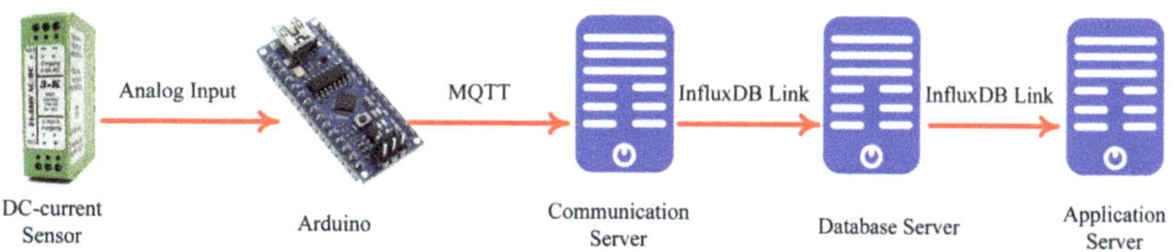



**Figure 8.** Data acquisition process for the DC-Current sensors deploying smart sensors. The micro-controller uses MQTT to send data to the database.

There are also three hybrid smart inverters, named SMA Sunny Island 8.0h, in this network that are connected to the battery banks that make the energy storage system. These inverters also communicate with another communication server via Modbus TCP/IP.

These inverters not only provide data but also receive data—commands—via Modbus TCP/IP in order to enable the user to control the reactive and active power flow of the grid by managing the charge and discharge of the battery banks. The inverters are capable of providing data regarding the measurements of the grid (AC-side) and data regarding the measurement of whichever battery bank is connected to it (DC-side); it is important to note that each inverter is connected to a phase conductor and they operate separately and do not communicate amongst themselves.

These data must be sent to the main communication server in order to post it to the application server to be incorporated into the supervisory system. The data must be sent via an Internet connection since the energy storage system is placed in a separate grid. It is then necessary to provide a secure communication protocol by applying SSL certificates. This security layer is applied to the lightweight MQTT protocol (MQTTS), together with password authentication, to send the data of the hybrid inverters to the main communication server. This solution cannot be seen as definitive, as research related to cybersecurity is advancing, mainly towards blockchain and the use of a well-structured sensor network [49].

In this type of architecture, there are devices that generate and publish data (publishers) and other devices that consume these data (subscribers). This architecture demands an extra entity that acts as a centralizer in the data exchange, the Message Broker. The devices that generate data publish it to the broker which organizes the data in topics. The devices that consume the data subscribe to the desired topic and are able to retrieve the data assigned to the topic. Every instant new information reaches the topic in the broker, this information is automatically sent to the device that subscribed to the topic. The process is summarized in Figure 9. Node-Red environment was used to implement all protocols due to its steady and reliable communication characteristics.

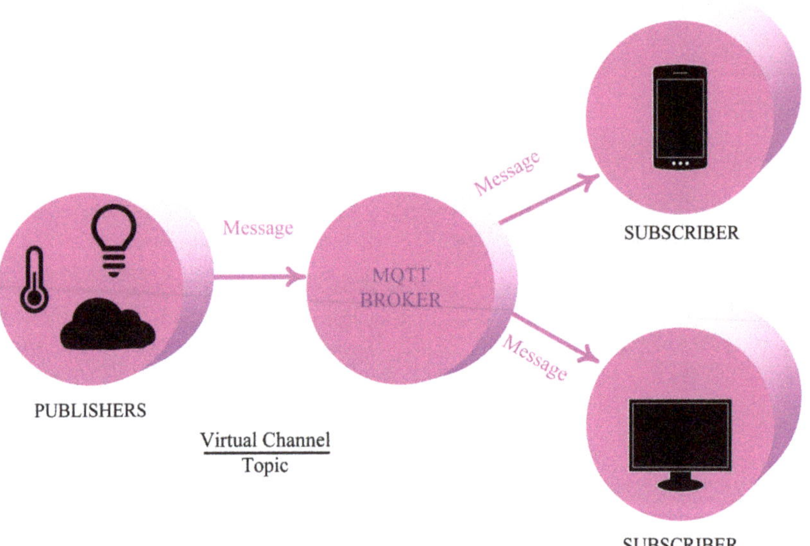


**Figure 9.** Schematic of the process involved in the MQTT protocol: publishers deliver the data to the broker and then the subscribers make a request to the broker for the desired data.

In order for the application server to access the data from the field devices, it is necessary to transmit the data from the communication server to the database server. For that, the InfluxDB database is used and the Node-Red environment already has built-in nodes that enable an easy configuration of the data transmission. InfluxDB is a password-protected time-series database, allowing the storage of sensor measurements with their timestamps. This database has a built-in time service that ensures the synchronization of time throughout the system. The advantages of using this particular tool are related to the processing and storage speed due to its simplified structure. Once the data are stored in the database server, it is possible to configure the link between it and the application server that houses Grafana, using its own databases configuration tool. Since these servers (communication, database, and application) are all on the same secure network, it is not necessary to implement another layer of security for the established links.

In order to summarize the architecture for the communication links among devices and servers, Figure 10 is shown. A flowchart with the steps of the work and processes along with the communication protocols is presented in Figure 11.

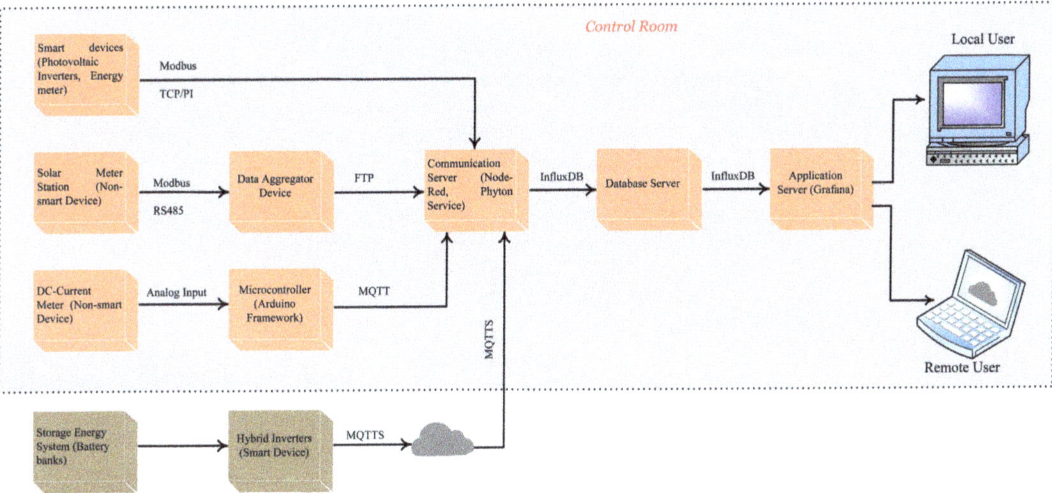

**Figure 10.** Schematic of the architecture of the data acquisition and data monitoring system, showcasing the communication protocols.

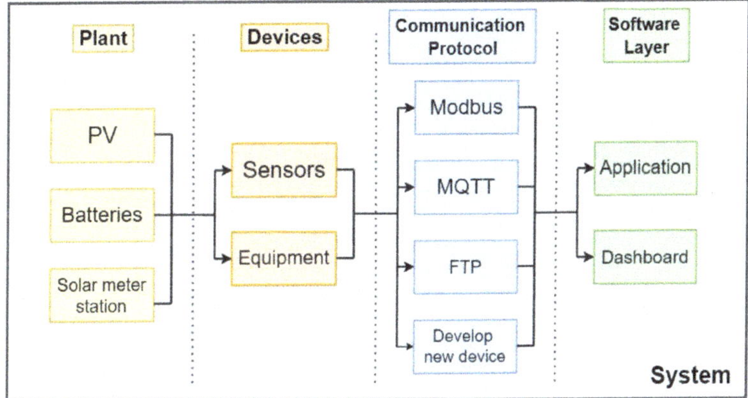

**Figure 11.** Flowchart of the development of the supervisory system including the steps necessary for the configuration of the communication protocols.

*3.2. Application Server*

In order to develop the screens of the supervisory system, the data are transmitted from the database server to the Grafana environment. Due to the features of this platform, it is possible to build any custom dashboards, meaning that this tool is capable of fulfilling the flexibility requirement to generate any dashboard necessary.

In order to retrieve the data, the application allows the configuration of the InfluxDB databases created in the database server. For this application, there are six databases and all of them must be configured in Grafana.

The users need an overview of the measurements of the micro-grid in real-time. Thus, it is interesting to provide the single-line diagram of the system with live measurements and the status of the phovoltaic power generation. The new supervisory system should also allow for the plotting of real-time and historical electrical and environmental data for further analysis.

The development of the screens of the supervisory system is conducted solely in the Grafana environment. This platform provides a user-friendly interface to build a custom-made supervisory system, presenting features and gadgets that allow customization. In addition, due to its built-in InfluxDB tool for the configuration on data transmission, the platform is easily expandable.

Grafana also allows the creation of multiple users. The user can be set up to have different access level rights. One user can be configured as the administrator; another can be a guest and only have viewing access.

The versatility of Grafana is one of the main advantages when comparing it to the provided current supervisory system. For the latter, new features and/or variables are cumbersome to be added; these inclusions have to be performed by the third party company. In addition, the structure of the current supervisory system is more rigid, not allowing customization.

## 4. Case Study

The case study is a photovoltaic (PV) plant (Tesla Laboratory Experimental Photovoltaic Power Plant) connected to a battery energy storage system and to the loads. The solar plant is 37 kWp, located in Belo Horizonte, Brazil. Figure 12 shows the single-line diagram, showcasing some of the PV panels, the control room, and the battery banks.

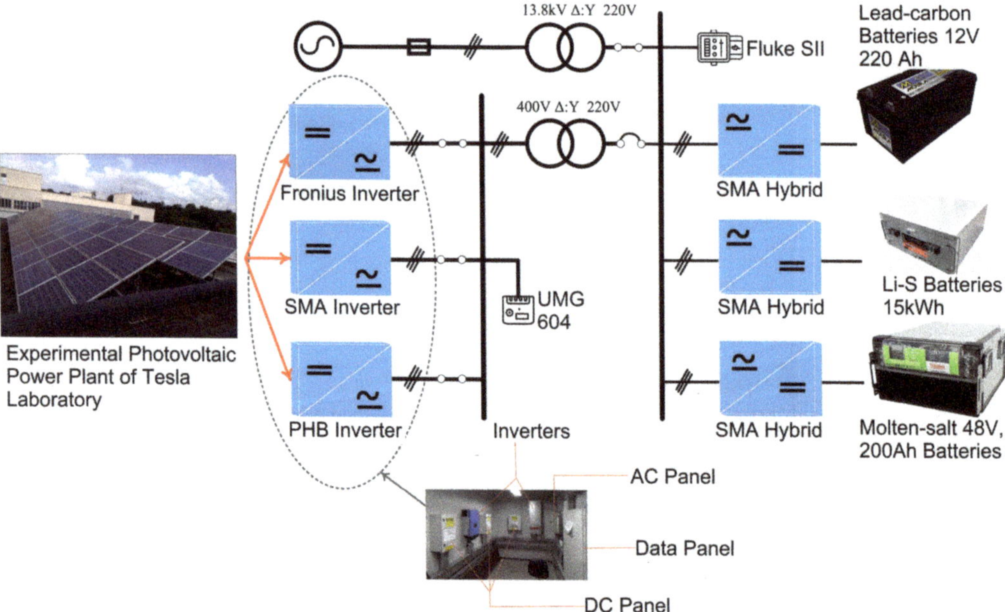


**Figure 12.** Single-line diagram of the solar power plant with the energy storage system.

The power plant has three power inverters: Fronius IG Plus 150V-3 (10 kW), SMA Sunny Tripower 12000TL (12 kW), and PHB20K-DT (20 kW), as seen in Figure 12—the last one was installed in May/2021. Since this project is a prototype, inverters from different manufacturers were installed in the power plant in order to investigate their differences. The photovoltaic plant is formed by 152 PV panels (Yingli 245P-32b, 245W), connected via inverters through a 45 kVA 400/220 V transformer in the university electrical grid.

The energy storage system is made of three independent battery banks and three independent single-phase hybrid inverters (SMA Sunny Island 8.0h). The latter are connected to the first, enabling the flow of data from/to the battery management system to/from the control room via local or remote commands/monitoring. This connection makes it possible to monitor the battery banks and to control the power flow between the solar power plant and the battery banks. The inverters are able to send commands to them to charge/discharge, adjusting the power flow.

Each battery banks presents a unique electrochemical technology, comprised by: 24 lead-carbon batteries pack, 6p4s of 12 V and 220 Ah, a molten-salt battery bank, 2p1s of 48 V and 200 Ah, and lithium-iron 48V battery bank, 5p1s of 48 V and 100 Ah. All the batteries are shown in Figure 12.

In order to obtain the measurements from the AC-side, a Janitza UMG 604 is connected to the grid. It is an energy meter that is able to provide various electric information: power, energy, frequency, current, and voltage. This device measures the output electric variables of the power plant, before the transformer. A Fluke SII (energy meter) is also installed. This one is connected in the 220 V busbar before the 13.8 k/220 V transformer.

## 5. Results and Analysis

The commercial supervisory system main screen could be viewed in Figure 13, demonstrating real-time main electrical, environmental data of the PV plant, and monetary value of the generated energy. The structure of the current supervisory system is stiff, not allowing expansions and alterations in the data format. There is also an evaluation page, seen in Figure 14, in which it is possible to plot any available variable along a specified period of time.

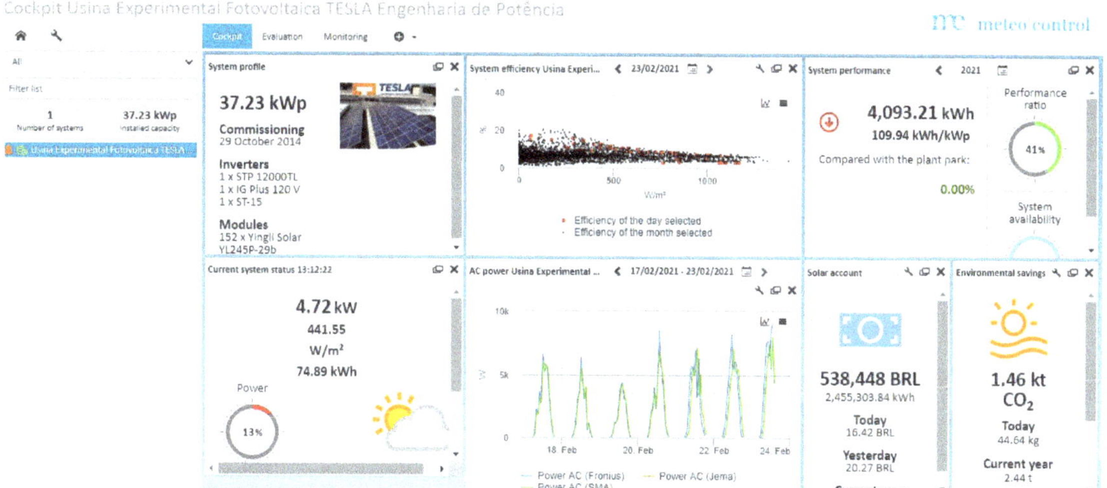

**Figure 13.** Main screen of the commercial supervisory system.

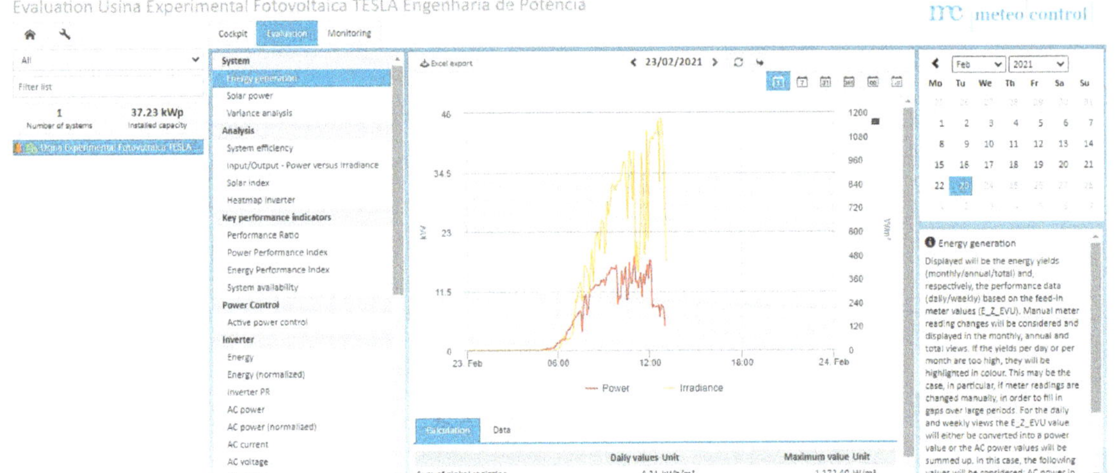

**Figure 14.** Graphs page of the current supervisory system.

The new supervisory system has been operating since Jan/2021. The box that contains the servers, the router, and the communication links can be seen in Figure 15.

**Figure 15.** Supervisory system hardware box.

The new supervisory system shows every electric measurement from the AC (from the energy meter and inverters) and DC buses (from the inverters) and the environmental variables made available by the solar meter station. Figures 16–18 are the new supervisory system. The new dashboards have a flexible and expandable structure in contrast with the old one. Any data and data format can be shown on the supervisory system for any user (all users and administrators). The developer is allowed to build any screen without restriction. The screens can be updated at a user-defined rate, allowing it to be more accurate with regard to the sampling rate of each measurement.

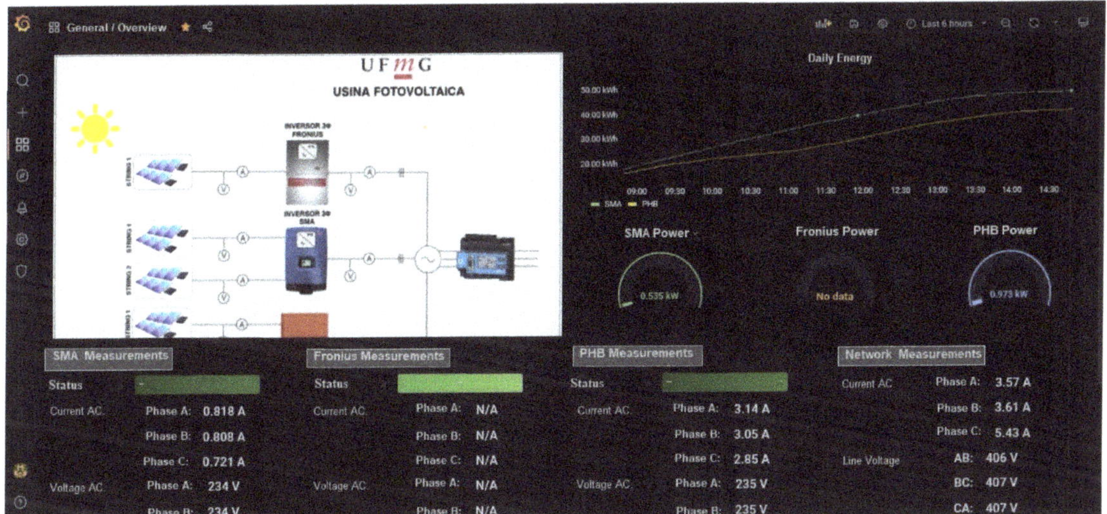

**Figure 16.** New supervisory system's main screen: PV plant overview.

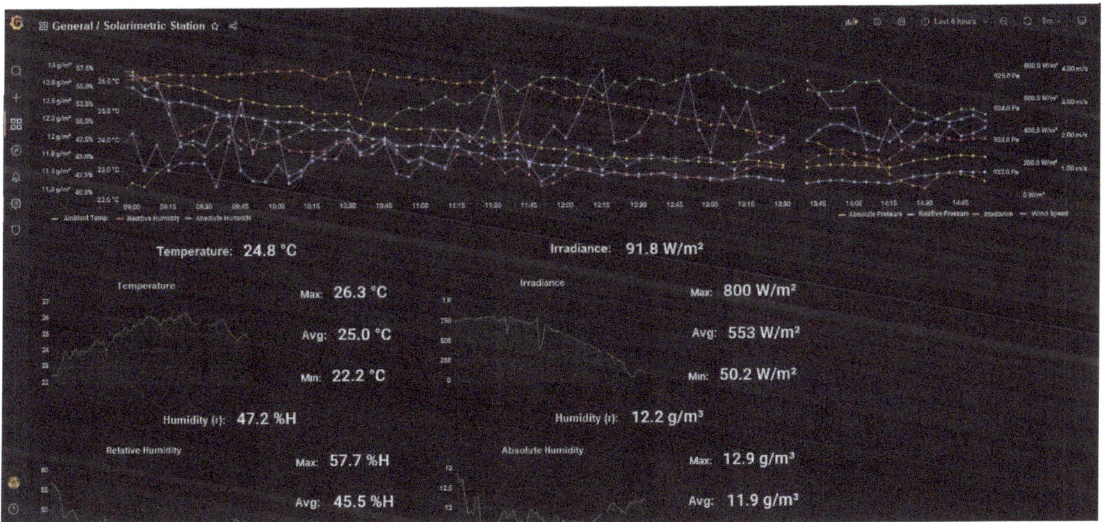

**Figure 17.** Screen for the measurements of the solar meter station.

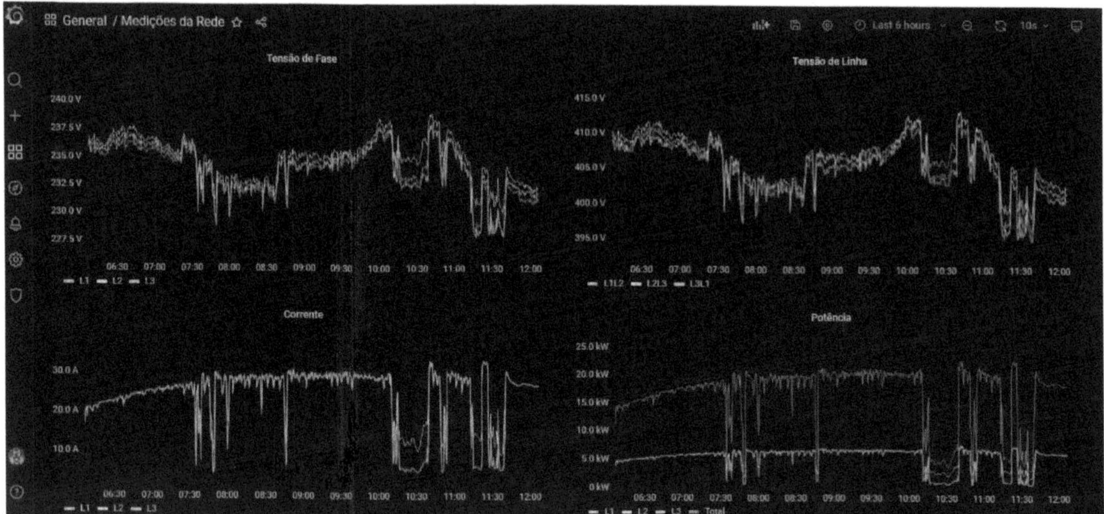

**Figure 18.** AC-Grid measurements for the solar power plant.

The system can also be accessed in a Read-Only capacity for all users, including the ones that do not have administrator capabilities. In addition, all the users are able to download any data in a CSV file for further analysis. All users can choose the sampling rate and the period of time for the file. The administrator user is the only one that can modify the dashboards and the communication with the databases.

Is it possible to see that the new supervisory system is able to display real-time current measurements, also with the ability to display peaks, lows, and averages as seen in the latter. This supervisory system is also capable of presenting historical data and the user (any access level) can select any period of time that is needed.

The new presented system is not only more flexible and more adaptable than the current supervisory system, but the developer and the user can also select a finer resolution for the display of measurements (historical and real-time). It is possible to establish a sampling rate of 5 s if desired; the current supervisory system has a fixed sampling rate of 5 min for every variable. The longer polling time could mean data loss especially on days in which there is a lot of weather variation due to the low inertia of the PV panels (cloudy days). Figure 19 gives a more detailed view in order to show the behavior of the active power during the morning of a cloudy day to highlight the data loss of the original system due to the polling time. The new system detects higher power peaks and lower power valleys since its polling time is lower: 10 s versus 5 min.

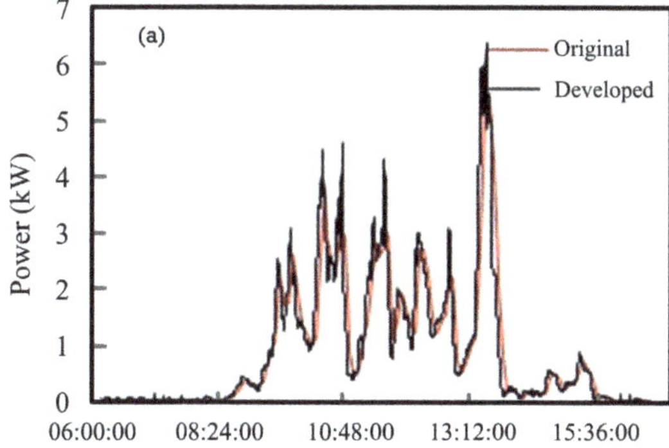
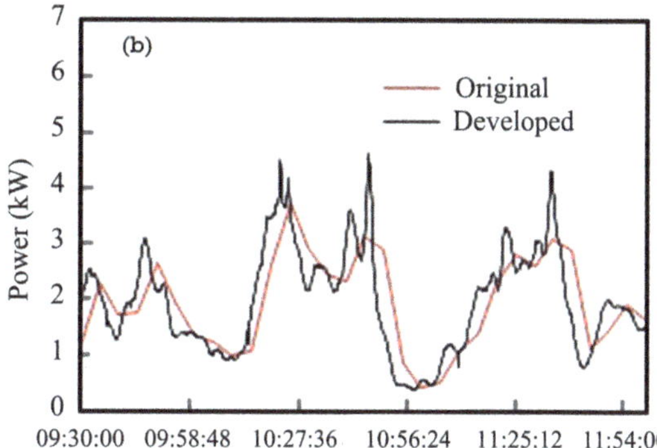

**Figure 19.** lActive AC power measured at the SMA Inverter. (**a**) Example with a long sampling; (**b**) Example with a short sampling enabling detailed measurements.

Besides the technical advantages, the comparison between equipment costs can also be used to illustrate other advantages. While the developed system had a total cost of around USD 550, the commercial one was USD 4500, plus USD 25 for monthly subscription. The main reason for this difference is that the commercial equipment can be used for other variety of applications, which has no use for the solar system. In other words, the commercial solution provides ADC inputs and communication interfaces that are not compatible with the solar system.

To sum up the main comparisons between the two supervisory systems, Table 1 is shown.

**Table 1.** Highlights of the comparison between the previous and the current supervisory systems.

| Feature | Previous Supervisory | Current Supervisory |
|---|---|---|
| Flexibility | Low | High |
| Price | USD 4500 + USD 25 monthly | USD 550 |
| Sampling Rate | Fixed at 5 min | Customizable for every variable |
| Compatibility | "Pseudo" Plug & Play | Requires configuration |

As it can be seen, the current supervisory system has advantages in three of the four considered features. The new system is more cost-effective, and the flexibility is considerably higher than the previous one, being more adaptable, and it is possible to expand more easily. The current supervisory system has a customizable sampling rate, which enables a better analysis of the behavior of each variable, making it more suitable for photovoltaic (PV) power plants since PV panels have low inertia. The feature that the previous supervisory system has over the current one is the compatibility since it is practically a system that only has to be installed to initiate the monitoring. The current one needed some configuration and adaptation.

## 6. Conclusions

The goal was to demonstrate the development steps of a supervisory system of a power plant and the energy storage system using IoT concepts, employing smart devices (network ready) and adapting the non-smart devices. The knowledge and techniques implemented for the case study can be transferred for the development of any supervisory system for any industrial/residential plant providing the means to establish a network.

Using smart devices and adapting the non-smart devices, every piece of information available from the micro-grid could be collected in the new supervisory system. The communication, database, and application servers could be built in order to develop the supervisory system.

The new supervisory system is a more flexible solution. This system could be modified to integrate any modifications that may be necessary due to field changes in the power plant or in the energy storage system or due to new monitoring requirements. The developer/user of the new system could configure polling times more appropriate to each electrical and environmental variable according to one's goal.

When comparing this study to other papers in this field, it is possible to see that the other research works did not compare traditional supervisory systems with supervisory systems that implemented IoT devices. This paper did stress the advantages technically, analytically, and financially of the current system when confronted with traditional systems and with the previous supervisory.

To further improve the new system, it would be interesting to implement a prediction feature of the daily energy generation. This would allow for the scheduling of the operation of the energy storage system. Adding this feature, the power output of the power plant would be predicted based on the weather forecast for the coming day and the monitoring system could command the field devices to control the charge/discharge of the battery banks based on this forecast in order to supplement the generation when necessary, fine-tuning it in real time.

An optimization tool for self-management of the plant would also be interesting. The goal of the tool could be to maximize the efficiency of the process and lower costs and losses.

**Author Contributions:** Conceptualization, I.A.P. and T.A.C.M.; Writing—original draft, F.M.Q.S.; Writing—review & editing, M.B.E.K., I.A.P. and T.A.C.M. All authors have read and agreed to the published version of the manuscript.

**Funding:** This research received no external funding.

**Institutional Review Board Statement:** Not applicable.

**Informed Consent Statement:** Not applicable.

**Data Availability Statement:** Not applicable.

**Acknowledgments:** The authors would like to acknowledge the help and the support of CAPES, CNPq, FAPEMIG, Pró-Reitoria de Pesquisa (PRPq)—UFMG, and Pró-Reitoria de Pós-Graduação (PRPG)—UFMG.

**Conflicts of Interest:** The authors declare no conflict of interest.

## References

1. Camacho, E.F.; Samad, T.; Garcia-Sanz, M.; Hiskens, I. Control for renewable energy and smart grids. *Impact Control. Technol. Control. Syst. Soc.* **2011**, *4*, 69–88.
2. Montaña, D.A.M.; Rodriguez, D.F.C.; Rey, D.I.C.; Ramos, G. Hardware and Software Integration as a Realist SCADA Environment to Test Protective Relaying Control. *IEEE Trans. Ind. Appl.* **2018**, *54*, 1208–1217. [CrossRef]
3. Sičanica, Z.; Sučić, S.; Milašinović, B. Architecture of an Artificial Intelligence Model Manager for Event-Driven Component-Based SCADA Systems. *IEEE Access* **2022**, *10*, 30414–30426. [CrossRef]
4. Forero, N.; Hernández, J.; Gordillo, G. Development of a monitoring system for a PV solar plant. *Energy Convers. Manag.* **2006**, *47*, 2329–2336. [CrossRef]
5. Correia, D.; Tomé, P.; Costa, P.M.; Marques, L. Monitoring system for small sized photovoltaic power plants. In Proceedings of the 11th Iberian Conference on Information Systems and Technologies (CISTI), Gran Canaria, Spain, 15–18 June 2016; pp. 1–6. [CrossRef]
6. Lorenzo, G.D.; Araneo, R.; Mitolo, M.; Niccolai, A.; Grimaccia, F. Review of O&M Practices in PV Plants: Failures, Solutions, Remote Control, and Monitoring Tools. *IEEE J. Photovoltaics* **2020**, *10*, 914–926. [CrossRef]
7. Pereira, R.I.S.; Jucá, S.C.S.; Carvalho, P.C.M.; Souza, C.P. IoT Network and Sensor Signal Conditioning for Meteorological Data and Photovoltaic Module Temperature Monitoring. *IEEE Lat. Am. Trans.* **2019**, *17*, 937–944. [CrossRef]
8. Adhya, S.; Saha, D.; Das, A.; Jana, J.; Saha, H. An IoT based smart solar photovoltaic remote monitoring and control unit. In Proceedings of the 2nd International Conference on Control, Instrumentation, Energy & Communication (CIEC), Kolkata, India, 28–30 January 2016; pp. 432–436. [CrossRef]

9.   Samosir, A.S.; Rozie, A.F.; Purwiyanti, S.; Gusmedi, H.; Susanto, M. Development of an IoT Based Monitoring System for Solar PV Power Plant Application. In Proceedings of the International Conference on Converging Technology in Electrical and Information Engineering (ICCTEIE), Bandar Lampung, Indonesia, 27–28 October 2021; pp. 82–86. [CrossRef]

10.  Manoharan, H.; Teekaraman, Y.; Kirpichnikova, I.; Kuppusamy, R.; Nikolovski, S.; Baghaee, H.R. Smart Grid Monitoring by Wireless Sensors Using Binary Logistic Regression. *Energies* **2020**, *13*, 3974. [CrossRef]

11.  Motlagh, N.H.; Mohammadrezaei, M.; Hunt, J.; Zakeri, B. Internet of Things (IoT) and the Energy Sector. *Energies* **2020**, *13*, 494. [CrossRef]

12.  Silva, F.M.Q.; Filho, B.J.C.; Pires, I.A.; Maia, T.A.C. Design of a SCADA System Based on Open-Source Tools. In Proceedings of the 14th IEEE International Conference on Industry Applications (INDUSCON), São Paulo, Brazil, 15–18 August 2021; pp. 1323–1328. [CrossRef]

13.  Osaretin, C.A.; Zamanlou, M.; Iqbal, M.T.; Butt, S. Open Source IoT-Based SCADA System for Remote Oil Facilities Using Node-RED and Arduino Microcontrollers. In Proceedings of the 11th IEEE Annual Information Technology, Electronics and Mobile Communication Conference (IEMCON), Vancouver, BC, Canada, 4–7 November 2020; pp. 571–575. [CrossRef]

14.  Mukaro, R.; Carelse, X.F. A microcontroller-based data acquisition system for solar radiation and environmental monitoring. *IEEE Trans. Instrum. Meas.* **1999**, *48*, 1232–1238. [CrossRef]

15.  Şahin, S.; İşler, Y. Microcontroller-Based Robotics and SCADA Experiments. *IEEE Trans. Educ.* **2013**, *56*, 424–429. [CrossRef]

16.  Benghanem, M.; Maafi, A. Data acquisition system for photovoltaic systems performance monitoring. *IEEE Trans. Instrum. Meas.* **1998**, *47*, 30–33. [CrossRef]

17.  Zedak, C.; Lekbich, A.; Belfqih, A.; Boukherouaa, J.; Haidi, T.; el Mariami, F. A proposed secure remote data acquisition architecture of photovoltaic systems based on the Internet of Things. In Proceedings of the 6th International Conference on Multimedia Computing and Systems (ICMCS), Rabat, Morocco, 10–12 May 2018; pp. 1–5. [CrossRef]

18.  Liu, Y. Research of Automatic Monitoring and Control Strategy of Photovoltaic Power Generation System. In Proceedings of the International Conference on Virtual Reality and Intelligent Systems (ICVRIS), Hunan, China, 10–11 August 2018; pp. 343–347. [CrossRef]

19.  Dumitru, C.-D.; Gligor, A. SCADA Based Software for Renewable Energy Management System. *Procedia Econ. Financ.* **2012**, *3*, 262–267. [CrossRef]

20.  Leahy, K.; Gallagher, C.; O'Donovan, P.; O'Sullivan, D.T. Issues with data quality for wind turbine condition monitoring and reliability analyses. *Energies* **2019**, *12*, 201. [CrossRef]

21.  Papageorgas, P.; Piromalis, D.; Antonakoglou, K.; Vokas, G.; Tseles, D.; Arvanitis, K.G. Smart Solar Panels: In-situ Monitoring of Photovoltaic Panels based on Wired and Wireless Sensor Networks. *Energy Procedia* **2013**, *36*, 535–545. [CrossRef]

22.  Gubbi, J.; Buyya, R.; Marusic, S.; Palaniswami, M. Internet of Things (IoT): A vision, architectural elements, and future directions. *Future Gener. Comput. Syst.* **2013**, *29*, 1645–1660. [CrossRef]

23.  Vashi, S.; Ram, J.; Modi, J.; Verma, S.; Prakash, C. Internet of Things (IoT): A vision, architectural elements, and security issues. In Proceedings of the International Conference on I-SMAC (IoT in Social, Mobile, Analytics and Cloud) (I-SMAC), Palladam, India, 10–11 February 2017; pp. 492–496. [CrossRef]

24.  Khan, L.U.; Yaqoob, I.; Imran, M.; Han, Z.; Hong, C.S. 6G Wireless Systems: A Vision, Architectural Elements, and Future Directions. *IEEE Access* **2020**, *8*, 147029–147044. [CrossRef]

25.  Weber, R. Internet of Things—New security and privacy challenges. *Comput. Law Secur. Rev.* **2010**, *26*, 23–30. [CrossRef]

26.  Lee, I.; Lee, K. The Internet of Things (IoT): Applications, investments, and challenges for enterprises. *Bus. Horizons* **2015**, *58*, 431–440. [CrossRef]

27.  Swamy, S.N.; Kota, S.R. An Empirical Study on System Level Aspects of Internet of Things (IoT). *IEEE Access* **2020**, *8*, 188082–188134. [CrossRef]

28.  Khan, R.; Khan, S.U.; Zaheer, R.; Khan, S. Future Internet: The Internet of Things Architecture, Possible Applications and Key Challenges. In Proceedings of the 10th International Conference on Frontiers of Information Technology, Islamabad, Pakistan, 17–19 December 2012; pp. 257–260. [CrossRef]

29.  Bertoldo, S.; Carosso, L.; Marchetta, E.; Paredes, M.; Allegretti, M. Feasibility analysis of a LoRa-based WSN using public transport. *Appl. Syst. Innov.* **2018**, *1*, 49. [CrossRef]

30.  Petajajarvi, J.; Mikhaylov, K.; Roivainen, A.; Hanninen, T.; Pettissalo, M. On the coverage of LPWANs: range evaluation and channel attenuation model for LoRa technology. In Proceedings of the 14th International Conference on ITS Telecommunications (ITST), Copenhagen, Denmark, 2–4 December 2015; pp. 55–59. [CrossRef]

31.  Sidorov, M.; Nhut, P.V.; Matsumoto, Y.; Ohmura, R. LoRa-Based Precision Wireless Structural Health Monitoring System for Bolted Joints in a Smart City Environment. *IEEE Access* **2019**, *7*, 179235–179251. [CrossRef]

32.  Pasolini, G.; Buratti, C.; Feltrin, L.; Zabini, F.; de Castro, C.; Verdone, R.; Andrisano, O. Smart city pilot projects using LoRa and IEEE802.15.4 technologies. *Sensors* **2018**, *18*, 1118. [CrossRef] [PubMed]

33.  Xu, W.; Kim, J.Y.; Huang, W.; Kanhere, S.S.; Jha, S.K.; Hu, W. Measurement, Characterization, and Modeling of LoRa Technology in Multifloor Buildings. *IEEE Internet Things J.* **2020**, *7*, 298–310. [CrossRef]

34.  Heller, A. The sensing internet—A discussion on its impact on rural areas. *Future Internet* **2015**, *7*, 363–371. [CrossRef]

35.  Munir, M.S.; Bajwa, I.S.; Naeem, M.A.; Ramzan, B. Design and Implementation of an IoT System for Smart Energy Consumption and Smart Irrigation in Tunnel Farming. *Energies* **2018**, *11*, 3427. [CrossRef]

36. Alsayyari, A.; Kostanic, I.; Otero, C.E. An empirical path loss model for Wireless Sensor Network deployment in a concrete surface environment. In Proceedings of the IEEE 16th Annual Wireless and Microwave Technology Conference (WAMICON), Cocoa Beach, FL, USA, 13–15 April 2015; pp. 1–6. [CrossRef]
37. Bezerra, N.S.; Åhlund, C.; Saguna, S.; de Sousa, V.A. Temperature impact in LoRaWAN—A case study in Northern Sweden. *Sensors* **2019**, *19*, 4414. [CrossRef]
38. Silverio-Fernández, M.; Renukappa, S.; Suresh, S. What is a smart device?-a conceptualisation within the paradigm of the internet of things. *Vis. Eng.* **2018**, *6*, 3. [CrossRef]
39. Jiménez-García, J.-L.; Baselga-Masia, D.; Poza-Luján, J.-L.; Munera, E.; Posadas-Yagüe, J.-L.; Simó-Ten, J.-E. Smart device definition and application on embedded system: performance and optimi-zation on a RGBD sensor. *ADCAIJ Adv. Distrib. Comput. Artif. Intell. J.* **2014**, *3*, 46–55. [CrossRef]
40. Tightiz, L.; Yang, H. A Comprehensive Review on IoT Protocols' Features in Smart Grid Communication. *Energies* **2020**, *13*, 2762. [CrossRef]
41. Vaca, N.; Garcia-Loro, F.; Martin, S.; Rodriguez-Artacho, M. Raspberry Pi Applications in Electronics and Control Laboratories. In Proceedings of the IEEE Global Engineering Education Conference (EDUCON), Tunis, Tunisia, 28–31 March 2022; pp. 1709–1713. [CrossRef]
42. Martin, S.; Fernandez-Pacheco, A.; Ruipérez-Valiente, J.A.; Carro, G.; Castro, M. Remote Experimentation Through Arduino-Based Remote Laboratories. *IEEE Rev. Iberoam. Tecnol. Del Aprendiz.* **2021**, *16*, 180–186. [CrossRef]
43. Ariza, J.Á.; Gil, S.G. RaspyLab: A Low-Cost Remote Laboratory to Learn Programming and Physical Computing Through Python and Raspberry Pi. *IEEE Rev. Iberoam. Tecnol. Del Aprendiz.* **2022**, *17*, 140–149. [CrossRef]
44. Mandanici, A.; Sarà, S.A.; Fiumara, G.; Mandaglio, G. Studying Physics, Getting to Know Python: RC Circuit, Simple Experiments, Coding, and Data Analysis With Raspberry Pi. *Comput. Sci. Eng.* **2021**, *23*, 93–96. [CrossRef]
45. Programming Tool-Node-Red. Available online: https://nodered.org/ (accessed on 5 July 2022).
46. Sumangali, K.; Kiran, P. Comparative study of open source tools for internet of things. *Int. J. Pharm. Technol.* **2016**, *8*, 26235–26241.
47. Sicari, S.; Rizzardi, A.; Coen-Porisini, A. How to evaluate an Internet of Things system: Models, case studies, and real developments. *Softw. Pract. Exp.* **2019**, *49*, 1663–1685. [CrossRef]
48. Grafana Labs. Available online: https://grafana.com/ (accessed on 5 July 2022).
49. Raimundo, R.J.; Rosário, A.T. Cybersecurity in the Internet of Things in Industrial Management. *Appl. Sci.* **2022**, *12*, 1598. [CrossRef]

*Article*

# Condition Assessment and Analysis of Bearing of Doubly Fed Wind Turbines Using Machine Learning Technique

Aiman Abbas Mahar [1], Nayyar Hussain Mirjat [2,*], Bhawani S. Chowdhry [3], Laveet Kumar [4], Quynh T. Tran [5] and Gaetano Zizzo [6,*]

[1] Department of Electronic Engineering, Mehran University of Engineering & Technology, Jamshoro 76062, Pakistan
[2] Department of Electrical Engineering, Mehran University of Engineering & Technology, Jamshoro 76062, Pakistan
[3] NCRA-HHRCMS Lab, Mehran University of Engineering and Technology, Jamshoro 76062, Pakistan
[4] Department of Mechanical Engineering, Mehran University of Engineering & Technology, Jamshoro 76062, Pakistan
[5] Institute of Energy Science, Vietnam Academy of Science and Technology, Hanoi 10000-04, Vietnam
[6] Department of Engineering, University of Palermo, 90128 Palermo, Italy
* Correspondence: nayyar.hussain@faculty.muet.edu.pk (N.H.M.); gaetano.zizzo@unipa.it (G.Z.)

**Abstract:** Condition monitoring of wind turbines is progressively increasing to maintain the continuity of clean energy supply to power grids. This issue is of great importance since it prevents wind turbines from failing and overheating, as most wind turbines with doubly fed induction generators (DFIG) are overheated due to faults in generator bearings. Bearing fault detection has become a main topic targeting the optimum operation, unscheduled downtime, and maintenance cost of turbine generators. Wind turbines are equipped with condition monitoring devices. However, effective and reliable fault detection still faces significant difficulties. As the majority of health monitoring techniques are primarily focused on a single operating condition, they are unable to effectively determine the health condition of turbines, which results in unwanted downtimes. New and reliable strategies for data analysis were incorporated into this research, given the large amount and variety of data. The development of a new model of the temperature of the DFIG bearing versus wind speed to identify false alarms is the key innovation of this work. This research aims to analyze the parameters for condition monitoring of DFIG bearings using SCADA data for k-means clustering training. The variables of k are obtained by the elbow method that revealed three classes of k (k = 0, 1, and 2). Box plot visualization is used to quantify data points. The average rotation speed and average temperature measurement of the DFIG bearings are found to be primary indicators to characterize normal or irregular operating conditions. In order to evaluate the performance of the clustering model, an analysis of the assessment indices is also executed. The ultimate goal of the study is to be able to use SCADA-recorded data to provide advance warning of failures or performance issues.

**Keywords:** bearings; condition monitoring; DFIG; K means; SCADA data; wind turbine

**Citation:** Mahar, A.A.; Mirjat, N.H.; Chowdhry, B.S.; Kumar, L.; Tran, Q.T.; Zizzo, G. Condition Assessment and Analysis of Bearing of Doubly Fed Wind Turbines Using Machine Learning Technique. *Energies* **2023**, *16*, 2367. https://doi.org/10.3390/en16052367

Academic Editors: Davide Astolfi and Abu-Siada Ahmed

Received: 2 January 2023
Revised: 6 February 2023
Accepted: 27 February 2023
Published: 1 March 2023



## 1. Introduction

In recent years, renewable energy sources have attracted considerable interest on a global scale, becoming a viable option due to technological development, cost reduction, and increasing demand, especially in developing countries [1,2]. Wind energy conversion is one of the most promising renewable energy technologies that has developed rapidly in recent years and provides a substantial share of electricity in an increasing number of countries.

The cumulative global installed capacity of wind energy is about 93.6 GW of the additional wind generating capacity installed globally in 2021, as shown in Figure 1.

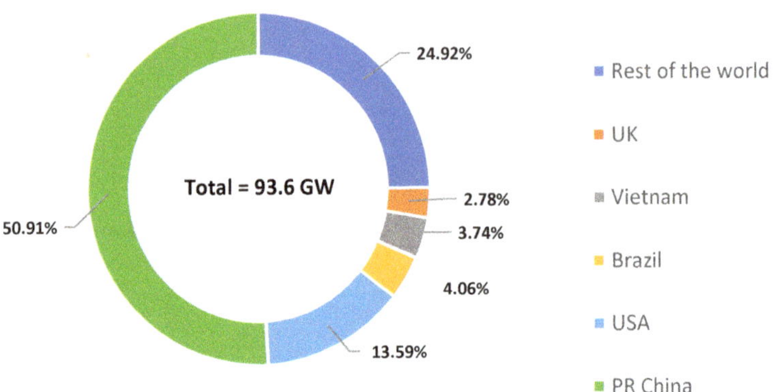



**Figure 1.** Global wind energy installation.

China, the U.S., Brazil, Vietnam, and the United Kingdom were the top five global markets for wind power installations in 2021 [3].

According to the report in [4], during 2021–2022, twelve wind power projects in Pakistan, with a cumulative installed capacity of 610 MW, achieved and started the supply of electricity to the national grid. As evidenced in Figure 2, currently, Pakistan has around 4% share of wind energy among other resources [5] and has a deployment of renewable energy throughout the land, particularly in two provinces, Sindh and Baluchistan [6], and aims to achieve 30% of its electricity generation from renewables by 2030 [7].

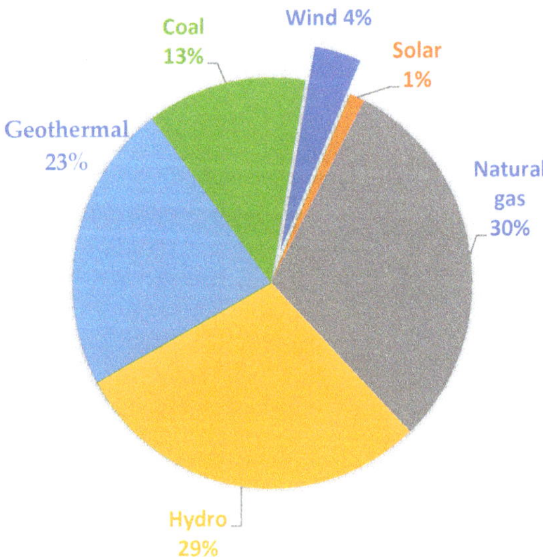



**Figure 2.** Pakistan's energy matrix.

The generator of a wind turbine is one of the most failure-prone assemblies due to the variable loads [8]. Continuous operations in all environmental conditions contribute to failures of wind turbine components, assemblies, and systems. Bearing failures account for more than 40% of the overall wind turbine generator failures leading to unexpected energy losses [9]. In Figure 3, a fire in a wind turbine is represented. According to [10], the turbine may have failed because of the generator's high bearing temperature due to a shutdown that led the generator to set on fire. The high temperature, combined with the leakage of oil from the rotating manifold nearby, probably triggered the fire.

**Figure 3.** Extend of damage due to wind turbine failure during October 2022 at Sapphire Wind Power Plant at Jhimpir.

Hence, a solution for effective condition monitoring of generator bearings and early identification of failure symptoms is needed. From actual maintenance practices of wind turbine generators, it was found that there is a high nonlinear relationship between the turbine fault and relevant factors [11]. Therefore, without proper monitoring, replacement of the damaged bearing will not solve the problem and will cause damage again. In order to avert such failures, it is critical to address the main causes of these failures in order to prevent them and create possible solutions to decrease the occurrence of damage in components. The development of condition monitoring techniques for rotating machinery, particularly the bearings, has received much attention during the past few decades. Since bearing failures results in prolonged downtime and wind turbine systems operate in adverse conditions with widely variable speeds, loads, and temperatures, they appear to be a solid option for model-based condition monitoring system [12,13]. However, cutting tools are the final executive component in the machining process and come into close contact with the product. This causes them to wear out quickly, which in turn impacts the workpiece's surface quality. According to [14], tool wear and damage are the primary factors causing the failure of the machining process. The resulting downtime accounts for 7–20% of the total downtime of the machining process, and the cost of tool and tool change accounts for 3–12% of the total machining cost.

A lack of information precisely describing primary bearing breakdowns in terms of frequency and damage types, as well as the most common bearing modeling and analysis setups, divided into dynamic and quasi-static categories, are presented in [15]. The study examines wind turbines' dynamic reliability under various control strategies and external conditions. The survival signature and fault tree analysis (FTA) were used to examine the system reliability level of wind turbine drive trains under various wind conditions in Ref. [16]. To maintain maximum power output, a doubly fed induction generator (DFIG) was chosen since it includes more than 50% of all major onshore wind power plants worldwide [17]. Doubly fed wind turbines are vulnerable to various types of generator losses; these failures lead to excess vibrations that might damage other components, such as bearing failures, which produce non-stationary vibrations [18,19]. Because the consequences of a failure are catastrophic for both business and customers, it is crucial to prevent the error more precisely [18]. Due to the longtime operation of the wind turbine in poor conditions, such as heavy loads, corrosion or failure in gearboxes can occur, such as misalignment, looseness, and contamination of the bearings [20]. Contaminants include dust in the air, dirt in the bearing, and any abrasive substance that gets into the bearing. Studies conducted on the reliability of wind turbines show that the drivetrain system accounts for about 20% of overall problems and accounts for nearly 30% of wind turbine outages when using DFIG [21].

With the use of data from the supervisory control and data acquisition system and the idea of energy saving, this study enhances a condition monitoring method for wind turbine main bearings. The objective of the current research was to employ model parameters as health indicators. The method was applied to find main bearing degradations over a two-year period in a wind farm with more than 100 WTGs. The method was evaluated because of the history of bearing failures that were known [22]. The two different types of datasets were used to test the newly created trigonometric entropy measure based on variational mode decomposition (VMD). One comes from the Centre for Intelligent Maintenance Systems, and the other from the XJTU-SY Bearing Databases. The suggested method has the ability to raise the alert about the beginnings of faults relatively at an early stage [23].

One of the research studies suggests an artificial neural network (ANN)-based defect detection approach for wind turbine main bearings based on current SCADA data. In this study, SCADA data from the Nord-Trndelag Elektrisitetsverk-owned Hundhammerfjellet wind farm were used. Turbine rear bearing temperature, a main shaft rear bearing characteristic in the SCDA data, provides an indication of how hot the bearings are operating and hence provides the opportunity to identify rear bearing overheating [24]. Another study was developed in such a way that 10 min average operating (SCADA) data for a group of 14 wind turbines are available, and a subset of 10 of those 14 turbines undergo monthly grease checks. The suggested method is completely hybrid and intended to combine data-driven and physics-informed layers in deep neural networks. The bearing damage of a wind turbine using recurrent neural networks was studied. It was specifically suggested that grease damage increments through a multi-layer perceptron and bearing fatigue damage through equations frequently employed in bearing reliability design [25]. Therefore, this research focuses on analyzing these faulty bearings in DFIG. The conditioning monitoring-based system was suggested to detect faulty bearings, and a machine learning approach was used for the detection of intensity/type of fault in DFIG. In this paper, system identification methods were applied to SCADA data to develop a condition monitoring model, which can be used to predict the generator's indices that affect the bearings in terms of wind speed, generator temperature, generator rotation speed, etc. The SCADA data were analyzed to assess wind turbine operating conditions using the developed model that effectively identifies potential failures or breakdowns.

The need for condition monitoring of wind turbines is growing as the size and location of modern wind turbines make their technical availability essential. Due to the limited accessibility of some remote-controlled wind farms on mountain and offshore wind farms, unexpected failures, especially of large and important components, might result in unnecessary delay and cost. Therefore, the goal of our research was to obtain real-time SCADA data for acquisition while minimizing downtime through condition monitoring. Many methods employing these data for early failure detection have recently been developed since CM using SCADA data is a potentially low-cost solution requiring no additional sensors. From all the parameters, it was found that the prediction of the temperature variation trend is crucial in order to provide an overheating warning and detect an optimization problem. In order to solve the optimization problem of the proposed model, the corresponding objective function was derived in a more tractable form, and an alternative update algorithm is presented, which is based on the identification of new concepts in unlabeled data. The method is used to achieve improved predictive performance in terms of improved predictive accuracy.

The rest of the paper is structured as follows: Section 2 presents the research methodology describing the data-based approach and reporting the research flow of the study. Section 3 provides the results and analysis of the study. Techniques and methods (K-means clustering approach, elbow method, boxplot visualization) are discussed and applied to the SCADA data where faults are known to have occurred, and conclusions and future recommendations are drawn.

## 2. Methodology

Wind turbine condition monitoring can be used to improve safety or to lower the cost of the existing level of safety, anticipating or detecting emerging major faults [26]. The study proposes a methodology based on wind turbine condition monitoring that, through a supervisory, control, and data acquisition SCADA system, records data from an inertial measurement unit (IMU) sensor mounted on the DFIG's bearings that detects system signals. The SCADA system provides historical signals, fault information, environmental condition parameters, and operational factors related to the DFIGs and their equipment in wind turbines. The collected data are then used by proposing a new method of bearing failure detection in the machines.

The method used in this study to compare the bearing status of the DFIG in the SCADA system is characterized by its feasibility and cost-effectiveness, and its flowchart is shown in Figure 4. Table 1 lists the key parameters of interest.

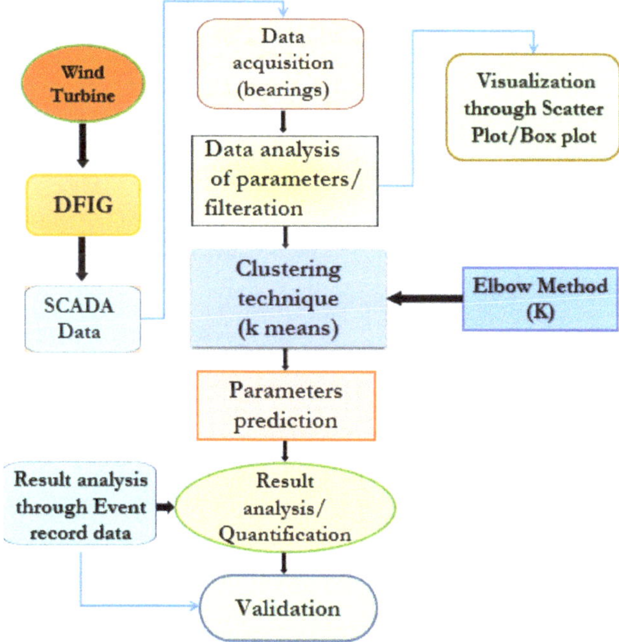


**Figure 4.** Research Methodology Flow chart.

Unlike supervised learning, where human-labeled data are necessary, unsupervised learning uses unlabeled data and provides an insight into the probable classes present in the dataset, which is then used to broadly classify the data. The main benefit of unsupervised learning lies in the fact that it does not require any human-labeled data, which saves much manual effort that is otherwise used to label every data point in a particular dataset individually.

**Table 1.** Assessment indices of DFIG bearings.

| S.no. | SCADA Parameters | Unit |
|---|---|---|
| 1 | Average Active Power | (kW) |
| 2 | Average Wind Speed | (m/s) |
| 3 | Average Reactive Power | (kVar) |
| 4 | Wind Turbine Energy yield | (kWh) |
| 5 | Average Ambient Temp | (°C) |
| 6 | Average Gen Rotation Speed | (rpm) |
| 7 | Average Maximum generator temp | (°C) |

Advanced condition monitoring systems are used by wind turbines, which are complex systems, to assess their state of health. Due to its high failure rate and downtime, the generator is one of the most important components. In wind farms, SCADA are used for real-time condition monitoring and control. New and reliable strategies for data analysis are needed because of the large amount and variety of the data. The development of a new model of the temperature of the DFIG bearing versus wind speed to identify false alarms was the key innovation of this work. A box plot was utilized to depict the distribution of the data, and a data partitioning strategy was used.

The model for determining failures in DFIG bearings is based on machine learning (ML) techniques for classification and parameter analysis for early diagnosis and predictive maintenance. The K-means clustering technique is used for the evaluation and condition analysis of DFIG bearings, and the elbow method is used to determine the K value for the classes in accordance with the clustering technique. Reducing the size and complexity of the dataset is the primary goal of the clustering technique. Compared to the original data, clustered groups of points occupy much less storage space and are easier to control, which is why this method is proposed. Once the clusters were determined, scatterplot and boxplot visualization was used for the statistical visualization of the predicted data. In the last step, validation of the predictive performance of the results was analyzed with the latest data obtained.

**K-means Clustering**

The unsupervised learning approach known as K-means clustering divides the unlabeled dataset into many clusters. The K-means algorithm was employed for the condition assessment of the DFIG bearings. The K-means clustering technique was applied to design the prediction model for the assessment indices that are based on DFIG bearings, such as temperature, rotation speed, and wind speed. The K-means model was used to identify the clusters in the available dataset. Here, the value of K (1, 2, 3, 4, 5) was determined.

The model is trained through the machine learning algorithm using python programming in Jupyter in Figure 5.

The steps in the K-means clustering algorithm are the following:

Let $X = \{x_1, x_2, x_3, \ldots, x_n\}$ be the set of data points and $V = \{v_1, v_2, \ldots, v_n\}$ be the set of centers:

(1)  Select "c" cluster centers;
(2)  Determine the Euclidean distance between the cluster centers and each data point;
(3)  Assign the cluster with the shortest distance from all the other cluster centers;
(4)  Recalculate the new cluster center using the following:

$$v_i = (1/c_i) \, \text{sum} \, (j = 1, c_i) \, (x_i)$$

where "$c_i$" denotes the no. of data points in the $i$th cluster;
(5)  Measure again the separation between each data point and the newly discovered cluster centers;
(6)  Stop if no data point was moved; otherwise, go back to step (3).

For the progression, keep changing the value until the optimal clusters can be achieved.

$$J(V) = \sum_{i=1}^{c} \sum_{j=1}^{ci} \left( \|x_i - v_j\| \right)^2$$

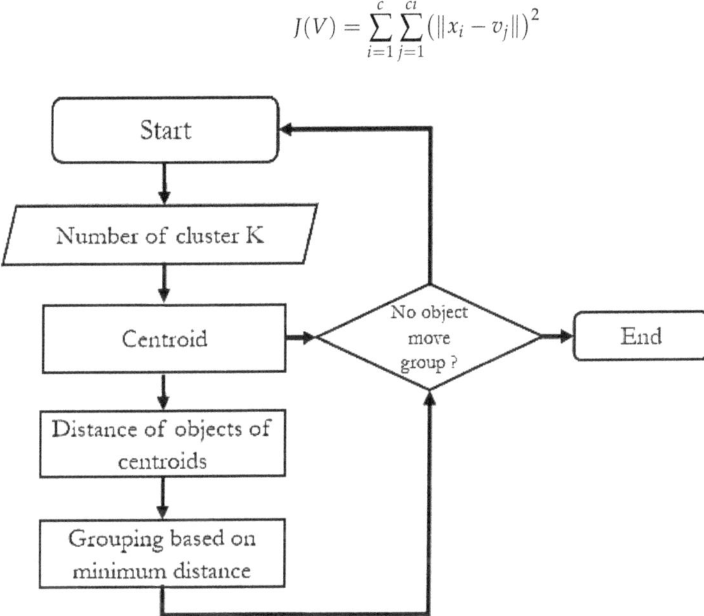


**Figure 5.** Flowchart of K-means technique.

The Euclidean distance between $x_i$ and $v_j$ is given as "$\|x_i - v_j\|$".

Prediction errors in the SCADA data of the model follow a normal distribution. Thus, it can be calculated based on the probability density function of normal distribution.

**Elbow method for optimizing K value:**

The elbow method is an effective method for cluster optimization and is used for clustering analysis since it is simple to implement and yields useful results. The elbow technique is a visual way to verify the consistency of the optimal number of clusters by analyzing the difference in the sum of square errors (SSE) of each cluster. The most severe difference creating the angle of the elbow indicates the optimum cluster number [27].

In this method, the optimal value of k using the elbow method is found by calculating the sum of square error (SSE) to evaluate K-means clustering using the elbow criterion. The idea of the elbow criteria approach is performed to select the k (number of clusters) at which the SSE decreases significantly. The elbow rule's fundamental concept is to employ a square of the distance between each cluster's centroid and sample points to generate a range of K values. As a performance measure, the sum of squared errors (SSE) is employed. SSE is calculated by iterating over the K-value. Smaller numbers represent more converging clusters. SSE displays a sharp reduction when the number of clusters is adjusted to be close to the number of actual clusters. When there are more clusters than there are actual clusters, SSE still decreases, but it does so more slowly. It is used to choose the best clusters.

**Data Visualization**

Large amounts of numerical data can be displayed using graphs, which can be used to demonstrate the relationships between the numerical values of various variables and to derive quantitative relationships between them. One of the most powerful and popular methods for visual data analysis is the scatter plot [28]. Different data points are positioned between an $x$- and $y$-axis to display these data. Each of these data points appears to be "scattered" across the graph. By using the generalized scatter plot technique, big datasets can be completely represented in the figure without any overlap. The primary concept is to provide the analyst with the flexibility to adjust the quantity of overlap and distortion to produce the optimal view. The model offers the option to smoothly zoom between the conventional and generalized scatter plots to enable effective usage. An optimization function considers overlap and visualization distortion.

**Performance Analysis System**

SCADA was utilized to create the system for analyzing model performance. Once a csv file is loaded with the dataset comprising the measurements, the model builds its predictions. Then, predictions are made using the seven parameters listed in Table 1. In the same sequence as the data received from the input file, the predictions are made and written to a csv file. By comparing the predicted label with the actual label, the generated csv files can be examined to determine how well the model performed. The software is used to analyze and visualize the performance of the model implemented in Jupyter Notebook using Python programming.

## 3. Results and Analysis

### 3.1. Case Study

In this study, SCADA data were collected and analyzed from an onshore wind farm with 66 wind turbines with DFIG generators with rated power of 1.5 MW each, located in the Jhimpir wind corridor. The data were collected in the form of excel sheets at 10 min intervals at a bearing speed of 1800 rpm. The database covers one month of operation, from January 2021 to February 2021. Up to 300 datasets were recorded for analysis and used to train the model.

### 3.2. Data Visualization

After data pre-processing, visualization of the data according to its correlation with all filtered parameters can be observed in Figure 5. Once the visualization of the data is acquired, the condition changes in the DFIG parameters summarized in Table 1 can be easily determined.

In Figure 6, a good correlation is observed between the average generator rotation speed and the average generator maximum temperature. Therefore, with the change in speed, the temperature also increases.

In order to analyze the trend and cycles, studies of the parameters with regard to time were devised, as shown in Figure 7. Based on the number of notices and downtime hours, Figure 6 displays the monthly report of the evaluation indices for a certain DFIG. One-month duration is displayed on the *x*-axis, and the left *y*-axis displays the ranges and downtime hours. The two temperature parameters exhibit nonlinear and progressive variation tendencies, and the wind speed varies significantly.

It is evident from the figure that as the wind speed increases, the generator's rotational speed and power generation also increase accordingly. The average temperature of the turbine also increases due to an increase in turbine speed. It can also be observed that the effect of ambient temperature on generator temperature is rather low compared to the effect of rotation speed. The turbine starts generation at 5 m/s and tries to maintain its speed at 17 m/s for maximum power generation. As wind speed reduces to less than 5 m/s, the turbine cannot maintain generation and reduces the generator speed to zero, as can be seen from the figure.

### 3.3. K-Means Clustering with Elbow Method

The K-means algorithm partitions the collected data into k clusters, in which each point belongs to the cluster with the smallest distance. K determines the default clusters to be generated during the process, with K = 2 creating two clusters and K = 3 creating three clusters. This allows for the analysis of a close connection between generator rotation speed and generator temperature. The technique is applied by using the elbow method algorithm, which provides a quick and intuitive response.

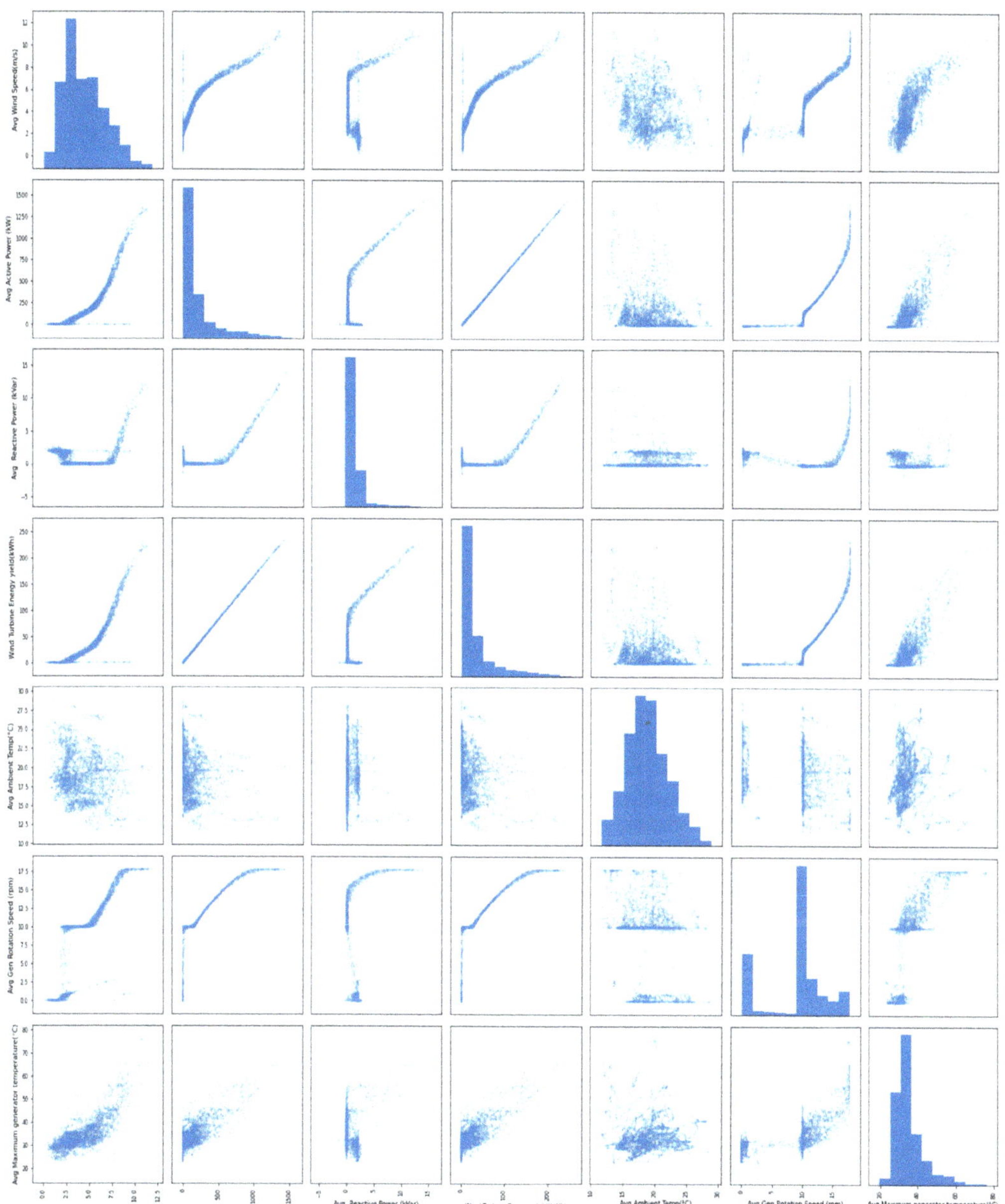

**Figure 6.** Scatterplot of units divided into clusters obtained according to its correlated parameters.

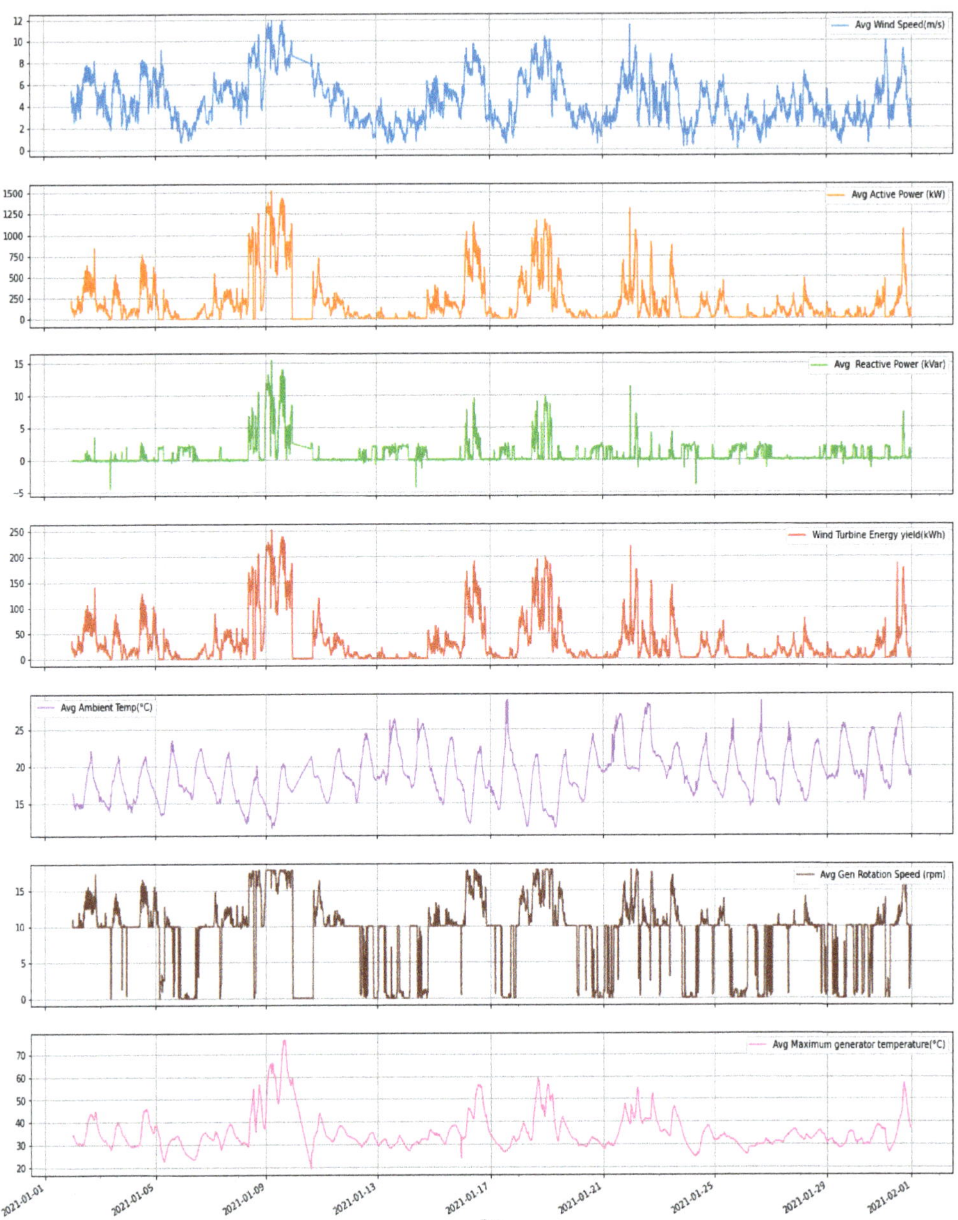

**Figure 7.** Time series graph of parameters for a given timeline.

In the elbow graph in Figure 8, it is seen that there is a sharp decrease in the SSE until the third cluster. Therefore, the optimal value of k = 3 is obtained.

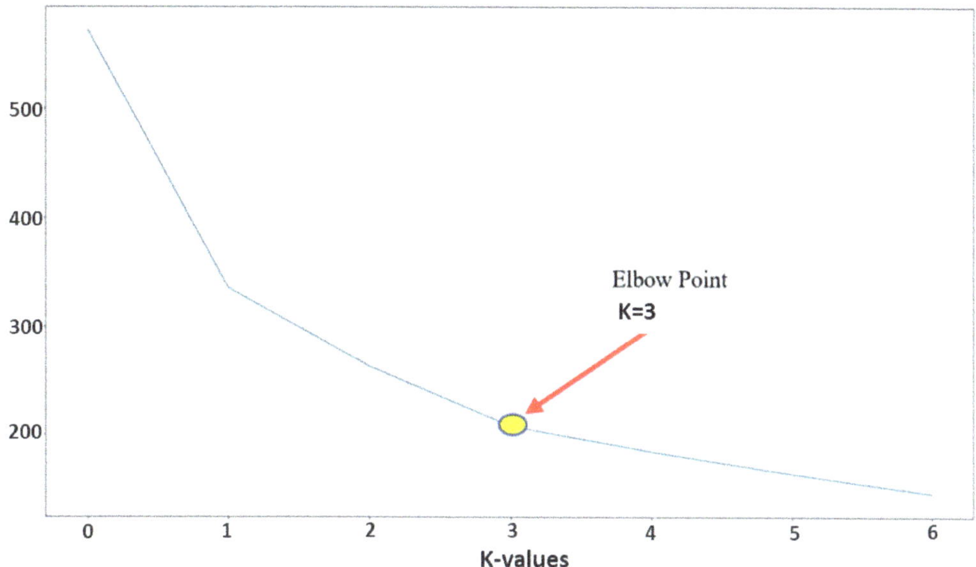


**Figure 8.** Determination of the best cluster using the elbow method: sum of squared errors (SSE) versus K values.

The centroid of a cluster and each individual observation allocated to that cluster are separated by the total within-cluster variation. The following centroids are chosen using the original max–min criterion, which involves picking the point that is farthest from its nearest centroid. When the cluster sizes are seriously out of balance, the max–min approach is extremely helpful in preventing the worst-case behavior of the random centroids [29].

The clusters are more clearly defined and confined the closer together these distances are by adjusting the k-value to 3 and observing the clusters again with respect to all parameters. The grouping of data points does not give a good correlation between all parameters except the average generator temperature and the average generator rotation speed, which are clearly grouped in Figure 9. However, a linear relationship (orange line) can be observed between the temperature and the rotation speed of the DFIG bearings.

The red dots showing the mean of each cluster's points, orange line shows the linearity and clusters are referred to as classes K = 0, 1, and 2, (3 colors = 3 clusters or k = 3) in Figure 10.

*3.4. Box Plot Visualization*

The relationship function between the assessment indices and their influencing factors was established using a boxplot. The condition assessment index of DFIG bearings was generated using a boxplot representation of k-means clustering algorithms. The boxplot visualization of the k-means cluster analysis groups individuals as average maximum generator temperature (°C) and average gen rotation speed (rpm).

The partition of units in clusters is:

Cluster 1: 0, for low temperature, low speed;
Cluster 2: 1, for medium temperature, moderate speed;
Cluster 3: 2, for severe temperature, high speed;

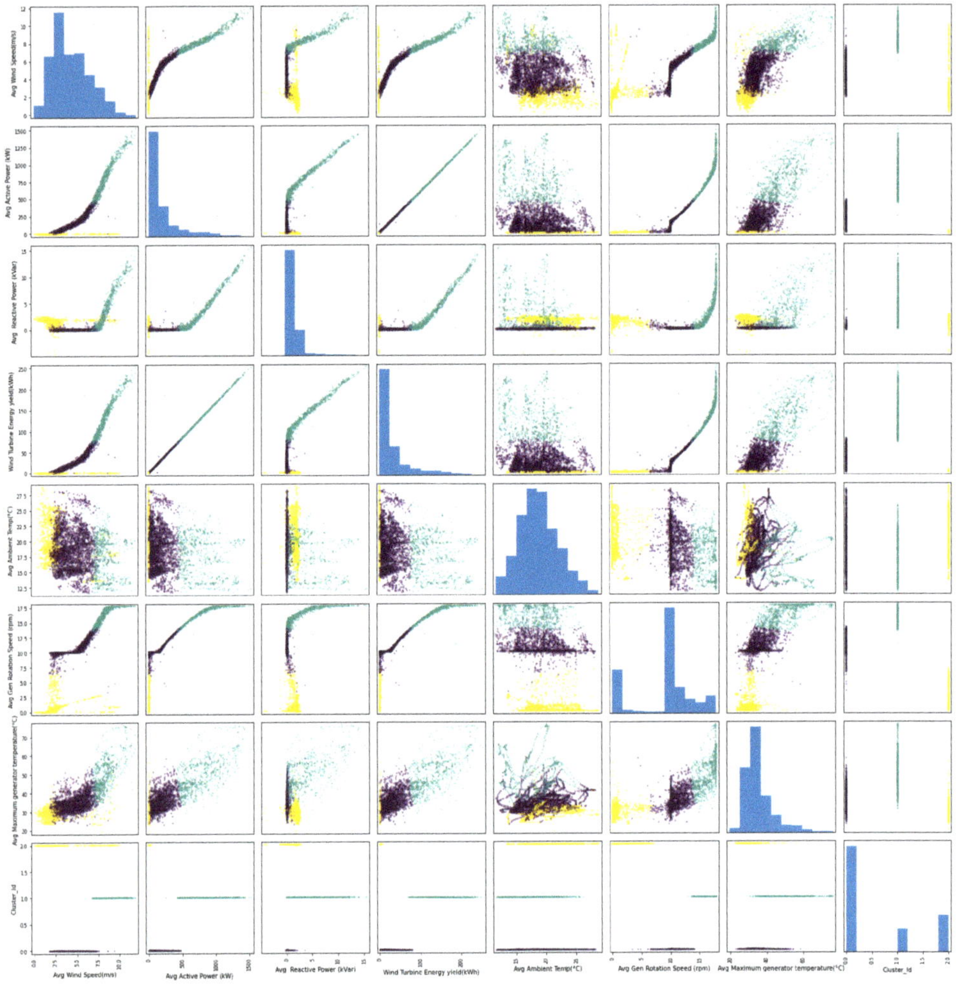

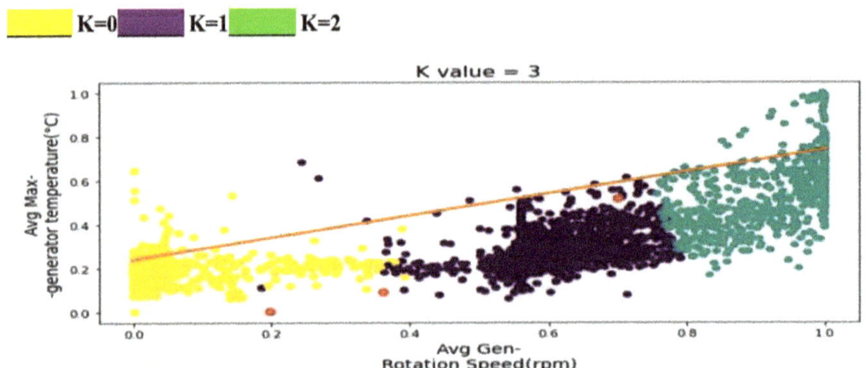

**Figure 9.** Data visualization according to its correlated parameters.



**Figure 10.** Data visualization of generator temperature and rotation speed according to K-means clustering.

In Figure 11, the relationship between rotation speed and temperature is analyzed. As the rotation increases, the temperature also increases. Therefore, three classes are extracted after the clustering method.

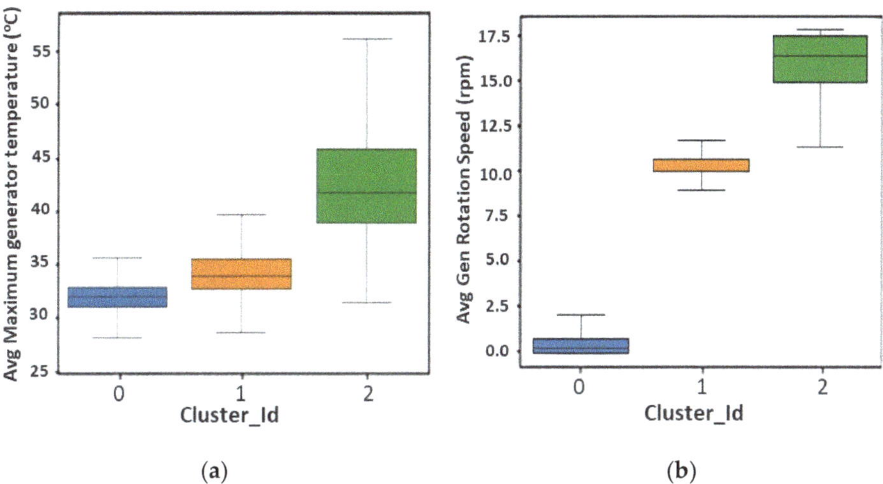

(a)  (b)

**Figure 11.** Boxplot variables: (**a**) rotation speed vs. cluster id; (**b**) temperature vs. cluster id.

### 3.5. Validation of Results

Validation of the predictive performance of the results with the latest data was obtained. After processing the model again for getting qualitative performance, SCADA operational data were recorded continuously at 10 min intervals. This can take the form of the average, minimum, maximum, or standard deviation of live values recorded by the controller in the previous 10 min period. Signals such as the turbine power output, wind speed, temperatures of various components, electrical signals, and environmental conditions such as anemometer-measured wind speed and ambient temperature were recorded. The flow chart of validation is shown in Figure 12.

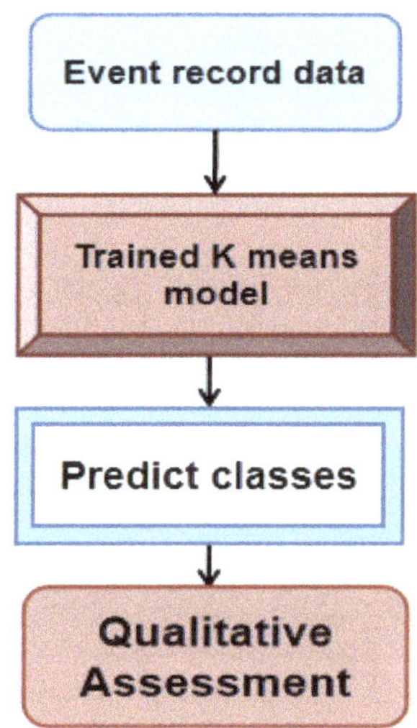


**Figure 12.** Flow chart of validation.

The comparison of average generator rotation speed and average generator temperature in Figure 13 shows their performance with their pros and cons in a particular scenario. The suggested evaluation approach can effectively forecast the change in operating circumstances prior to fault occurrences and can provide early warning of developing faults in DFIG bearings, according to the validation results.

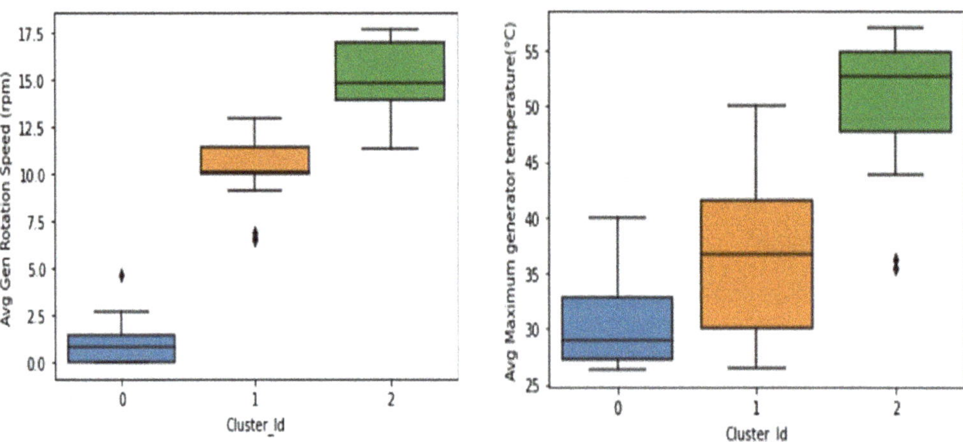

**Figure 13.** Validation of predicted results.

Validation of this research shows the change in speed and temperature. If the temperature exceeds greater than the range of 25 °C to 75 °C, it may fault the DFIG. After applying the machine learning technique, it was found that the rotation of the generator changes its speed, the high temperature could also be exceeded, and it can affect DFIG and could damage the bearings. Therefore, this research provides early fault detection of DFIG bearings.

## 4. Conclusions, Social Impact, and Recommendations

The approach was applied in a field study using one month of SCADA data from a wind farm consisting of 66 1.5 MW turbines in order to develop an effective condition monitoring system for early diagnosis and prognoses the conditions of the wind turbine's drive train to investigate bearing failures in DFIG. The box plot shows the visual graph of the affected parameters. In order to achieve forecasting with high accuracy, this paper proposes a novel model for the bearings in DFIG and a machine learning method for predictive maintenance. Therefore, forecasting the trend of temperature change is critical for overheating warnings. In order to evaluate the performance of the clustering model, experimental analysis was carried out. By combining the condition parameters in a scatter matrix, the linear elbow technique revealed the relationships between temperature and related variables. The results of comparative studies and early fault diagnosis show that the proposed method has better performance for temperature forecasting and average rotation speed of the main bearing of large-scale DFIG bearings. If these data can be used to identify potential failures or breakdowns, then this may well prove to be a very cost-effective means of condition monitoring; the SCADA data parameters prove to be a cost-effective method of condition monitoring that can be used for potential faults or malfunctions. The method will enable reducing downtime and monetary losses due to maintenance and replacement of various wind turbine components. The study provides an adaptable but reliable framework for the early identification of developing wind turbine damage, reducing wind turbine outages, and raising wind turbine dependability and revenue through operational improvement. Forecasting the trend of temperature change is essential for issuing overheating alerts. The outcomes of comparative studies and early fault diagnosis demonstrate that the proposed method has improved performance for temperature and rotation speed for forecasting the bearing of DFIG bearings. This research suggests a novel model for the bearings in DFIG and a machine learning method for preventive maintenance to achieve forecasts with high accuracy.

The sustainability of wind energy generation is ensured by using the approach suggested in this paper. Additionally, wind turbines may lessen the amount of power produced using fossil fuels, which lowers overall air pollution and carbon dioxide emissions. The potential for wind turbines to negatively impact wild animals through collisions as well as indirectly through noise pollution, habitat loss, and decreased survival or reproduction is a major issue for the business. Moreover, this study could be carried out to analyze the viability of this method for more than seven parameters of DFIG wind turbines in the future. With some changes, our project can also be used for health/condition monitoring of other equipment such as wind turbine gearboxes, medical sectors, helicopters, cars, etc.

**Author Contributions:** Conceptualization, A.A.M., N.H.M., B.S.C., L.K. and Q.T.T.; validation, A.A.M., N.H.M., B.S.C., L.K. and G.Z.; formal analysis, A.A.M., N.H.M., B.S.C., L.K. and G.Z.; writing—original draft preparation, A.A.M., N.H.M., B.S.C. and L.K.; writing—review and editing, A.A.M., N.H.M., B.S.C., L.K., Q.T.T. and G.Z. All authors have read and agreed to the published version of the manuscript.

**Funding:** The first author of the article has received funding/stipend from ICT endowment of Mehran University of Engineering and Technology, Jamshoro; Pakistan.

**Acknowledgments:** This research work is carried out at Mehran University of Engineering and Technology and is supported by NCRA established at the IICT building, Mehran UET, Jamshoro. Gaetano Zizzo is supported by the Research Fund for the Italian Electrical System under the Contract Agreement "Accordo di Programma 2022-2024" between ENEA and the Ministry of Ecological Transition.

**Conflicts of Interest:** The authors declare no conflict of interest.

# References

1.  Soomro, M.A.; Memon, Z.A.; Kumar, M.; Baloch, M.H. Wind energy integration: Dynamic modeling and control of DFIG based on super twisting fractional order terminal sliding mode controller. *Energy Rep.* **2021**, *7*, 6031–6043. [CrossRef]
2.  Khatri, S.A.; Harijan, K.; Uqaili, M.A.; Shah, S.F.; Mirjat, N.H.; Kumar, L. A Logistic Modelling Analysis for Wind Energy Potential Assessment and Forecasting its Diffusion in Pakistan. *Front. Energy Res.* **2022**, *10*, 860092. [CrossRef]
3.  *Global Wind Report 2022*; Global Wind Energy Council: Brussels, Belgium, 2022.
4.  Annual Report 2021-22. Available online: https://www.nepra.org.pk/publications/Annual%20Reports/Annual%20Report%20 2021-22.pdf (accessed on 6 February 2023).
5.  Turi, J.A.; Rosak-Szyrocka, J.; Mansoor, M.; Asif, H.; Nazir, A.; Balsalobre-Lorente, D. Assessing Wind Energy Projects Potential in Pakistan: Challenges and Way Forward. *Energies* **2022**, *15*, 9014. [CrossRef]
6.  Asghar, R.; Ullah, Z.; Azeem, B.; Aslam, S.; Hashmi, M.H.; Rasool, E.; Shaker, B.; Anwar, M.J.; Mustafa, K. Wind Energy Potential in Pakistan: A Feasibility Study in Sindh Province. *Energies* **2022**, *15*, 8333. [CrossRef]
7.  Solangi, Y.A.; Tan, Q.; Khan, M.W.A.; Mirjat, N.H.; Ahmed, I. The selection of wind power project location in the Southeastern Corridor of Pakistan: A factor analysis, AHP, and fuzzy-TOPSIS application. *Energies* **2018**, *11*, 1940. [CrossRef]
8.  Kusiak, A.; Verma, A. Analyzing bearing faults in wind turbines: A data-mining approach. *Renew Energy* **2012**, *48*, 110–116. [CrossRef]
9.  Tavner, P.J.; Greenwood, D.M.; Whittle, M.W.G.; Gindele, R.; Faulstich, S.; Hahn, B. Study of weather and location effects on wind turbine failure rates. *Wind Energy* **2013**, *16*, 175–187. [CrossRef]
10. Islam, M.M.; Sohag, K.; Alam, M.M. Mineral import demand and clean energy transitions in the top mineral-importing countries. *Resour. Policy* **2022**, *78*, 102893. [CrossRef]
11. Zhang, D.; Yuan, J.; Zhu, J.; Ji, Q.; Zhang, X.; Liu, H. Fault Diagnosis Strategy for Wind Turbine Generator Based on the Gaussian Process Metamodel. *Math. Probl. Eng.* **2020**, *2020*, 4295093. [CrossRef]
12. Garlick, W.G.; Dixon, R.; Watson, S.J.; Garlick, W.G.; Dixon, R.; Watson, S.J. A Model-Based Approach to Wind Turbine Condition Monitoring Using SCADA A Model-Based Approach to Wind Turbine Condition Monitoring Using SCADA Data Data A Model-based Approach to Wind Turbine Condition Monitoring Using SCADA Data. Available online: https://hdl.handle.net/2134/5488 (accessed on 6 February 2023).
13. Elasha, F.; Shanbr, S.; Li, X.; Mba, D. Prognosis of a wind turbine gearbox bearing using supervised machine learning. *Sensors* **2019**, *19*, 3092. [CrossRef] [PubMed]

14. Zhou, Y.; Zhi, G.; Chen, W.; Qian, Q.; He, D.; Sun, B.; Sun, W. A new tool wear condition monitoring method based on deep learning under small samples. *Measurement* **2022**, *189*, 110622. [CrossRef]
15. Hart, E.; Clarke, B.; Nicholas, G.; Kazemi Amiri, A.; Stirling, J.; Carroll, J.; Dwyer-Joyce, R.; McDonald, A.; Long, H. A review of wind turbine main bearings: Design, operation, modelling, damage mechanisms and fault detection. *Wind. Energy Sci.* **2020**, *5*, 105–124. [CrossRef]
16. Li, Y.; Zhu, C.; Chen, X.; Tan, J. Fatigue reliability analysis of wind turbine drivetrain considering strength degradation and load sharing using survival signature and FTA. *Energies* **2020**, *13*, 2108. [CrossRef]
17. Soomro, M.A.; Memon, Z.A.; Kumar, M. PWM Based VSC for Power Quality Assessment of Grid Integrated DFIG-WECS. *Int. J. Integr. Eng.* **2020**, *12*, 239–252. [CrossRef]
18. Liu, W.Y.; Tang, B.P.; Han, J.G.; Lu, X.N.; Hu, N.N.; He, Z.Z. The structure healthy condition monitoring and fault diagnosis methods in wind turbines: A review. *Renew. Sustain. Energy Rev.* **2015**, *44*, 466–472. [CrossRef]
19. Han, W.; Kim, J.; Kim, B. Effects of contamination and erosion at the leading edge of blade tip airfoils on the annual energy production of wind turbines. *Renew Energy* **2018**, *115*, 817–823. [CrossRef]
20. Qian, P.; Ma, X.; Zhang, D. Estimating health condition of the wind turbine drivetrain system. *Energies* **2017**, *10*, 1583. [CrossRef]
21. Gu, H.; Liu, W.; Gao, Q.; Zhang, Y. A review on wind turbines gearbox fault diagnosis methods. *J. Vibroeng.* **2021**, *23*, 26–43. [CrossRef]
22. de Oliveira-Filho, A.M.; Cambron, P.; Tahan, A. Condition Monitoring of Wind Turbine Main Bearing Using SCADA Data and Informed by the Principle of Energy Conservation. In Proceedings of the 2022 Prognostics and Health Management Conference, PHM-London 2022, London, UK, 27–29 May 2022; pp. 276–282. [CrossRef]
23. Kumar, A.; Gandhi, C.P.; Vashishtha, G.; Kundu, P.; Tang, H.; Glowacz, A.; Shukla, R.K.; Xiang, J. VMD based trigonometric entropy measure: A simple and effective tool for dynamic degradation monitoring of rolling element bearing. *Meas. Sci. Technol.* **2022**, *33*, 014005. [CrossRef]
24. Zhang, Z.Y.; Wang, K.S. Wind turbine fault detection based on SCADA data analysis using ANN. *Adv. Manuf.* **2014**, *2*, 70–78. [CrossRef]
25. Yucesan, Y.A.; Viana, F.A.C. A Physics-informed Neural Network for Wind Turbine Main Bearing Fatigue. *Int. J. Progn. Health Manag.* **2020**, *11*. [CrossRef]
26. Entezami, M. Wind Turbine Condition Monitoring System Wayside Acoustic Monitoring for Train Bearings View Project EC FP7-NIMO View Project. 2010. Available online: https://www.researchgate.net/publication/242568720 (accessed on 6 February 2023).
27. Umargono, E.; Suseno, J.E.; Gunawan, S. *K-Means Clustering Optimization Using the Elbow Method and Early Centroid Determination Based-On Mean and Median*; Atlantis Press: Paris, France, 2020; pp. 234–240. [CrossRef]
28. Keim, D.A.; Hao, M.C.; Dayal, U.; Janetzko, H.; Bak, P. Generalized scatter plots. *Inf. Vis.* **2010**, *9*, 301–311. [CrossRef]
29. Fränti, P.; Sieranoja, S. How much can k-means be improved by using better initialization and repeats? *Pattern Recognit* **2019**, *93*, 95–112. [CrossRef]

*energies*

*Article*

# Effective Utilization of Distributed Power Sources under Power Mismatch Conditions in Islanded Distribution Networks

Zohaib Hussain Leghari [1,2,3,*], Mohammad Yusri Hassan [1,2], Dalila Mat Said [1,2], Laveet Kumar [4], Mahesh Kumar [3], Quynh T. Tran [5,*] and Eleonora Riva Sanseverino [6,7,*]

1    Centre of Electrical Energy Systems (CEES), Institute of Future Energy (IFE), Universiti Teknologi Malaysia (UTM), Skudai, Johor Bahru 81310, Johor, Malaysia
2    School of Electrical Engineering (SKE), Faculty of Engineering, Universiti Teknologi Malaysia (UTM), Johor Bahru 81310, Johor, Malaysia
3    Department of Electrical Engineering, Mehran University of Engineering and Technology (MUET), Jamshoro 76062, Sindh, Pakistan
4    Department of Mechanical Engineering, Mehran University of Engineering and Technology (MUET), Jamshoro 76062, Sindh, Pakistan
5    Institute of Energy Science, Vietnam Academy of Science and Technology, 18 Hoang Quoc Viet, Cau Giay, Hanoi 10072, Vietnam
6    Department of Engineering, University of Palermo, 90128 Palermo, Italy
7    EnSiEL, Consorzio Interuniversitario Nazionale "Energia e Sistemi Elettrici", 16145 Genova, Italy
*    Correspondence: zohaib.leghari@faculty.muet.edu.pk (Z.H.L.); tranttq@hawaii.edu (Q.T.T.); eleonora.rivasanseverino@unipa.it (E.R.S.)

Citation: Leghari, Z.H.; Hassan, M.Y.; Said, D.M.; Kumar, L.; Kumar, M.; Tran, Q.T.; Sanseverino, E.R. Effective Utilization of Distributed Power Sources under Power Mismatch Conditions in Islanded Distribution Networks. *Energies* 2023, *16*, 2659. https://doi.org/10.3390/en16062659

Academic Editor: Anca D. Hansen

Received: 24 January 2023
Revised: 6 March 2023
Accepted: 9 March 2023
Published: 12 March 2023

**Abstract:** The integration of distributed generation (DG) into a power distribution network allows the establishment of a microgrid (MG) system when the main grid experiences a malfunction or is undergoing maintenance. In this case, the power-generating capacity of distributed generators may be less than the load demand. This study presents a strategy for the effective utilization of deployed active and reactive power sources under power mismatch conditions in the islanded distribution networks. Initially, the DGs' and capacitors' optimal placement and capacity were identified using the Jaya algorithm (JA) with the aim to reduce power losses in the grid-connected mode. Later, the DG and capacitor combination's optimal power factor was determined to withstand the islanded distribution network's highest possible power demand in the event of a power mismatch. To assess the optimal value of the DG–capacitor pair's operating power factor ($pf_{source}$) for the islanded operation, an analytical approach has been proposed that determines the best trade-off between power losses and the under-utilization of accessible generation. The test results on 33-bus and 69-bus IEEE distribution networks demonstrate that holding the islanded network's load power factor ($pf_{load}$) equal to $pf_{source}$ during the power imbalance conditions allows the installed distributed sources to effectively operate at full capacity. As expected, the proposed strategy will assist the utility companies in designing efficient energy management or load shedding schemes to effectively cope with the power mismatch conditions.

**Keywords:** capacitors; distributed generation; distribution network; islanded operation; microgrid; power supply–demand imbalance

## 1. Introduction

Obtaining favorable results by introducing distributed generation (DG) and capacitors into distribution networks necessitates a well-thought-out design strategy. The optimal positioning and sizing of capacitors and DGs potentially result in improved voltage stability, reliability, power quality, reduced power losses, and eliminating or deferring the upgrades of the electrical power networks. In addition, DGs' and capacitors' joint presence allows the distribution networks to be operated as autonomous grids whenever the primary grid faces a fault or is being serviced. In the recent literature, heuristic and meta-heuristic techniques have been commonly used to optimize the capacity and placement of DGs and capacitor units in the distribution networks. However, the literature focuses on identifying the DGs' and capacitors' simultaneous allocation for grid-integrated distribution networks.

Among the examined papers, two studies [1,2] in the literature have retrieved the pivotal option of microgrid (MG) formation, while allocating the DGs and capacitors in on-grid radial distribution networks. Gholami et al. [1] optimally allocated the capacitor units in DG-integrated power distribution networks for both on-grid and off-grid operation modes. Under islanded operation, the authors assumed that a section of the distribution system was operated with accessible energy. Nevertheless, they did not entail any method for fully operating the installed sources. Wang and Zhong [2] optimally allocate the capacitor and DG units both on the grid and in the islanded mode of the distribution system in order to boost voltage levels at network buses. However, the authors chose different DG and capacitor banks for both operation modes, and their placements changed when the operation mode switched, which is an unrealistic strategy. In one study, Yazdavar et al. [3] optimize the capacitor and DG sizes, locations, and types for a standalone MG that is not the part of central grid. Thus, the authors ignored both the grid-integrated and islanded modes of operation. In addition, while optimally allocating the DGs alone, a method for determining the best DG locations and sizes for islanded networks has been reported in a few studies [4–6]. However, the mentioned studies hypothesized that the DG units' capacity was more than the total energy consumption (power losses and demand) and hence used the isolated operation approach to address islanded distribution networks.

Although islanded and isolated networks have nearly identical control and operational needs, they differ in planning, owing to the islanded mode's short interval of MG operation [7]. The on-grid distribution networks typically operate in autonomous mode for a brief time when the primary grid experiences a fault or is undergoing maintenance. Therefore, the installation of larger DGs can ensure the sustainable operation of the electric grid. However, it will raise the overall cost of the power system, making the electric grid more complicated. Therefore, a mechanism must be developed to utilize the same installed DGs as a standby power source till the grid's supply is restored. It will be possible if the installed DGs supply power to a specific distribution network section by dividing the entire network into several zones. Alternatively, if distribution system operators (DSOs) can persuade consumers to limit their electricity consumption, then at least a portion of each consumer's load demand can be met with available energy. Hence, it is necessary to create a mechanism to fully utilize the installed DG and capacitor units to serve a significant part of the islanded network's total load demand under supply–demand mismatch conditions. Such an approach will also allow the utilities to design effective energy conservation, load shedding, and demand-side management (DSM) schemes to improve the reliability of the islanded distribution networks when there is a supply–demand disparity (Figure 1). The systematic literature findings have been compiled in Table 1. In addition, in [8], the authors of this paper presented a detailed critical review of the studies that have been especially undertaken in the domain of simultaneous DG and capacitor allocation in the power distribution networks.

Furthermore, in the literature, various studies [9,10] emphasize the risk of a total blackout and suggest restoration techniques that leverage distributed generation to restore critical loads, especially since shedding load is a critical method for handling power imbalance conditions when the power supply is less than the demand. Load shedding enables DSOs to allocate available energy to critical loads and discard the remaining load. However, an analysis of how the distributed sources' available power generation capacity can be optimized to serve the maximum possible share of the total network load while minimizing load shedding is necessary to make an informed decision.

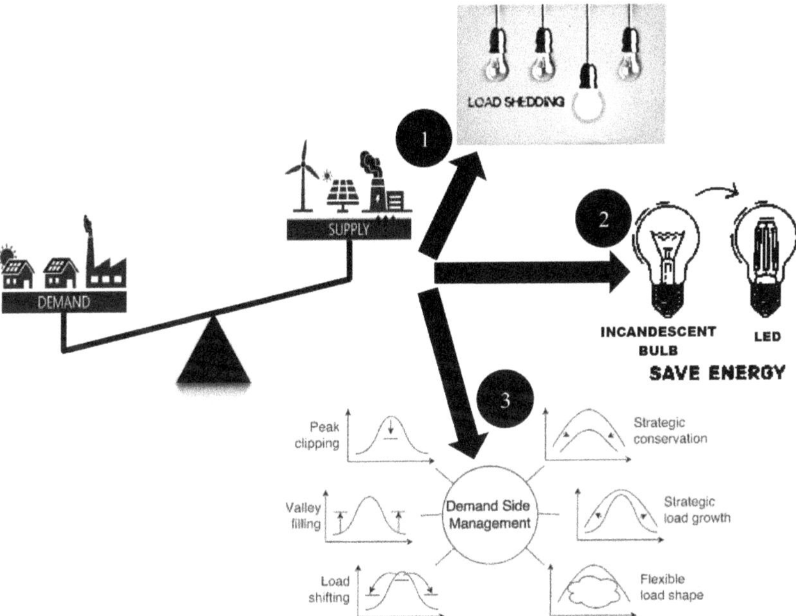


**Figure 1.** Strategies to handle the supply–demand imbalance ($P_{demand} \geq P_{supply}$) condition.

Considering this fact, this study proposes a dual-stage strategy for optimally placing DGs and capacitors during the grid-integrated mode of distribution networks and efficiently operating them for islanded operation. The optimal DG and capacitor allocation for grid-integrated operation is determined in the first part using the Jaya algorithm (JA) to decrease losses while keeping voltage deviation (*VD*) at buses within safe ranges. The second part determines the DG and capacitor combination's optimal operating point for carrying the maximum power demand of the islanded distribution network in the event of power deficiency conditions. The second part of this study proposes an analytical approach to determine the optimum power factor of the DG–capacitor combination, evaluating power losses and under-utilization of the mounted distributed power sources.

The main contributions of this study are listed below:

i.     A methodology correlating the effective utilization of the DG and capacitor units under autonomous operation mode is proposed for the scenario where the power supply is less than the power demand;

ii.    A bi-objective minimization function, incorporating the accessible power generation's under-utilization and active power loss reduction, is established to optimize the islanded distribution network's operation during power supply and demand imbalance events.

The rest of the article is arranged in four sections: The optimization problem is formulated in Section 2. The proposed planning strategy is discussed in Section 3. Section 4 presents the results and discussion, and the conclusion is drawn in Section 5.

**Table 1.** A literature review summary on optimal DG and capacitor placements in distribution systems.

| Ref# | Author(s)/Year | Objective Function(s) | Optimization Technique(s) | Operation Mode(s) |
|---|---|---|---|---|
| [11] | Ahmad Eid (2022) | Active and reactive power loss, voltage deviation, and stability | Jellyfish search algorithm | Grid-connected |
| [12] | Mouwaf et al. (2022) | Active power loss, voltage deviation, and stability | Chaotic bat algorithm | Grid-connected |
| [13] | Leghari et al. (2022) | Active power loss and voltage deviation | Best–worst optimizers | Grid-connected |
| [14] | Naderipour et al. (2021) | Costs of energy loss, installation, and maintenance | Spotted hyena optimizer | Grid-connected |
| [15] | Malik et al. (2020) | Active power loss, voltage deviation, and voltage stability index | Multiobjective particle swarm optimization | Grid-connected |
| [16] | Tolabi et al. (2020) | Active power loss, voltage stability, and operational cost | Thief and police algorithm | Grid-connected |
| [17] | Almabsout et al. (2020) | Active power loss | Enhanced genetic algorithm | Grid-connected |
| [18] | Manikanta et al. (2019) | Active power loss | Quantum-inspired evaluation algorithm | Grid-connected |
| [19] | Sambaiah and Jayabarathi (2019) | Active power loss, voltage deviation, voltage stability index, installation and maintenance costs of DGs and capacitors, and gas emissions | Salp swarm algorithm | Grid-connected |
| [20] | Lotfi et al. (2018) | Active power loss | Particle swarm optimization and genetic algorithm | Grid-connected |
| [21] | Mehmood et al. (2018) | Energy loss index, voltage enhancement index, and investment cost index | Elitist speciation-based genetic algorithm | Grid-connected |
| [22] | Dixit et al. (2017) | Active power loss | Gbest-guided artificial bee colony | Grid-connected |
| [23] | Biswas et al. (2017) | Active and reactive power loss | Multiobjective evolutionary algorithm based on the decomposition | Grid-connected |
| [24] | Ghanegaonkar and Pande (2017) | Active power loss, energy loss, and capacitor switching events | Particle swarm optimization | Grid-connected |
| [25] | Kumar et al. (2017) | Active power loss, voltage deviation, and voltage stability index | Multiobjective particle swarm optimization | Grid-connected |
| [26] | Muthukumar and Jayalalitha (2016) | Active power loss | Hybrid harmony search—particle artificial bee colony algorithm | Grid-connected |
| [27] | Khodabakhshian and Andisghae (2016) | Cost of losses | Intersect mutation differential evolution | Grid-connected |
| [28] | Jannat and Savic (2016) | Voltage deviation and installed reactive power capacity | Non–dominated sorting genetic algorithm | Grid-connected |
| [29] | Lalitha et al. (2016) | Active power loss and voltage deviation | Symbiotic organisms search | Grid-connected |
| [30] | Andebili (2016) | Investment and maintenance costs of DGs and capacitors, cost of energy loss, and risk cost | Genetic algorithm | Grid-connected |
| [31] | Ghaffarzadeh and Sadeghi (2016) | The benefit of reductions in active power loss, reactive power loss, and power purchased from the grid | Biogeography-based optimization algorithm | Grid-connected |
| [32] | Pereira et al. (2016) | Investment costs of DGs and capacitors and system's operation costs | Hybrid Tabu search-Chu-Beasly genetic algorithm | Grid-connected |
| [33] | Kayal and Chanda (2016) | Active power loss, voltage stability factor, network security index, economic index, and annual carbon dioxide emission | Non-dominated sorting multiobjective particle swarm optimization | Grid-connected |
| [34] | Khan et al. (2015) | Active power loss and voltage deviation | Binary collective animal behavior optimization algorithm | Grid-connected |
| [35] | Zeinalzadeh et al. (2015) | Active power loss, voltage stability index, and sections current index | Genetic algorithm | Grid-connected |
| [1] | Gholami et al. (2015) | Costs of energy loss, peak power loss, and capacitors | Genetic algorithm | Grid-connected and islanded |
| [36] | Jain et al. (2014) | Active power loss, reactive power loss, voltage profile, and gas emissions | Modified particle swarm optimization | Grid-connected |
| [37] | Mahari and Mahari (2014) | Active power loss | Discrete imperialistic competition algorithm | Grid-connected |
| [38] | Syed and Injeti (2014) | Active power loss | Backtracking search algorithm | Grid-connected |
| [39] | Hosseinzadehdehkordi et al. (2014) | Investment and operation costs of capacitors and cost of power/energy loss | Differential evolution | Grid-connected |
| [40] | Aman et al. (2013) | Active power loss | Particle swarm optimization | Grid-connected |
| [41] | Musa et al. (2013) | Active power loss | Particle swarm optimization | Grid-connected |
| [42] | Manafi et al. (2013) | Active power loss | Differential evolution and particle swarm optimization | Grid-connected |
| [43] | Karimi et al. (2012) | Investment and operation costs of capacitors and cost of power/energy loss | Differential evolution | Grid-connected |
| [2] | Wang and Zhong (2011) | Voltage profile | Optimal power flow | Grid-connected and islanded |
| [44] | Zou et al. (2009) | Investment costs for DG and capacitors | Particle swarm optimization | Grid-connected |
| [45] | Zou et al. (2008) | Costs of DG units, capacitors, energy loss, and distribution system reliability | Particle swarm optimization | Grid-connected |

## 2. Problem Formulation

The primary goal of this study was to efficiently utilize the installed DGs and capacitors to their full capacity so that the mounted devices can carry the highest potential share of the networks' entire load during the autonomous operation. In this context, a multi-criterion function was formulated considering the objectives of power loss minimization and reduced under-utilized capacity of the accessible active–reactive power generation. The details of the proposed objective functions are presented below.

### 2.1. Power Loss

Because of the greater line resistance to reactance ratio, high current flows, and radial configuration, the distribution system is the most unreliable. It has the highest power losses among the three components of the power system: generation, transmission, and distribution [46]. It has been noted that about 13 percent of the total generated power is wasted as the real power loss ($I^2 \times R$) in the distribution networks [47,48]. As per the statistics, the distribution system is responsible for over 70 percent of the total power system losses [49]. According to the study [50], these losses range between 33.7 percent and 64.9 percent. In this condition, the inappropriate selection of the DGs and capacitor sizes and locations will further add to the system losses. Therefore, this study's first objective is to minimize the power loss with simultaneous DG and capacitor placements.

Consider the single-line diagram of a simple two-bus radial distribution system depicted in Figure 2. Radial distribution systems use series impedances to represent power distribution lines and constant loads to establish a symmetrical system. The power flow solution for such networks may be computed by using Equations (1)–(3).

$$P_{br} = P_{b2} + P_{br,loss} \tag{1}$$

$$Q_{br} = Q_{b2} + Q_{br,loss} \tag{2}$$

$$V_{b2} = V_{b1} + I_{br}(R_{br} + jX_{br}) \tag{3}$$

where $P_{br}$ and $Q_{br}$ are the $br$ branch's active and reactive powers that flows between buses $b1$ and $b2$; $P_{b2}$ and $Q_{b2}$ are the real and reactive loads connected at bus $b2$; $P_{br,loss}$ and $Q_{br,loss}$ are the real and reactive power losses of branch $br$; $V_{b1}$ and $V_{b2}$ are the voltages across $b1$ and $b2$ buses, respectively; $R_{br}$ and $X_{br}$ are the branch resistance and reactance; and $I_{br}$ is the branch current flowing between buses $b1$ to $b2$.

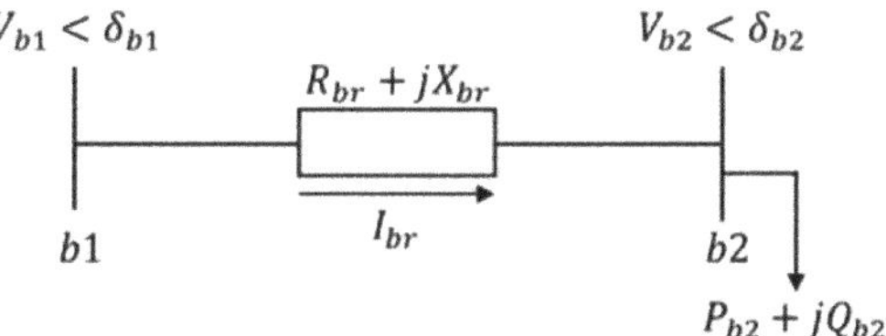


**Figure 2.** Single-line diagram of two-bus distribution network.

Equations (4) and (5) can be used to compute power losses encountered across each distribution system's branch, whereas Equation (6) can be used to calculate the amount of power dissipated over the entire distribution system.

$$P_{br,loss} = R_{br} \times \frac{(P_{b2}^2 + Q_{b2}^2)}{|V_{b2}^2|} \forall br \in (1, 2, \ldots, nb - 1) \tag{4}$$

$$Q_{br,loss} = X_{br} \times \frac{(P_{b2}^2 + Q_{b2}^2)}{|V_{b2}^2|} \forall br \in (1, 2, \ldots, nb - 1) \tag{5}$$

$$T_{loss} = \sum_{br=1}^{nb-1} P_{br,loss} + j \sum_{br=1}^{nb-1} Q_{br,loss} \tag{6}$$

where $T_{loss}$ is the distribution network's total loss, including both active and reactive power losses, and $nb$ represents the total number of buses in the distribution network.

The power flow for the radial distribution networks incorporated with DG–capacitor units in Figure 3 can be computed using Equations (7) and (8). The active and reactive power flows at the terminal node of the $b + 1$th branch can be mathematically stated as

$$P_{b+1} = \left[ P_{b,b+1} - \left( R_{b,b+1} \frac{P_{b,b+1}^2 + Q_{b,b+1}^2}{|V_b|^2} \right) - P_{b+1}^L + \alpha_{PDG} P_{b+1}^{DG} \right] \tag{7}$$

$$Q_{b+1} = \left[ Q_{b,b+1} - \left( X_{b,b+1} \frac{P_{b,b+1}^2 + Q_{b,b+1}^2}{|V_b|^2} \right) - Q_{b+1}^L + \alpha_{QDG} Q_{b+1}^{DG} + \alpha_{QCB} Q_{b+1}^{CB} \right] \tag{8}$$

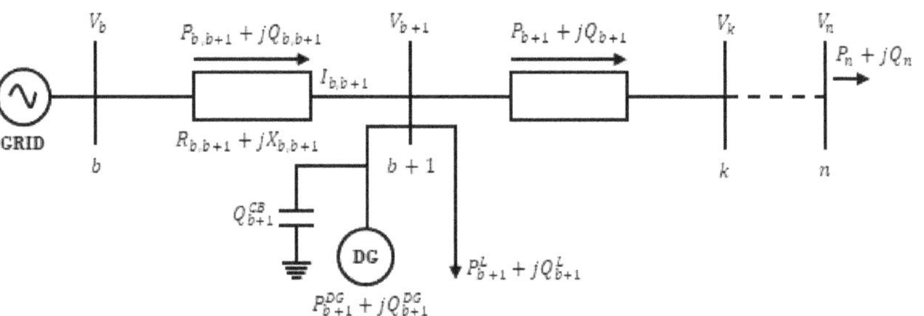


**Figure 3.** Radial distribution system with DG and capacitor banks equipped on a bus.

Since the resistance to reactance ratio in distribution networks is so high, active power losses dominate in these networks. For this reason, the active power loss minimization is chosen as a first objective for this study rather than the total loss function. In a distribution network, the branch connecting buses $b$ and $b + 1$ results in a power loss that can be computed as given in Equation (9).

$$P_{lossb,b+1} = \left| I_{b,b+1} \right|^2 R_{b,b+1} \tag{9}$$

The current flow through that branch (between buses $b$ and $b + 1$) can be calculated as Equation (10).

$$I_{b,b+1} = \sqrt{\frac{P_{b,b+1}^2 + Q_{b,b+1}^2}{|V_b|^2}} \tag{10}$$

The cumulative real power loss of distribution system is the sum of power losses across branches of the distribution network, as shown in Equation (11).

The mathematical expression of the first objective of the considered optimization problem minimizing the power loss function is presented in Equation (12).

$$P_{lossT} = \sum_{br=1}^{nb-1} P_{lossb,b+1} \tag{11}$$

$$f_1 = Min(P_{lossT}) \tag{12}$$

### 2.2. Accessible Generation Capacity's Utilization

In the event of a grid failure, DGs can fulfill the distribution network's energy needs. The deployed capacitors can aid DGs in fulfilling the reactive power demand of load. Since the installed power-generating capacity of the DGs and capacitors is typically less than the load demand, the electricity injected into the autonomous distribution network is less than the energy needs. In this instance, the electric utility can address the supply–demand disparity by lowering the electricity consumption, shedding the load, or executing a demand-side management (DSM) scheme.

However, it is critical to figure out how much of the network's load share can be supplied with the available DG and capacitor powers to employ any of these solutions. The problem is to find out how the installed DG and capacitor units can be used so as to maximize the supplied load share during islanded distribution network operation.

Power-generating sources generally operate at high power factors to minimize power losses and optimize the distribution networks' utilization capacity. Furthermore, the utilization factor of the DGs, which is the ratio of the device's actual output to the maximum achievable output (or rated capacity), can alter significantly as the operational power factor varies. However, the amount of active and reactive power generated is usually determined by the load. To put it in another way, the load on the distribution network should be assigned a quantity that allows the coupled DGs and capacitors to run at the required power factor and serve the maximum amount of energy to the load with minimal system losses. Therefore, the second minimization function chosen for this study is the under-utilization of accessible power generation, which evaluates whether or not a resource is being used to its maximum potential. The proposed objective function's mathematical formulation is given in Equations (13)–(18).

$$S_{under-utilization} = S_{available} - S_{generated} \tag{13}$$

$$f_2 = min(S_{under-utilization}) \tag{14}$$

where $S_{generated}$ is the function of $pf_{source}$, which can be computed as Equation (15):

$$S_{generated}(pf_{source}) = \sqrt{P^2_{DG,generaed} + Q^2_{Cap,generated}} \tag{15}$$

The values of $P_{DG,generated}$ and $Q_{Cap,generated}$ at any value of source power factor $pf_{set}$ can be calculated as Equations (16) and (17), respectively.

$$P_{DG,generated} = \frac{Q_{Cap,available} \times pf_{set}}{\sqrt{1 - pf^2_{set}}} \tag{16}$$

$$Q_{Cap,generated} = \sqrt{\left(\frac{P_{DG,available}}{pf_{set}}\right)^2 - (P_{DG,available})^2} \tag{17}$$

Here the $S_{under-utilization}$ value might be anywhere between 0 and $S_{available}$. A value of 0 for $S_{under-utilization}$ implies that the installed DGs/capacitors are fully operational. In contrast, $S_{under-utilization} > 0$ indicates that the DGs–capacitors' power is less than their installed capacity. As a result, the mathematical description of the proposed multi-criteria optimization problem is given as Equation (18) using a weighted sum approach:

$$F = Min(f_1 + f_2) \tag{18}$$

### 2.3. Constraints

To evade undesired results in the proposed optimization problem, some constraints are imposed that must be satisfied in order to solve the problem successfully. As there is no power supply from the central grid during the islanded operation of the distribution network, only the mounted DGs and capacitors are responsible for meeting the load's energy requirements. Since the installed units' power output often is less than the load requirement, the first constraint posed for the islanded distribution network is the maximum active and reactive power flows across the network ($P_{island,max}$, $Q_{island,max}$) that must be equal to or less than the installed generation capacity of the installed DG and capacitor units (Equations (19) and (20)). In addition, each bus in the distribution network must have a voltage ($V_b$) that should be within $\pm 5\%$ of the rated voltage ($V_{rated}$) (Equation (21)). The assumed $V_{rated}$ for this study is 1 p.u.

$$P_{island\_max} \leq \sum P_{DG} \tag{19}$$

$$Q_{island\_max} \leq \sum Q_{Cap} \tag{20}$$

$$0.95\text{p.u.} \leq V_b \leq 1.05\text{p.u.} \tag{21}$$

*2.4. Decision Variables*

For the studied optimization problem, the chosen decision (design) variable is the power factor of the DG–capacitor combination ($pf_{source}$) at which they are functioning. The upper and lower bounds set for $pf_{source}$ are 0.8 and 0.93, respectively.

*2.5. Modeling of DGs' Power Output*

For this study, the DGs' power output is assumed as deterministic. The literature has classified the DGs into four types. Type 1 DG can inject only active power, including solar photovoltaic panels, fuel cells, and battery energy storage. DGs of type 2 only supply reactive power either using capacitors or D-STATCOM. Type 3 DG usually involves the synchronous generator, which can generate both active and reactive powers. Finally, type 4 DG includes an induction generator that can generate active power while absorbing the reactive power. This study assumes a type 1 DG, such as photovoltaic panels paired with batteries, to generate controlled active power output.

*2.6. Modeling of Capacitors' Power Output*

Capacitor banks are devices that produce reactive power. They are currently available in the markets as fixed discrete types. Therefore, this paper considers the capacitors of discrete size for the optimal simultaneous allocation of capacitor and DG units in the distribution system. The capacitors available in the market are smaller units (50 kVAR), which are further integer multiples of factor $k$. Hence, the required amount of capacitor size can be determined using Equation (22).

$$Q_{max} = k \times Q_o \tag{22}$$

where $k$ is an integer. The required amount of kVAR can be assessed such as $[Q_o, 2Q_o, 3Q_o, \ldots, kQ_o]$.

## 3. Proposed Methodological Framework to Optimize the Autonomous Network's Operation

It is worth recalling that the distribution networks operate in grid-connected mode except when the grid has a fault or is being maintained. For this reason, the sizing and placement of capacitors and DGs must be chosen considering a grid-connected operation. Therefore, the first part of the proposed methodology is to obtain the optimal capacity and placement of the DGs and capacitors for grid-integrated distribution networks. In order to determine the optimal location and sizing for the DG and capacitor units, the Jaya algorithm (JA) was used. The JA, developed by Rao [51], is a stochastic heuristic optimization algorithm that only involves a single recombination step. In addition, the JA is different from most population-based optimization approaches as it uses no parameters, chosen by the user, in its execution. Only the maximum number of iterations (MaxItr) and population size (nPop) are to be specified for the JA [52,53]. In the recent literature, the efficiency of the JA has been demonstrated in various studies [54–58].

To mathematically describe the JA's implementation cycle, let $z$ be the real valued vector of decision variables composing one solution. The $z^k_{j,best}$ and $z^k_{j,worst}$ indicate the best and worst so far solutions. The $i$th candidate's $j$th decision variable in the $k$th iteration is represented as $z_{i,j}^k$. Then, at the $k$th iteration, the numerical equation used to update the candidate solution is given as represented in Equation (23).

$$Z_{i,j}^k = z_{i,j}^k + r_{1,j}^k\left(z_{j,best}^k - \left|z_{i,j}^k\right|\right) - r_{2,j}^k\left(z_{j,worst}^k - \left|z_{i,j}^k\right|\right) \tag{23}$$

where $r_{1,j}^k$ and $r_{2,j}^k$ are random numbers selected with uniform probability between 0 and 1 at the $k$th iteration. The objective function value decides whether the updated solution $Z_{i,j}^k$ can be preferred or not over the current solution $z_{i,j}^k$.

The update procedure is repeated for each solution vector of the current population at iteration k. The outputs of all updates are the input population to the subsequent iteration. The termination conditions can be either a maximum number of iterations or a convergence to the desired outcome. The flowchart presented in Figure 4 elaborates the execution cycle of the JA.

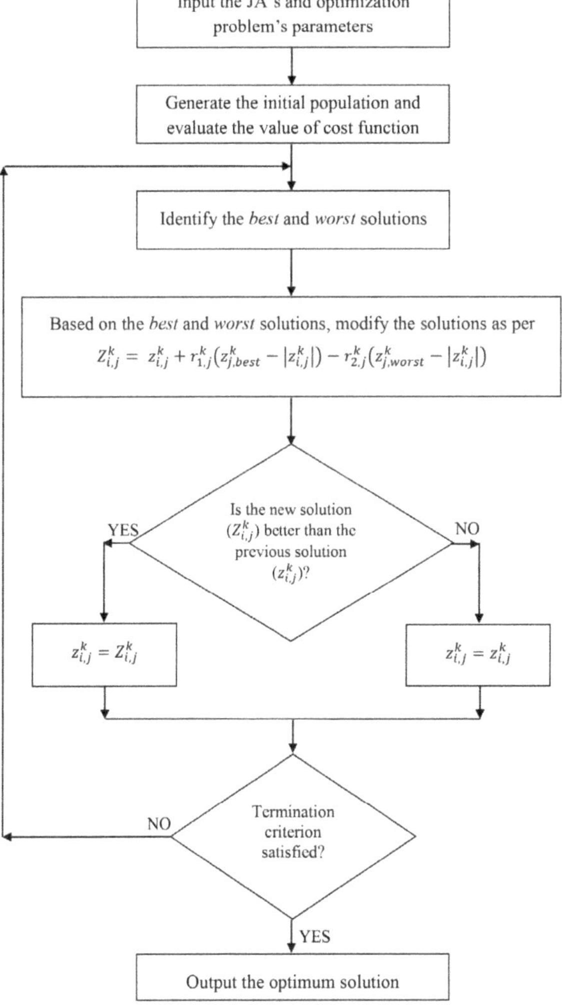


**Figure 4.** Execution cycle of the Jaya algorithm.

Once the DGs and capacitors have been installed for the on-grid distribution network operation, replacing or relocating them for the short interval of the network's autonomous operation is not viable. Hence, it is not feasible to repeatedly optimize the power-generating equipment's size and location to enhance the functioning of distribution network under the autonomous operation mode. Therefore, this study considers the DG–capacitor combination's power factor ($pf_{source}$) as the decision variable to achieve the desired goal of efficient and maximum utilization of the mounted devices under the distribution networks' independent operation.

To evaluate the effect of $pf_{source}$ on the autonomous distribution network's functioning, a detailed framework is presented whose stepwise elaboration is explained as follows.

**Step 1.** Define the base power, base voltage, load data, and line data for the selected distribution network.

**Step 2.** Calculate the starting values of the objective functions, the active power loss in this case, by running the base case load flow for all the solutions of the starting population.

**Step 3.** Set the JA's parameters, $nPop$ and $MaxItr$, and the parameters of the optimization problem, $n$ (number of design variables) and $U_d$ and $L_d$ (upper and lower bounds).

**Step 4.** Initialize the starting population with random values of the design variables.

**Step 5.** Execute the power flow to compute the value of the objective function for each search agent of the starting population.

**Step 6.** Find out the cost function values to determine the best and worst solutions.

**Step 7.** Update the solutions of the current population, based on known best and worst solutions, as per Equation (23).

**Step 8.** Carry out the power flow for each new solution vector and determine the cost function's updated values.

**Step 9.** Compare the new updated cost function values with the previous values for each solution. Adopt the new solution if it is superior to the old one; else, stick with the old solution. Create the new population replacing the old one.

**Step 10.** Stop the optimization process if the maximum iteration count is completed. Otherwise, repeat steps 6 to 9. Finally, report the obtained final optimum solutions of DG–capacitor sizes and locations.

**Step 11.** Disconnect the distribution network from the grid and identify the available maximum active and reactive power generations for the autonomous distribution network.

**Step 12.** Specify a value for the DG–capacitor combination's working power factor, $pf_{set}$ (Equation (24)).

**Step 13.** Gradually increase the active and reactive power demands of the load while keeping the source power factor constant at $pf_{set}$. Let $P_{o,i}$, $Q_{o,i}$ be the initial active and reactive power demands of load connected at bus $i$, which are assumed as 50% of $P_{DG,available}$ and $Q_{Cap,available}$.

Raise the load's active and reactive power demands gradually, $\Delta P_{load,i}$ and $\Delta Q_{load,i}$, s.t. Equation (25) holds:

$$pf_{source} = \frac{P_{DG,generated}}{\sqrt{P^2_{DG,generated} + Q^2_{Cap,generated}}} = pf_{set} \qquad (24)$$

$$\Rightarrow \forall i \in (1, 2, \ldots, nb - 1) \; \exists \; \Delta P_{load,i} \text{ and } \Delta Q_{load,i}:$$

$$\frac{\Delta P_{load,i}}{P_{load,i}} = \frac{\Delta P_{load,i+1}}{P_{load,i+1}} \text{ and } \frac{\Delta Q_{load,i}}{Q_{load,i}} = \frac{\Delta Q_{load,i+1}}{Q_{load,i+1}} \qquad (25)$$

where $P_{DG,generated}$ and $Q_{Cap,generated}$ are the active and reactive powers generated by the DG and capacitor, respectively.

**Step 14.** Stop adding to the load demand, if

$$P_{DG,generated} = P_{DG,available} \;\Big|\; Q_{Cap,generated} = Q_{Cap,available} \;\Big|\; VD_b > 5\%$$

where $VD$ is the voltage deviation ($|V_{rated} - V_b|$) at any bus $b$.

Output the values of $P_{DG,generated}$ and $Q_{Cap,generated}$:

**Step 15.** Compute the cost function value as Equation (18).

**Step 16.** For the next $pf_{set}$ value, repeat steps 12 to 15.

**Step 17.** Compare the values of the cost function acquired at each $pf_{set}$ and display the best solution value of $pf_{set}$.

A summary of the proposed methodological framework is presented in Figure 5.

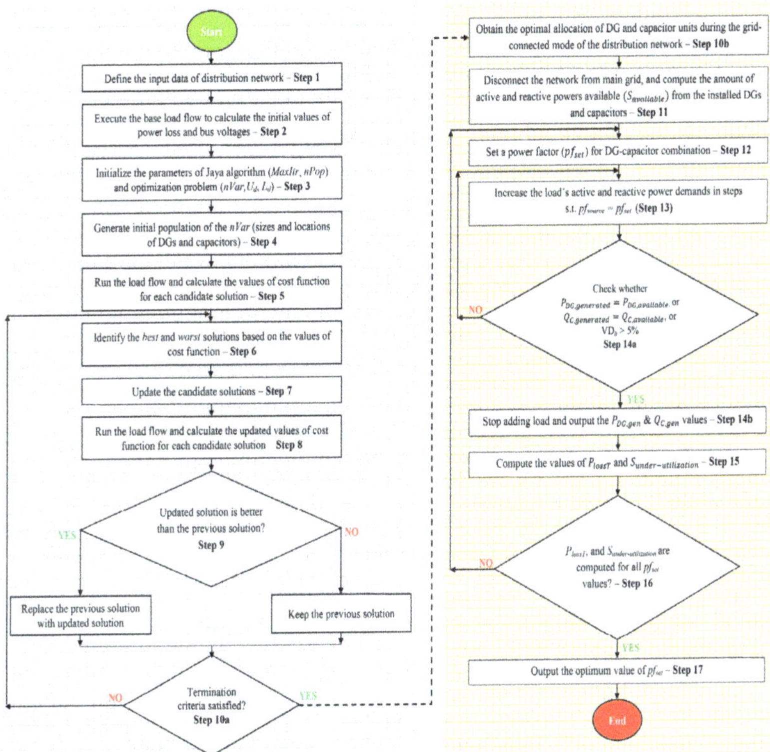


**Figure 5.** The proposed methodology to efficiently utilize the mounted distributed sources in islanded distribution networks under the power mismatch condition.

## 4. Results and Discussion

For this study, the IEEE 33- and 69-bus networks were used to implement the proposed methodological framework. Due to their limited size, these test systems have been widely employed in the literature. The detailed data of the 33-bus and 69-bus test systems are provided in [40,59]. The $pf_{source}$ value was calibrated between 0.8 and 0.93 to investigate the effect of the source power factor on the operation of the off-grid islanded distribution network. However, the comprehensive examination of the power factor effect has been provided for four specific cases, as follows:

Case 1:  DG–capacitor couple supplying power at a power factor of 0.93 (i.e., at the maximum bound).

Case 1:  DG–capacitor couple supplying power at $pf_{source}$ (also termed as $pf_{DG-Cap}$).

Case 3:  DG–capacitor couple supplying power to the load at the load power factor ($pf_{load}$).

Case 4:  DG–capacitor couple supplying power at a power factor of 0.8 (i.e., at the minimum bound).

*4.1. Optimal DG and Capacitor Unit Allocation for Grid-Integrated 33-Bus and 69-Bus Distribution Networks*

The first part of the proposed methodology is to optimize the siting and sizing of active and reactive power sources in the grid-integrated distribution networks. The obtained results for the simultaneous DG–capacitor allotment in the 33-bus and 69-bus systems are presented in Table 2. After running JA 30 times, starting from randomly generated populations, the best optimal sizes and locations of the DG and capacitor units were determined and reported. In the 33-bus test system, the optimal allocations for the DG and capacitor units were found to be 2.54 MW at bus 6 and 1.26 MVAR at bus 30. Similarly, for the 69-bus test system, bus 61 was identified as the optimal location for both the DG and capacitor units, with a capacity of 1.8285 MW and 1.3 MVAR.

**Table 2.** The results of simultaneous DG–capacitor allocation in 33-bus and 69-bus test systems.

| Parameters | 33-Bus Test System | 69-Bus Test System |
| --- | --- | --- |
| DG size in MW (bus location) | 2.54 (bus 6) | 1.8285 (bus 61) |
| Capacitor size in MVAR (bus location) | 1.26 (bus 30) | 1.3 (bus 61) |
| Power losses before DG and capacitor integration, MW | 211 | 225 |
| Power losses after DG and capacitor integration, MW | 58.452 | 23.171 |
| Minimum bus voltage (p.u.) before DG and capacitor integration, @ bus | 0.9038 (bus 18) | 0.9092 (bus 65) |
| Minimum bus voltage (p.u.) after DG and capacitor integration, @ bus | 0.9538 (bus 18) | 0.9725 (bus 27) |
| Available power generation from DG and capacitor in MVA | 2.835 | 2.244 |
| Distribution networks' total load demand in MVA | 4.369 | 4.660 |
| Available generation from distributed power units (percentage of network load) | 64.90% | 48.15% |

The active power loss minimization ($f_1$) is considered the primary objective function to attain the optimal DG and capacitor values (sizing and siting). The voltage magnitude at the network buses (Equation (21)) and the power flow limit through the wires (Equations (26) and (27)) are the non-equality and equality constraints posed for this optimization problem.

$$P_{SS} + \sum P_{DG} = \sum P_{load} + \sum P_{loss} \tag{26}$$

$$Q_{SS} + \sum Q_{Cap} = \sum Q_{load} + \sum Q_{loss} \tag{27}$$

where $P_{SS}$ and $Q_{SS}$ are the sub-station's active and reactive power supplies; $P_{DG}$ is the DG's active power injected into the distribution network; $Q_{Cap}$ is the capacitor's reactive injected into the distribution network; $P_{load}$ and $Q_{load}$ are the active and reactive power loads connected to each bus of the distribution network; and $P_{loss}$ and $Q_{loss}$ are the active and reactive power losses encountered across each network branch.

The four decision variables chosen for this complex optimization problem are DG size and location ($P_{DG}, N_{DG}$) and capacitor size and location ($Q_{Cap}, N_{Cap}$). These decision variables' lower and upper bounds are set as $P_{DG} = [0 \leq P_{DG} \leq P_{load}]$ and $Q_{Cap} = [0 \leq Q_{Cap} \leq Q_{load}]$, except for bus 1 (i.e., the slack bus), DGs and capacitors can be located on any of the network buses, $N_{DG} = [2 \leq N_{DG} \leq n_b]$, and $N_{Cap} = [2 \leq N_{Cap} \leq n_b]$.

*4.2. 33-Bus Autonomous Distribution Network*

The 33-bus network's total load is 4.369 MVA with a 0.85 power factor ($pf_{load}$), whereas the total power available from the mounted DG and capacitor is 2.835 MVA with a 0.896 power factor ($pf_{source}$). According to the data shown above, the available power of the attached DG and capacitor is 64.90 percent of the total need for electricity. It is important to note that the $pf_{source}$ value is dependent on the mounted DG and capacitor units' rated outputs; thus, it may differ accordingly. In the studies that focus solely on the optimal DG placement, the $pf_{source}$ value will be defined by the stated optimal real and reactive power values produced by DG alone. The optimal allocation of the DG and capacitor units in the off-grid 33-bus islanded distribution network is shown in Figure 6.

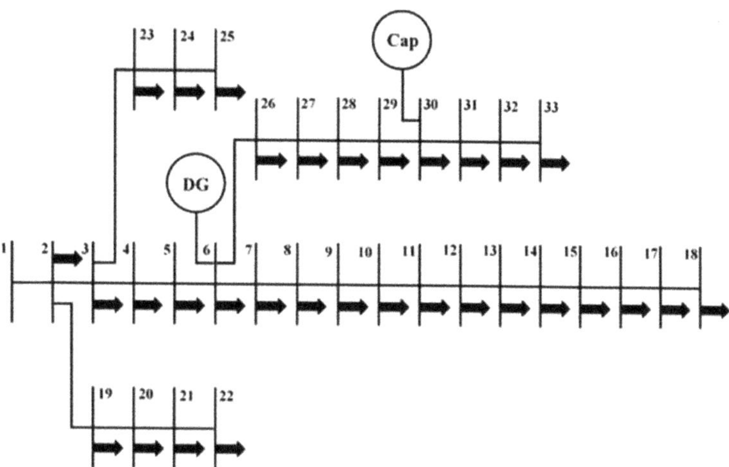

**Figure 6.** The 33-bus autonomous network with mounted DGs.

In the first case, the DG and capacitor combination's operating power factor is set to 0.93 ($> pf_{source}$). In this situation, the total load handled by the DG–capacitor pair is just 2.677 MVA or 61.27 percent of the distribution system's entire load demand. The aggregate active power losses of 0.047 MW are reported under this situation, with bus 18 having the lowest bus voltage of 0.975 p.u. The DG and capacitor collectively produce 2.731 MVA power, which is 62.3 percent of the total power demand of the 33-bus network. As a result, assuming the installed DG and capacitor run at a power factor of 0.93, they can produce enough energy to meet 62.3 percent of the network's total load in autonomous mode. The accessible generation capacity of 3.99 percent remains unutilized in this situation, whereas the operating efficiency evaluated in this case is 98.35 percent.

In case 2, the DG–capacitor combination's operating power factor is set to 0.896, based on the installed units' rated values (2.54 MW and 1.26 MVAR). The DG and capacitor operate at full capacity in this state, and no electricity generation is left over. The DG and the capacitor collectively carry 2.782 MVA of load, accounting for 63.68 percent of the total demand of the islanded network. With a minimum bus voltage of 0.974 p.u. obtained at bus 18, an active power loss of 0.045 MW is recorded in this case. The power output produced by the DG and capacitor in this scenario is 2.835 MVA or 64.90 percent of the total load. In this case, the network's operating efficiency is assessed to be 98.13 percent, which is somewhat higher than in the prior case. In addition, in case 2, the power output obtained from the mounted power-generating units is higher.

In the third case, the $pf_{source}$ was kept at 0.85, which was the same as $pf_{load}$. The load capacity collectively supplied by the DG and capacitor in this state is 2.352 MVA or 53.83 percent of the entire network load. In this circumstance, the power loss accounted in the islanded network is 0.034 MW, and bus 18 has the lowest voltage of 0.979 p.u. The total generated power is obtained as 2.392 MVA that is 54.75 percent of the 33-bus distribution network's peak demand. As a result, 15.63 percent of the generation capacity is under-utilized, with an operating efficiency of 98.33 percent in this situation. However, unlike prior cases, the installed units offer less power that is marginally more than 50% of the whole power demand.

For the proposed case 4, the $pf_{source}$ is tuned to the minimum 0.8 value that is lower than both $pf_{source}$ and $pf_{load}$. In this circumstance, the combined load handled by the DG and capacitor is 2.067 MVA or 47.31 percent of the overall load. With a minimum voltage of 0.981 p.u., this arrangement produces a real power loss of 0.028 MW. Bus 18 observed the minimal bus voltage once again, just like in previous cases. The total electricity produced by the DG and capacitor, in this case, is 2.1 MVA, 48.07 percent of the total demand, and the network's operating efficiency is 98.43 percent. It is the worst-case scenario among the four proposed cases, as the under-utilization of the power-generating devices reaches 25.93 percent, i.e., during the independent functioning of the distribution network, more than a quarter of the available power-producing capacity is left unused. Table 3 presents the statistical summary of the four cases examined; moreover, the graphical representation of the comparative analysis is provided in Figures 7 and 8.

**Table 3.** Results for the 33-bus autonomous distribution network.

| Quantity | Case 1 | Case 2 | Case 3 | Case 4 |
|---|---|---|---|---|
| Total power collectively produced by DG and capacitor in MVA | 2.731 | 2.835 | 2.392 | 2.100 |
| Load's total power consumption in MVA | 2.677 | 2.782 | 2.352 | 2.067 |
| Operating power factor of the DG–capacitor combination | 0.93 | 0.896 ($pf_{source}$) | 0.85 ($pf_{load}$) | 0.8 |
| Real power loss in MW | 0.047 | 0.045 | 0.034 | 0.028 |
| Operating efficiency of the islanded distribution network | 98.02% | 98.13% | 98.33% | 98.43% |
| Total power produced by DG and capacitor units (percentage of network load) | 62.30% | 64.89% | 54.75% | 48.07% |
| The load portion supplied with accessible power generation (percentage of network load) | 61.27% | 63.68% | 53.83% | 47.31% |
| Under-utilization of mounted distributed generation capacity (percentage of network load) | 2.59% | 0.0% | 10.14% | 16.82% |
| Under-utilization of mounted distributed generation capacity (percentage of available power generation) | 3.99% | 0.0% | 15.63% | 25.93% |
| Minimum voltage in p.u. (@ bus) | 0.975 (bus 18) | 0.974 (bus 18) | 0.979 (bus 18) | 0.981 (bus 18) |

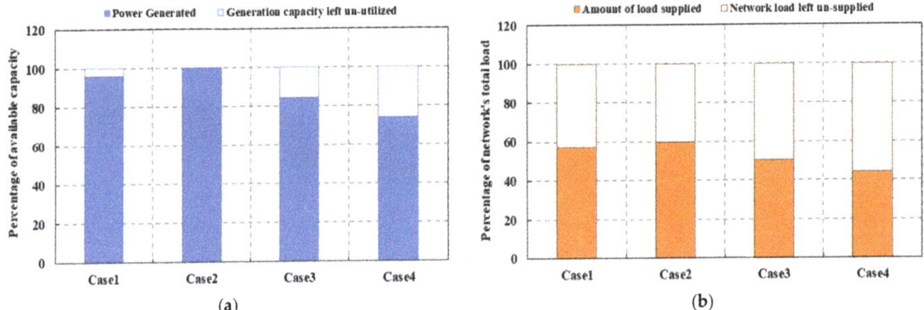


**Figure 7.** Performance comparison of the 33-bus autonomous distribution network at different pfsource. (**a**) Installed power generation capacity's % utilization and (**b**) network's % load share delivered with accessible power.

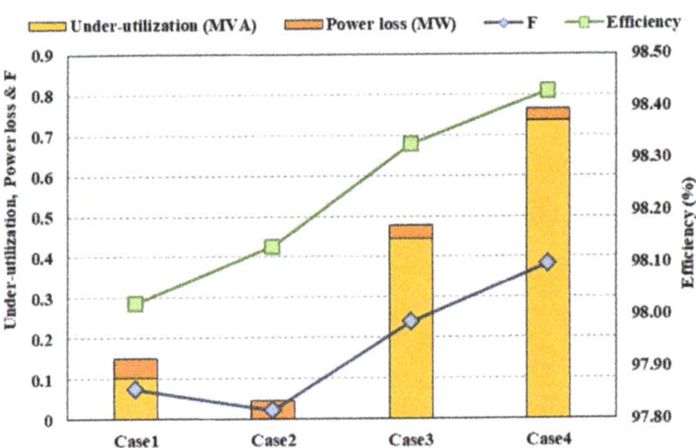


**Figure 8.** The observed efficiency (%), under-utilization (MVA), power losses (MW), and objective function (F) values in four cases for 33-bus autonomous distribution network.

The findings illustrate that keeping the DG–capacitor combination's power factor close to $pf_{source}$ is the most efficient approach to operate the islanded distribution networks.

The $pf_{source}$ value can be computed from the DG and capacitor's optimal ratings obtained during the grid-integrated network operation. At this power factor, both the DG and capacitor supply power to their full potentials. It can be seen from the results that the operating efficiency of the islanded distributed network achieved in cases 3 and 4 is marginally higher than that obtained in cases 1 and 2 because the mounted active and reactive power-generating components are not utilized to their full capability in the latter two cases. Therefore, the reduced power flows through the distribution lines cause a considerable reduction in power loss. While these cases have lower power losses than cases 1 and 2, the difference is insignificant compared with the additional power delivered by the mounted units in the first two cases. In addition to the distribution network's operating efficiency, the impact of the DG–capacitor pair's operational power factor on the voltage profile of the islanded distribution networks has also been analyzed, as shown in Figure 8. Since the distribution network is less loaded in cases 3 and 4 due to the opted $pf_{set}$ values, the bus voltages attained for these cases show a modest enhancement. In four cases, the minimum bus voltages for the 33-bus islanded network were recorded at bus 18, ranging from 0.974 to 0.981. Despite the fact that the power grid was not feeding energy during the islanded operation, the bus voltages remained within the defined limits even when the mounted devices were fully loaded. This is owing to the optimum DG and capacitor positions attained during on-grid operation, with the DG located at the central position of bus 6 and the capacitor's placement at bus 30 near the end point buses, which allows the voltage to be maintained within a set margin (i.e., ±5% of $V_{rated}$) on the weakest buses. Furthermore, even though the installed power-generating units can carry more load, the proposed strategic approach's operating mechanism prevents the islanded network from being loaded beyond the limit that drops the voltage below the predefined margin, as specified in step 14 of Section 3. For this reason, the end buses (buses 18, 25, and 33) have voltages that are well within the operational limits, as seen in Figure 9.

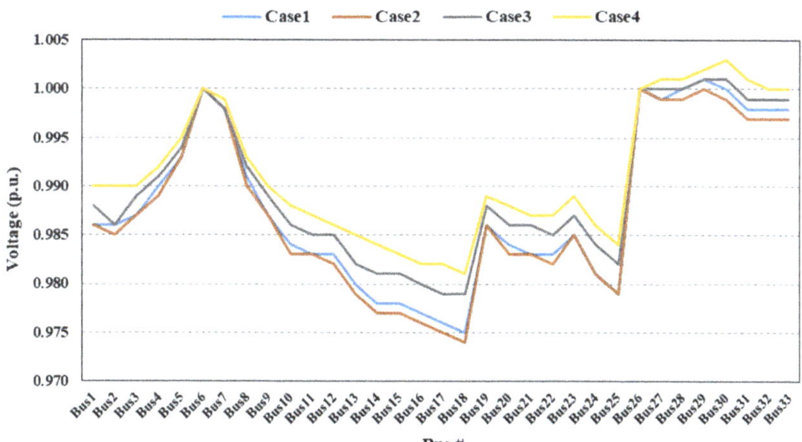


**Figure 9.** Voltage profiles of the 33-bus autonomous distribution network attained in each case.

### 4.3. 69-Bus Autonomous Distribution Network

For the 69-bus distribution network, the optimal siting and sizing of the single DG and single capacitor units determined during the grid-integrated mode using JA were 1.8285 MW (at bus 61) and 1.3 MVAR (at bus 61), respectively. Hence, the total accessible electricity from the mounted distributed resources is 2.244 MVA with a power factor of 0.815 ($pf_{source}$). The optimal allocation of the distributed active–reactive power units in the 69-bus autonomous distribution network is shown in Figure 10. The IEEE 69-bus radial distribution network's active power load is 3.80219 MW, and the reactive power load is j2.6946 MVAR (total 4.66 MVA) with a $pf_{load}$ of 0.816. The computed percentage of the power generation share is 48.15% of the total demand from the presented DG and capacitor values. To examine how this accessible power can be utilized to the maximum potential, the studied four power factor values for the 69-bus network are 0.93, 0.815 ($pf_{source}$), 0.816 ($pf_{load}$), and 0.8. It must be noted that unlike the 33-bus distribution network, the $pf_{source}$ value for the 69-bus network is less than the $pf_{load}$. Here the $pf_{source}$ value is close to the minimum bound of the opted range of power factors, i.e., 0.8. For the 33-bus network, the $pf_{source}$ value was close to the upper limit (0.93) of the selected range of the operating power factors. Hence, the 69-bus distribution network's power factor values also offer the sensitivity analysis of the proposed methodology for the autonomous distribution networks by examining their performance under uncertain input conditions. The quantities measured for the 69-bus distribution system are presented in Table 4.

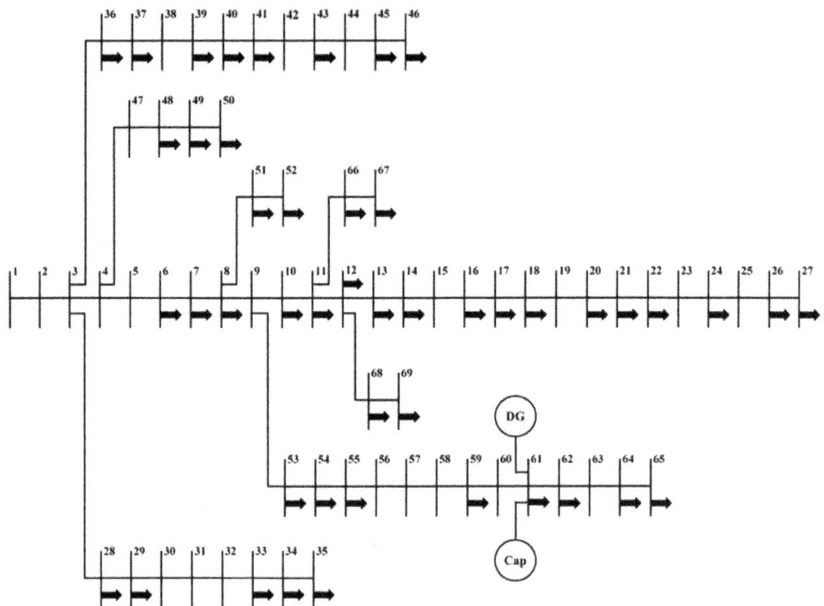

**Figure 10.** The 69-bus autonomous distribution network with installed distributed sources.

For the 69-bus autonomous network, although the DG–capacitor pair's operating rated power factor is low, the results revealed that the installed devices produce maximum power when operating at or close to the rated $pf_{source}$. Dissimilar to the 33-bus network, where the minimum power generation was observed in case4, the minimum amount of generated power is acquired in case 1 when the mounted DG and capacitor were generating power at the 0.93 $pf_{source}$ (Figure 11). This clearly shows that the best power factor for the autonomous distribution networks lies close to the $pf_{source}$. Any other power factor farther from this range will cause a significant dip in the under-utilization of the mounted devices' power-generating potential.

**Table 4.** Results for the 69-bus autonomous distribution network.

| Quantity | Case 1 | Case 2 | Case 3 | Case 4 |
|---|---|---|---|---|
| Total power collectively produced by DG and capacitor in MVA | 1.966 | 2.244 | 2.237 | 2.167 |
| Load's total power consumption in MVA | 1.923 | 2.190 | 2.182 | 2.117 |
| Operating power factor of the DG–capacitor combination | 0.93 | 0.815 ($pf_{source}$) | 0.816 ($pf_{load}$) | 0.8 |
| Real power loss in MW | 0.0409 | 0.0532 | 0.0531 | 0.0496 |
| Operating efficiency of the islanded distribution network | 97.81% | 97.59% | 97.59% | 97.69% |
| Total power produced by DG and capacitor units (percentage of network load) | 42.19% | 48.15% | 48.09% | 46.50% |
| The load portion supplied with accessible power generation (percentage of network load) | 41.27% | 47.00% | 46.93% | 45.43% |
| Under-utilization of mounted distributed generation capacity (percentage of network load) | 5.88% | 0.00% | 0.06% | 1.65% |
| Under-utilization of mounted distributed generation capacity (percentage of available power generation) | 12.23% | 0.00% | 0.13% | 3.43% |
| Minimum voltage in p.u. (@ bus) | 0.958 (buses 17–27) | 0.954 (buses 19–27) | 0.954 (buses 20–27) | 0.956 (buses 19–27) |

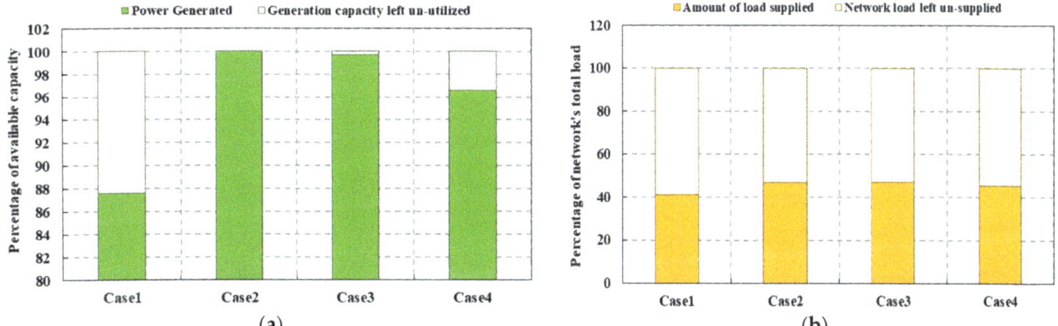


**Figure 11.** Performance comparison of the 69-bus autonomous distribution network at different pfsource. (**a**) Installed power generation capacity's percent utilization. (**b**) Network's percent load share delivered with accessible power.

Conversely, for the 33-bus and 69-bus independent distribution networks, the highest power losses were produced at the power factors close to the $pf_{source}$. Undoubtedly, this is because of the maximum power that flows at the $pf_{source}$ values. The graphical illustration of the obtained results in the studied four cases for the 69-bus autonomous distribution network is presented in Figure 12. In addition, the graphical illustration of the network's voltage profiles obtained in four cases is presented in Figure 13. The voltage profiles show that the voltages at bus 61 and neighboring buses lie close to the rated value of 1 p.u., which are due to the DG and capacitor units' allocation at bus 61. In addition, because of the tactically placed DG–capacitor modules and proposed mechanism of the methodological framework, the voltage profiles lie within acceptable bounds in all cases, even if the DG–capacitor operates to its full capacity. Unlike the 33-bus distribution network, where the lowest voltage value was observed at a single bus, the minimum bus voltages for the 69-bus islanded network were observed across several buses (from bus 17 to bus 27), ranging from 0.954 to 0.958 in four cases.

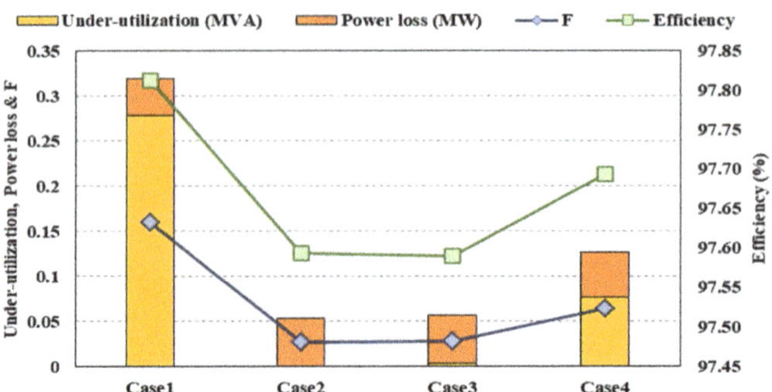

**Figure 12.** The observed efficiency (%), under-utilization (MVA), power losses (MW), and function F values in four cases for the 69-bus autonomous distribution network.

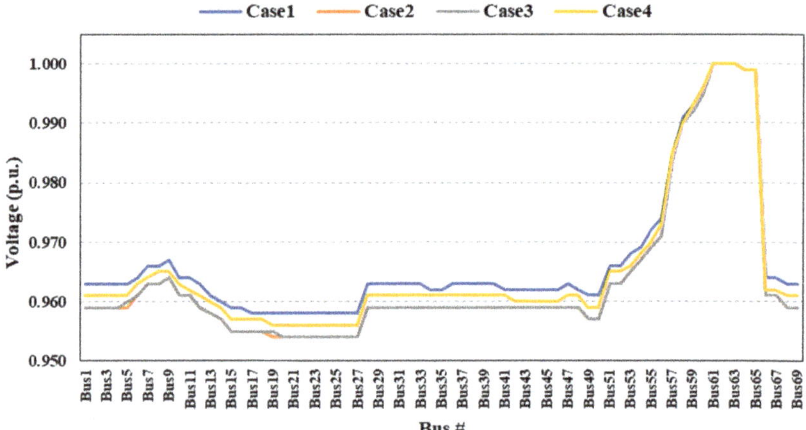

**Figure 13.** Voltage profiles of autonomous 69-bus distribution network.

## 5. Conclusions and Future Work

This paper introduces a strategic planning methodology for the efficient utilization of installed distributed generation units in distribution networks' during their islanded operation, due to a fault. Unlike the existing studies, the event of power supply–demand mismatch has been considered to analyze the extent to which the energy needs of the islanded network can be fulfilled by mounting dispersed power-generating units.

A multi-criterion bi-objective function has been developed, including minimization objectives of active power loss reduction and decrement in accessible power generation's under-utilization.

Four cases were devised to see how much of the available DG-capacitor capacity was under-utilized at the different operational power factor values of the DG-capacitor combination.

The study found that the non-utilization of available power generation capacity in the 33-bus distribution network can vary by up to 25.93% across power factors ranging from 0.8 to 0.93. Similarly, the under-utilization in the 69-bus distribution network can deviate up to 12.23%. Furthermore, the obtained results show that deployed DGs and capacitors operating at a source power factor (i.e., power factor incurred from their optimal capacities) can meet the energy requirements of a larger portion of the network load.

This is because the load demand handled by DGs and capacitors is less than the highest possible share at any other power factor. Conversely, operating the DG-capacitor combination close to the source power factor causes an increase in power loss. However, compared with the increased load served with the available generation, the increase in power loss is negligible.

The efficient employment of the active and reactive power sources in autonomous mode during power deficit events necessitates developing effective demand-side management (DSM) schemes for the electrical power systems. The establishment of any such comprehensive plan is beyond the scope of this study. However, this research will pave the way for such future investigations. Henceforward, the proposed methodological framework can also be applied to different DGs, evaluating the uncertainties of power generation.

**Author Contributions:** Conceptualization, Z.H.L., M.Y.H. and D.M.S.; data curation, Z.H.L., M.Y.H., D.M.S. and L.K.; writing—original draft preparation, Z.H.L., M.Y.H., D.M.S., M.K., Q.T.T. and E.R.S.; writing—review and editing, Z.H.L., M.Y.H., D.M.S., M.K., Q.T.T. and E.R.S.; supervision, M.Y.H. and D.M.S. All authors have read and agreed to the published version of the manuscript.

**Funding:** This research received no external funding.

**Data Availability Statement:** Data is contained within the article in the form of results provided in tables. No new data were created or analyzed in this study. Therefore, data sharing is not applicable to this article.

**Acknowledgments:** The first author wishes to express his gratitude to the HEC (Higher Education Commission), Pakistan, and MUET (Mehran University of Engineering & Technology), Jamshoro, Sindh, Pakistan, for providing financial assistance in pursuing higher education. Furthermore, all authors wish to extend their appreciation to the UTM (Universiti Teknologi Malaysia) for its technical services.

**Conflicts of Interest:** The authors declare no conflict of interest.

# References

1.  Gholami, R.; Shahabi, M.; Haghifam, M. An efficient optimal capacitor allocation in DG embedded distribution networks with islanding operation capability of micro-grid using a new genetic based algorithm. *Int. J. Electr. Power Energy Syst.* **2015**, *71*, 335–343. [CrossRef]
2.  Wang, M.; Zhong, J. A novel method for distributed generation and capacitor optimal placement considering voltage profiles. In Proceedings of the 2011 IEEE Power and Energy Society General Meeting, Detroit, MI, USA, 24–28 July 2011; pp. 1–6.
3.  Yazdavar, A.H.; Shaaban, M.F.; El-Saadany, E.F.; Salama, M.M.A.; Zeineldin, H.H. Optimal planning of distributed generators and shunt capacitors in isolated microgrids with nonlinear loads. *IEEE Trans. Sustain. Energy* **2020**, *11*, 2732–2744. [CrossRef]
4.  Kirthiga, M.V.; Daniel, S.A.; Gurunathan, S. A methodology for transforming an existing distribution network into a sustainable autonomous micro-grid. *IEEE Trans. Sustain. Energy* **2013**, *4*, 31–41. [CrossRef]
5.  Anand, M.P.; Ongsakul, W.; Singh, J.G.; Sudhesh, K.M. Optimal allocation and sizing of distributed generators in autonomous microgrids based on LSF and PSO. In Proceedings of the 2015 International Conference on Energy Economics and Environment (ICEEE), Greater Noida, India, 27–28 March 2015; pp. 1–6.
6.  Jamian, J.J.; Mustafa, M.W.; Mokhlis, H.; Baharudin, M.A.; Abdilahi, A.M. Gravitational search algorithm for optimal distributed generation operation in autonomous network. *Arab. J. Sci. Eng.* **2014**, *39*, 7183–7188. [CrossRef]
7.  Farag, H.E.Z.; El-Saadany, E.F. Optimum shunt capacitor placement in multimicrogrid systems with consideration of islanded mode of operation. *IEEE Trans. Sustain. Energy* **2015**, *6*, 1435–1446. [CrossRef]
8.  Leghari, Z.H.; Kumar, M.; Shaikh, P.H.; Kumar, L.; Tran, Q.T. A Critical Review of Optimization Strategies for Simultaneous Integration of Distributed Generation and Capacitor Banks in Power Distribution Networks. *Energies.* **2022**, *15*, 8258. [CrossRef]
9.  Vita, V.; Fotis, G.; Pavlatos, C.; Mladenov, V. A New Restoration Strategy in Microgrids after a Blackout with Priority in Critical Loads. *Sustainability* **2023**, *15*, 1974. [CrossRef]
10.  Fotis, G.; Vita, V.; Maris, T.I. Risks in the European Transmission System and a Novel Restoration Strategy for a Power System after a Major Blackout. *Appl. Sci.* **2023**, *13*, 83. [CrossRef]
11.  Eid, A. Cost-based analysis and optimization of distributed generations and shunt capacitors incorporated into distribution systems with nonlinear demand modeling. *Expert Syst. Appl.* **2022**, *198*, 116844. [CrossRef]
12.  Mouwafi, M.T.; El-sehiemy, R.A.; El-ela, A.A.A. A two-stage method for optimal placement of distribution generation units and capacitors in distribution systems. *Appl. Energy* **2022**, *307*, 118188. [CrossRef]
13.  Leghari, Z.H.; Hussain, S.; Memon, A.; Memon, A.H.; Baloch, A.A. Parameter-Free Improved Best-Worst Optimizers and Their Application for Simultaneous Distributed Generation and Shunt Capacitors Allocation in Distribution Networks. *Int. Trans. Electr. Energy Syst.* **2022**, *2022*, 1–31. [CrossRef]
14.  Naderipour, A.; Malek, Z.A.; Hajivand, M. Spotted hyena optimizer algorithm for capacitor allocation in radial distribution system with distributed generation and microgrid operation considering different load types. *Sci. Rep.* **2021**, *11*, 2728. [CrossRef] [PubMed]

15. Malik, M.Z.; Kumar, M.; Soomro, A.M.; Baloch, M.; Gul, M.; Farhan, M.; Kaloi, G.S. Strategic planning of renewable distributed generation in radial distribution system using advanced MOPSO method. *Energy Rep.* **2020**, *6*, 2872–2886. [CrossRef]
16. Tolabi, H.B.; Ara, A.L.; Hosseini, R. A new thief and police algorithm and its application in simultaneous reconfiguration with optimal allocation of capacitor and distributed generation units. *Energy* **2020**, *203*, 117911. [CrossRef]
17. Almabsout, E.; El-Sehiemy, R.; An, O.; Bayat, O. A hybrid local search-genetic algorithm for simultaneous placement of DG units and shunt capacitors in radial distribution systems. *IEEE Access* **2020**, *8*, 54465–54481. [CrossRef]
18. Manikanta, G.; Mani, A.; Singh, H.P.; Chaturvedi, D.K. Simultaneous placement and sizing of DG and capacitor to minimize the power losses in radial distribution network. In *Soft Computing: Theories and Applications*; Springer: Berlin/Heidelberg, Germany, 2019; pp. 605–618.
19. Sambaiah, K.S.; Jayabarathi, T. Optimal allocation of renewable distributed generation and capacitor banks in distribution systems using salp swarm algorithm. *Int. J. Renew. Energy Res.* **2019**, *9*, 96–107.
20. Lotfi, H.; Elmi, M.B.; Saghravanian, S. Simultaneous placement of capacitor and DG in distribution networks using particle swarm optimization algorithm. *Int. J. Smart Electr. Eng.* **2018**, *7*, 35–41.
21. Mehmood, K.K.; Kim, C.-H.; Khan, S.U.; Haider, Z. Unified Planning of Wind Generators and Switched Capacitor Banks: A Multiagent Clustering-Based Distributed Approach. *IEEE Trans. Power Syst.* **2018**, *33*, 6978–6988. [CrossRef]
22. Dixit, M.; Kundu, P.; Jariwala, H.R. Incorporation of distributed generation and shunt capacitor in radial distribution system for techno-economic benefits. *Eng. Sci. Technol. Int. J.* **2017**, *20*, 482–493. [CrossRef]
23. Biswas, P.P.; Mallipeddi, R.; Suganthan, P.N.; Amaratunga, G.A.J. A multiobjective approach for optimal placement and sizing of distributed generators and capacitors in distribution network. *Appl. Soft Comput. J.* **2017**, *60*, 268–280. [CrossRef]
24. Ghanegaonkar, S.P.; Pande, V.N. Optimal hourly scheduling of distributed generation and capacitors for minimisation of energy loss and reduction in capacitors switching operations. *IET Gener. Transm. Distrib.* **2017**, *11*, 2244–2250. [CrossRef]
25. Mahesh, K.; Nallagownden, P.; Elamvazuthi, I. Optimal placement and sizing of renewable distributed generations and capacitor banks into radial distribution systems. *Energies* **2017**, *10*, 1–24.
26. Muthukumar, K.; Jayalalitha, S. Integrated approach of network reconfiguration with distributed generation and shunt capacitors placement for power loss minimization in radial distribution networks. *Appl. Soft Comput. J.* **2017**, *52*, 1262–1284.
27. Khodabakhshian, A.; Andishgar, M.H. Simultaneous placement and sizing of DGs and shunt capacitors in distribution systems by using IMDE algorithm. *Int. J. Electr. Power Energy Syst.* **2016**, *82*, 599–607. [CrossRef]
28. Jannat, M.B.; Savić, A.S. Optimal capacitor placement in distribution networks regarding uncertainty in active power load and distributed generation units production. *IET Gener. Transm. Distrib.* **2016**, *10*, 3060–3067. [CrossRef]
29. Lalitha, M.P.; Babu, P.S.; Adivesh, B. Optimal distributed generation and capacitor placement for loss minimization and voltage profile improvement using symbiotic organisms search algorithm. *Int. J. Electr. Eng.* **2016**, *9*, 249–261.
30. Rahmani-Andebili, M. Simultaneous placement of DG and capacitor in distribution network. *Electr. Power Syst. Res.* **2016**, *131*, 1–10. [CrossRef]
31. Ghaffarzadeh, N.; Sadeghi, H. A new efficient BBO based method for simultaneous placement of inverter-based DG units and capacitors considering harmonic limits. *Int. J. Electr. Power Energy Syst.* **2016**, *80*, 37–45. [CrossRef]
32. Pereira, B.R.; Martins Da Costa, G.R.M.; Contreras, J.; Mantovani, J.R.S. Optimal distributed generation and reactive power allocation in electrical distribution systems. *IEEE Trans. Sustain. Energy* **2016**, *7*, 975–984. [CrossRef]
33. Kayal, P.; Chanda, C.K. Strategic approach for reinforcement of intermittent renewable energy sources and capacitor bank for sustainable electric power distribution system. *Int. J. Electr. Power Energy Syst.* **2016**, *83*, 335–351. [CrossRef]
34. Khan, N.; Ghoshal, S.; Ghosh, S. Optimal allocation of distributed generation and shunt capacitors for the reduction of total voltage deviation and total line loss in radial distribution systems using binary collective animal behavior optimization algorithm. *Electr. Power Compon. Syst.* **2015**, *43*, 119–133. [CrossRef]
35. Zeinalzadeh, A.; Mohammadi, Y.; Moradi, M.H. Optimal multi objective placement and sizing of multiple DGs and shunt capacitor banks simultaneously considering load uncertainty via MOPSO approach. *Int. J. Electr. Power Energy Syst.* **2015**, *67*, 336–349. [CrossRef]
36. Jain, N.; Singh, S.; Srivastava, S. PSO based placement of multiple wind DGs and capacitors utilizing probabilistic load flow model. *Swarm Evol. Comput.* **2014**, *19*, 15–24. [CrossRef]
37. Mahari, A.; Mahari, A. Optimal DG and capacitor allocation in distribution systems using DICA. *J. Eng. Sci. Technol.* **2014**, *9*, 641–656.
38. Syed, M.S.; Injeti, S. Simultaneous optimal placement of DGs and fixed capacitor banks in radial distribution systems using BSA optimization. *Int. J. Comput. Appl.* **2014**, *108*, 28–35.
39. Hosseinzadehdehkordi, R.; Nasab, M.E.; Hossein, S.; Karimi, M.; Farhadi, P. Optimal sizing and siting of shunt capacitor banks by a new improved differential evolutionary algorithm. *Int. Trans. Electr. Energy Syst.* **2013**, *24*, 1089–1102. [CrossRef]
40. Aman, M.M.; Jasmon, G.B.; Solangi, K.H.; Bakar, A.H.A.; Mokhlis, H. Optimum simultaneous DG and capacitor placement on the basis of minimization of power losses. *Int. J. Comput. Electr. Eng.* **2013**, *5*, 516–522. [CrossRef]
41. Musa, I.; Zahawi, B.; Gadoue, S.M. Integration of induction generator based distributed generation and shunt compensation capacitors in power distribution networks. In Proceedings of the 4th International Conference on Power Engineering, Energy and Electrical Drives, Istanbul, Turkey, 13–17 May 2013; pp. 1105–1109.

42. Manafi, H.; Ghadimi, N.; Ojaroudi, M.; Farhadi, P. Optimal Placement of Distributed Generations in Radial Distribution Systems Using Various PSO and DE Algorithms. *Electron. Electr. Eng.* **2013**, *19*, 53–57. [CrossRef]
43. Karimi, M.; Shayeghi, H.; Banki, T.; Farhadi, P.; Ghadimi, N. Solving optimal capacitor allocation problem using DE algorithm in practical distribution networks. *Prz. Elektrotechniczny* **2012**, *88*, 90–93.
44. Zou, K.; Agalgaonkar, A.P.; Muttaqi, K.M.; Perera, S. Voltage support by distributed generation units and shunt capacitors in distribution systems. In Proceedings of the 2009 IEEE Power & Energy Society General Meeting, Calgary, AB, Canada, 26–30 July 2009; pp. 1–8.
45. Zou, K.; Agalgaonkar, A.; Muttaqi, K.; Perera, S. Optimisation of Distributed Generation Units and shunt capacitors for economic operation of distribution systems. In Proceedings of the2008 Australasian Universities Power Engineering Conference, Sydney, NSW, Australia, 14–17 December 2008; pp. 1–7.
46. Sultana, S.; Roy, P.K. Optimal capacitor placement in radial distribution systems using teaching learning based optimization. *Int. J. Electr. Power Energy Syst.* **2014**, *54*, 387–398. [CrossRef]
47. Rashtchi, V.; Darabian, M.; Molaei, S. A robust technique for optimal placement of distribution generation. In Proceedings of the International Conference on Advanced Computer Science, Information Technology, Electronics Engineering & Communication, Hangzhou, China, 23–25 March 2012; pp. 31–35.
48. Nekooei, K.; Farsangi, M.M.; Nezamabadi-pour, H.; Lee, K. An improved multi-objective harmony search for optimal placement of DGs in distribution systems. *IEEE Trans. Smart Grid* **2013**, *4*, 557–567. [CrossRef]
49. Nojavan, S.; Jalali, M.; Zare, K. Optimal allocation of capacitors in radial/mesh distribution systems using mixed integer nonlinear programming approach. *Electr. Power Syst. Res.* **2014**, *107*, 119–124. [CrossRef]
50. Singh, H.; Hao, S.A. Papalexopoulos, Transmission congestion management in competitive electricity markets. *IEEE Trans. Power Syst.* **1998**, *13*, 672–680. [CrossRef]
51. Rao, R.V. Jaya: A simple and new optimization algorithm for solving constrained and unconstrained optimization problems. *Int. J. Ind. Eng. Comput.* **2016**, *7*, 19–34.
52. Demircali, A.; Koroglu, S. Jaya algorithm-based energy management system for battery- and ultracapacitor-powered ultralight electric vehicle. *Int. J. Energy Res.* **2020**, *44*, 4977–4985. [CrossRef]
53. Leghari, Z.H.; Hassan, M.; Said, D.; Memon, Z.A.; Hussain, S. An efficient framework for integrating distributed generation and capacitor units for simultaneous grid- connected and islanded network operations. *Int. J. Energy Res.* **2021**, *45*, 1–39. [CrossRef]
54. Jumani, T.A.; Mustafa, M.W.; Hussain, Z.; Rasid, M.M.; Saeed, M.S.; Memon, M.M.; Khan, K.S. Nisar, Jaya optimization algorithm for transient response and stability enhancement of a fractional-order PID based automatic voltage regulator system. *Alexandria Eng. J.* **2020**, *59*, 2429–2440. [CrossRef]
55. Abhishek, K.; Kumar, V.R.; Datta, S.; Mahapatra, S. Application of Jaya algorithm for the optimization of machining performance characteristics during the turning of CFRP (epoxy) composites: Comparison with TLBO, GA, and ICA. *Eng. Comput.* **2017**, *33*, 457–475. [CrossRef]
56. Du, D.C.; Vinh, H.H.; Trung, V.D.; Hong Quyen, N.T.; Trung, N.T. Efficiency of Jaya algorithm for solving the optimization-based structural damage identification problem based on a hybrid objective function. *Eng. Optim.* **2018**, *50*, 1233–1251. [CrossRef]
57. Rao, R.; Saroj, A. Economic optimization of shell-and-tube heat exchanger using Jaya algorithm with maintenance consideration. *Appl. Therm. Eng.* **2017**, *116*, 473–487. [CrossRef]
58. Jin, R.V.; Wang, L.; Huang, C.; Jiang, S. Wind turbine generation performance monitoring with Jaya algorithm. *Int. J. Energy Res.* **2019**, *43*, 1604–1611. [CrossRef]
59. Aman, M.M.; Jasmon, G.B.; Bakar, A.H.A.; Mokhlis, H. Optimum network reconfiguration based on maximization of system loadability using continuation power flow theorem. *Int. J. Electr. Power Energy Syst.* **2014**, *54*, 123–133. [CrossRef]

*energies*



*Article*

# Frequency Support Studies of a Diesel–Wind Generation System Using Snake Optimizer-Oriented PID with UC and RFB

Vikash Rameshar [1], Gulshan Sharma [1,*], Pitshou N. Bokoro [1] and Emre Çelik [2]

[1]   Department of Electrical Engineering Technology, University of Johannesburg,
     Johannesburg 2006, South Africa; vikashr@uj.ac.za (V.R.); pitshoub@uj.ac.za (P.N.B.)
[2]   Department of Electrical and Electronics Engineering, Engineering Faculty, Düzce University,
     Düzce 81620, Turkey; emrecelik@duzce.edu.tr
*    Correspondence: gulshans@uj.ac.za

**Abstract:** The present paper discusses the modeling and analysis of a diesel–wind generating system capable enough to cater to the electrical power requirements of a small consumer group or society. Due to high variations of the load demand or due to changes in the wind speed, the frequency of the diesel–wind system will be highly disturbed, and hence to regulate the frequency and power deviations of the wind turbine system, an effective controller design is a necessary requirement, and therefore this paper proposes a novel controller design based on PID scheme. The parameters of this controller is effectively optimized through a new snake optimizer (SO) in an offline manner to minimize frequency and power deviations of an isolated diesel–wind system. The performance of SO-PID for the diesel–wind system is evaluated by considering the integral of time multiplied absolute error (*ITAE*), integral absolute error (*IAE*), and integral of time multiplied square error (*ITSE*). The results were calculated for a step change in load, step change in wind speed, load change at different instants of time with diverse magnitude, and for random load patterns, and they were compared with some of the recently published results under similar working conditions. In addition, the effect of an ultracapacitor (UC) and redox flow battery (RFB) on SO-PID was investigated for the considered system, and the application results demonstrated the advantages of our proposal over other studied designs.

**Keywords:** RES; diesel engine generator; wind turbine generator; ultracapacitor; redox flow battery

---

**Citation:** Rameshar, V.; Sharma, G.; Bokoro, P.N.; Çelik, E. Frequency Support Studies of a Diesel–Wind Generation System Using Snake Optimizer-Oriented PID with UC and RFB. *Energies* **2023**, *16*, 3417. https://doi.org/10.3390/en16083417

Academic Editors: Saeed Sepasi and Quynh Thi Tu Tran

Received: 16 March 2023
Revised: 5 April 2023
Accepted: 10 April 2023
Published: 13 April 2023

## 1. Introduction

As a working citizen, every improvement in the public domain stems from enhancements made within the industrial sectors, and these changes require massive amounts of energy directly associated with electricity. The demand for electrical power has induced the need for electrical industries to bring out their competitive side. Although competition is encouraged among the industries, the fossil fuel (FF) units that produce the bulk of the electrical energy cannot meet the demand by utilizing only standard steam-based energy generation. This demand for electrical power can only be met by finding new energy-producing methods. One of the global development goals is for renewable energy sources (RESs) which, going into the future, would provide clean essential energy generation and would remove the reliance on fossil fuels [1]. Clean, renewable energy is produced and includes biomass, hydrogen, biofuels, hydro control, sun-powered radiation, wind pressure, geothermal assets, and ocean essentials which RESs have been utilizing. Technology towards sun-radiated, that is, photovoltaic (PV) and wind energy units, have been dominating and increasing due to the neighborhood-specific environments and advancements within the RESs. Additionally, it boosts positive finance and the development of communities [2]. Within these RES structures that promote growth, one has to adhere to the control and importance of the generated power and how it progresses if RESs are to be fully implemented within the energy system.

Concerns are increasing within the FF sector, including increasing costs and shortages, which enhances the positive move toward RESs to subsidize the control framework [3]. Sources within the renewable energy system cannot predict adequate control and efficiency, which in turn affects the planning of frequency control. This would insinuate that these significant issues could be resolved with the idea of a micro-grid in the grid-connected and isolated mode that could improve and control the frequency framework in immense commercial and provincial areas. Other methods of reliable power supply within the electrical control system can be substantiated through the use of diesel-generating and wind-based units, which have gathered momentum by way of green vitality in confined control systems. Wind-based units are susceptible to fluctuations in yield and the framework frequency due to variations in wind speed or load demand [4]. The change in frequency and power greatly influences the control technique and stability of locally linked networks which also influences the security of these networks, as discussed in [5]. Some studies such as [6] have examined the unconnected fusion models made up of a fuel cell, diesel engine generator (DEG), and wind turbine generator (WTG), while in [7], the basis of mixed PV–fuel cells for standalone applications was explored. Although it is safer to have large-capacity batteries at one location and not in motion, the control and interchange of PV–fuel cells and battery banks powering an electric vehicle (EV) were studied in [8]. The load demand of 3 kW within the EV was powered by a battery bank, a film fuel cell, and a PV-generating control system. In [9], an investigation of wind turbines, fuel cells, and solar PVs was combined with numerical modeling, and the battery bank control environment was a crucial part of the study. In [10], the control application of a hybrid model was explored by Aissou et al., which included wind generation apart from the primary battery and PV inverter scenario to cater to expanding loads. The outstanding part of the work by Tamalouzt et al. in [11] was the recreation of a micro-grid control system that utilized a battery bank, PV generator, fuel cell, doubly fed induction generator, and a wind turbine. In most cases, the RES depended on the climate circumstance at that instance in time. This, moderately said, would be the circumstance where the electricity demand has expanded over the supply ability. If the latter occurs, i.e., an excess of supply (an overflow), the RES systems would have to control its overflow with energy storage devices (ESDs). The ESD would be able to handle and control the overflow of supply for a duration as the load increases; the demand would use up the ESD as the system requires. Storage devices utilized for energy, consisting of a battery, capacitor, flywheel, and superconducting magnet energy storage (SMES), can accommodate an overflow of energy and convey it at the top of the load request [12–14]. ESDs utilized with a hydrogen generative aqua electrolyzer (HAE), which includes a fuel cell, are labelled as capable in comparison to other capacity ESDs due to their needed operation. RESs, which can accommodate diverse energy storage systems, could utilize the HAE structure to break down water or normal gas into oxygen and hydrogen; then, it is a matter of compression techniques to store the gas. This would entail transferring the gas to fuel the cells via pipelines. In an instance of higher load demand, the HAE has the ability and potential to stabilize the system state [14]. Another storage device that can supply a fast-acting capacity to the system in the event that the generator rotors need a headway due to control demand is the redox flow battery (RFB). The RFB has the ability to dampen electromechanical expansions of the control system by discharging through control systems that include inverter/rectifier networks [15]. This essentially allows the RFB to almost instantaneously adapt to load changes, thereby decreasing the demand. Furthermore, ultracapacitors (UC) are currently powerful and inexpensive storage devices and have been used by several researchers for various power system applications [16]. Be it a bigger or smaller controlling capacity such as an isolated diesel–wind system, frequency regularization is an essential feature of power delivery systems as required by clients that should be in the form of inexpensive, tenacious, and high-quality electrical energy. The strategy to be implemented for essential control in this paper will be based on the PID as suggested within the scope of [17].

Hybrid power system (HPS) applications have met with numerous evolutionary computational intelligence (CI) techniques which have to some extent, enhanced and developed the HPS system. In [18], genetic algorithms were applied to design a hybrid solar–wind generator with a battery bank as an ESD. In [19], the amalgamation of different energy resources and energy storage elements was utilized to minimize frequency deviations. This research explored a PID controller with optimization algorithms to improve the frequency deviation of the generating structures, which gathered valuable methodological information from [20], that utilized the particle swarm optimization (PSO) technique in the simulation studies of autonomous hybrid energy generation and energy storage structures. Although the PID controller has been optimized with different techniques, in [21], the design and optimization of the controller parameters were optimized using the JAYA algorithm to be specific to the self-adapting multi-population elitist optimization, which targeted the *ITAE* and *IAE* objective functions. In [22], the efficiency of the wind–solar generating structure utilizing hybrid energy storage was optimized by applying a combination of simulated annealing and PSO. The HPS model made up of solar–wind–fuel cell simulation in [23] used the artificial bee swarm algorithm technique that not only resulted in the optimization of the model but showed that the HPS design was also cost-effective. In [24], the studies of the simulated hybrid models utilized PV, fuel cells, DEGs, WTGs, battery energy storage, flywheel, aqua electrolyzer, and ultracapacitors, which were all optimized by a GA optimizer. A chaotic PSO and fractional order fuzzy control method was applied to an HPS with renewable energy generation, which was studied in [25]. A new and powerful quasi-opposition harmony search (QOHS) technique was used in [26] to solve the frequency regulation problem of various power system models. In the current field of optimization techniques, multi-area power systems (MAPs) of a two-area non-reheat thermal system were explored and optimized with an Artificial Rabbit Algorithm (ARA) [27]. The AR algorithm was employed in optimizing the PID controller parameters for the load frequency control. Another MAP exploration optimization technique is the Coyote optimization algorithm, utilized with cascading controllers [28]. This multi-area power system exploration also included PV panels as a RES. With the advancement in the field of metaheuristic techniques used for the optimization problem, there is still a possibility to explore new and powerful optimization techniques such as the snake optimizer (SO) [29] to solve power system problems in a much better and in an effective way and hence, this work set out to:

- Study and present the modeling of a diesel–wind isolated system in an interconnected mode for frequency regularization studies capable enough to provide uninterrupted electrical energy during a change in the load demand.
- Present the new control strategy for a diesel–wind isolated system to reduce or minimize the frequency and power deviations in the event of load change or in case of a change in wind speed. The proposed design is formulated by using PID and the gains of PID, which play an important role and are obtained via the new SO algorithm.
- Study and present the modeling of UC and RFB for diesel–wind generating systems.
- Investigate the performance of SO-PID for various working conditions and to compare the results with the Ziegler–Nichols method [1] and with the quasi-opposition harmony search (QOHS) technique [26].
- The results are matched for a step change in load, step change in wind speed, load change at the different instances of time, and for continuous load patterns and via calculating gains of the controller and through error values, i.e., integral of time multiplied absolute error (*ITAE*), integral absolute error (*IAE*), and integral of time multiplied square error (*ITSE*).
- The outcome of SO-PID was also compared via plotting frequency and power deviation response for the diesel–wind system.
- In addition, the effect of UC and RFB impact on the output of the diesel–wind system was also investigated and compared to determine the best control design within the system for diverse working conditions.

## 2. The Detailed Modeling of the Isolated Diesel–Wind System

The wind turbine and diesel engine generators are some of the most reliable energy sources to accomplish a continued supply of power. The DEG and WTG individual generating systems can provide a capacity of 150 kW. Due to the WTG being reliant on weather conditions, more accurately, wind speed, the DEG has an advantage in supplying uninterrupted power (UP). Figure 1 below is a representation of the linear model of the DEG. The system parameters are given in Appendix A in Table A1. The model is powered by a diesel engine and also includes a turbine that is restricted by a speed governor contrivance. Within the generator, the transfer function is as follows [26]:

$$\Delta P_{GD} = \frac{1}{(1 + sT_{D4})} \Delta P_{GT} \tag{1}$$

$$\Delta P_{GT} = \left( \frac{K_D(1 + sT_{D1})}{(1 + sT_{D2})(1 + sT_{D3})} \right) \Delta P_G \tag{2}$$

$$\Delta P_G = \Delta P_{CD} - \left( \frac{1}{R_D} \right) \Delta F \tag{3}$$

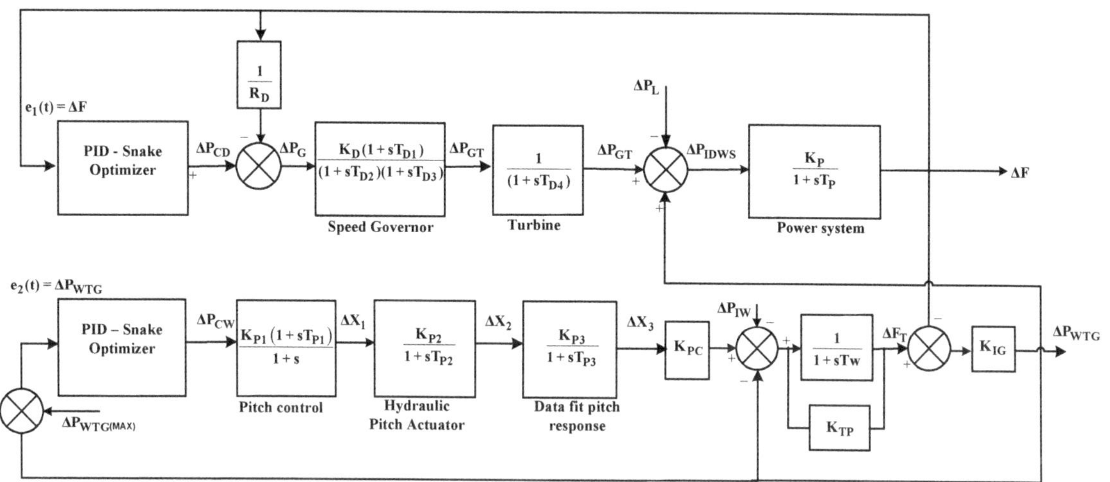


**Figure 1.** Diesel–wind isolated system.

Due to the speed governor being the eye of the model, the above equation equates to $\Delta P_{GT}$ providing an incremental change of the output of the speed governor (p.u.) while providing the control signal of the speed governor within the diesel system (p.u.). $\Delta P_G$ provides the control signal of the speed governor (p.u.). $\Delta F$ provides the frequency alteration, while $K_D$ provides the gain of the speed governor, and $T_{D1}$, $T_{D2}$, and $T_{D3}$ are the time constants in seconds. The turbine has a unity gain and a responding time $T_{D4}$, and a speed regulator of the diesel system represented by $R_D$.

The *WTG* system utilizes the dynamic energy of the wind through a process of mechanical excellence to finally produce electrical energy. This dynamic model of wind generation is featured in Figure 1. Within the *WTG* model, a transfer from the hydraulic connection to the actual speed must be a clear distinction between the turbine and generator frequency. The change in produced power is expressed in Equation (4) as described by [26].

$$\Delta P_{WTG} = K_{IG}[\Delta F_T - \Delta F] \tag{4}$$

$\Delta F_T$ signifies the wind generator's speed change, and $K_{IG}$ represents the gain obtained from the hydraulic connection. Equation (5) displays the induction generator speed:

$$\Delta F_T = \left(\frac{1}{1+sTw}\right)[K_{TP}\Delta F_T - \Delta P_{WTG} + K_{PC}\Delta X_3 + \Delta P_{IW}] \tag{5}$$

In most wind generation techniques, blade characteristics have to be considered. $K_{PC}$ in Equation (5) is considered the blade gain contribution, while $\Delta P_{IW}$ is considered the input wind control (in p.u.). The data fit pitch (DFP) yield is considered with $\Delta X_3$ within the system. With most systems, there is some lag that has to be compensated for, which has a function to match the gain characteristics of the micro-grid model. The DFP yield of the system is represented as:

$$\Delta X_3 = \Delta X_2\left[\frac{K_{P3}}{1+sT_{P3}}\right] \tag{6}$$

In Equation (6), the yield of the hydraulic pitch is represented with $\Delta X_2$, and the system also considers the responding time of the DFP with $T_{P3}$, also a gain indication with $K_{P3}$. The regulation of the pitch angle of the blades within the wind turbine is the function of the hydraulic pitch system (HPS) that has an output of:

$$\Delta X_2 = \Delta X_1\left[\frac{K_{P2}}{1+sT_{P2}}\right] \tag{7}$$

In Equation (7), $\Delta X_1$ is the yield of the pitch controller with the HPS gain indication with $K_{P2}$ and the time constant of the HPS represented by $T_{P2}$ (in seconds). The importance of the pitch angle has to be carefully considered due to the wind turbine achieving a maximum yield. Representation of the pitch angle systems is as follows:

$$\Delta X_1 = \Delta P_{CW}\left[\frac{K_{P1}(1+sT_{P1})}{1+s}\right] \tag{8}$$

Equation (8) provides a control signal in (p.u.) of the pitch angle system by $\Delta P_{CW}$ and $K_{P1}$ showing the gain of the pitch angle output. $T_{P1}$ allows the pitch angle mechanism to respond in seconds to provide control and gain to the system.

The quality of power delivered by the micro-grid calls for the restriction of the frequency in the system, which has to consider the load demand for the *DEG* and *WTG* system to generate total power generation. With that being said, any misinterpretation between the created control and load request may vary the micro-grid frequency, which has to be controlled as seamlessly as possible. As indicated in Equation (9), the micro-grid frequency response creates a first-order representation in the form of gain, transfer function, and time constant. $K_P$ is an indication of the gain, and $T_P$ is an indication of the micro-grids' time constant as shown below:

$$\Delta F = \Delta P_{IDWS}\left[\frac{K_P}{1+sT_P}\right] \tag{9}$$

## 3. Redox Flow Battery (RFB) and Ultracapacitor (UC) Details

One crucial energy storage device with a quick control reaction concerning frequency deviations is the redox flow battery (RFB). As a reaction speed indication, the RFB can control its operating time within seconds. The RFB storage device has a higher efficiency due to its robust charge and discharge capacity. This discussion of the RFB is illustrated completely in [15]. An interesting feature of the redox flow battery is that the electrolyte within the battery is stored in two different tanks. This feature allows no self-discharge of the RFB and hence enhances its life span, which is much higher than other batteries. The electrolytes comprise sulfuric acid solutions containing vanadium ions confined in a positive and negative storage tank. Furthermore, the RFB is relatively safe when operated at normal temperatures, although it has been known for its rapid charging and discharging abilities. Another feature of the redox flow battery is the unit's ability to efficiently reload after a load disturbance has occurred. This means that the RFB set value can be restored quickly enough so as to not to cause any delays within the next load disturbance. RFB set value and the power exchange to the system would depend on the deviated frequency signal, and its incorporation is shown in Figure 2. The system parameters are given in Appendix A in Table A1.

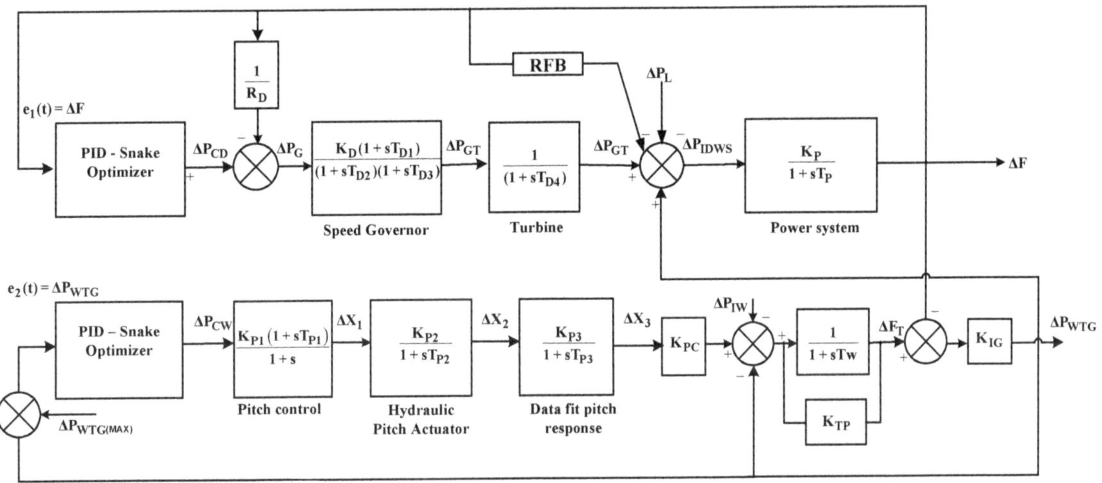

**Figure 2.** Diesel–wind isolated system with RFB.

Advanced technology in energy devices has surfaced within the electrochemical device range, namely ultracapacitors (UCs), which are also known as electric double-layer capacitors or supercapacitors. The technical aspects of a UC include a composition of potassium hydroxide as an electrolyte and two permeable anodes which have an ion-exchange layer segregating them. The difference from a standard capacitor is the vast surface area of the capacitor's permeable terminals and electrolytic fluid. UCs also stand out with an improvement in capacitance by the slightly thicker two-fold layer within the capacitor. These two factors increase the reliability of UCs from (100 to 1000 times) more than an ordinary electrolytic capacitor. UCs' compact design and high energy density provide an edge over standard capacitors and allow for massive amounts of energy storage. While most UCs offer an increased power capacity compared to batteries, they have a rating of high specific energy of 1–10 Wh/kg and a specific power energy rating of 1000–5000 W/kg. They can be discharged and charged quicker than a battery or normal capacitor, which is also maintenance-free and offers a longer life cycle. However, the qualities of a UC make it suitable for a wide range of LFCs, nonlinearities coaxed by various assumptions, and the validation of a UC can be indicated through a first-order TF. A control zone frequency deviation can work as an input to the UC and its TF is shown in Equation (10) below [16]:

$$\Delta P_{UC}(s) = \left\{ \frac{K_{UC}}{1 + sT_{UC}} \right\} \Delta F(s) \tag{10}$$

$K_{UC}$ is known as the gain pick time of a UC, with $T_{UC}$ being the time constant. The state of charge (SOC) is the state of charge that makes the gain $K_{UC}$ dependent on it. The gain of the UC is maintained steadily between 50% and 90% of the working SOC. The incorporation of a UC is shown in Figure 3.

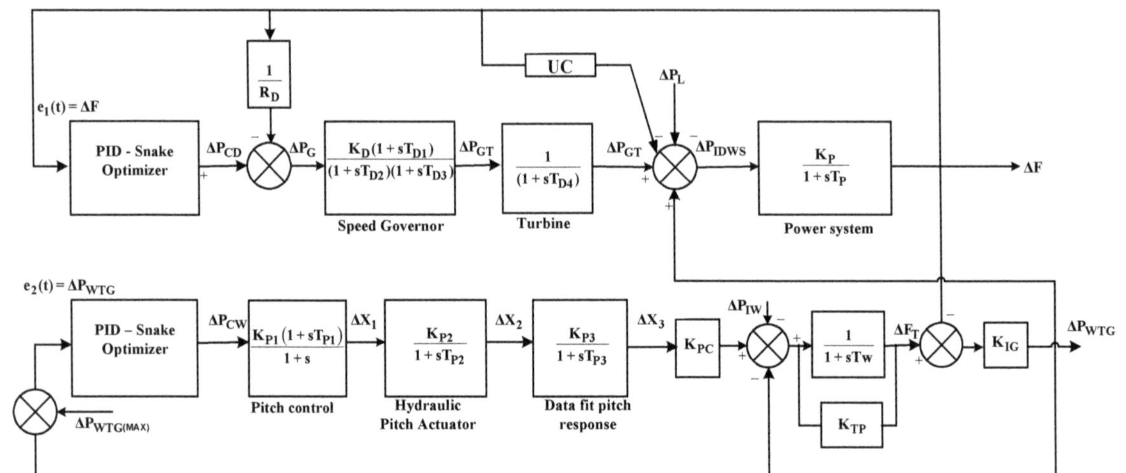


**Figure 3.** Diesel–wind isolated system with UC.

## 4. Snake Optimizer Details and Execution Steps

The last ten years have seen the development of many algorithms that provide solutions to significant data difficulties and analyses. Many of the difficulties stem from real-world engineering applications. Therefore, continuously solving these problems with different optimization techniques and algorithms is needed. One approach that utilizes a relationship between snakes has produced positive results. The SO technique is currently used in this research and a sequence of events around snakes outlined below [29].

### 4.1. Mating Behavior of Snakes

Male and female snakes mate under the influence of a few factors. These factors involve the temperature of the area, the availability of food, as well as the male and female dominance. Questions also arise if there is more than one dominant male: a competition will naturally emerge, in the form of a fight, to attract the female. If the female is attracted and mating does occur, the female would lay eggs and leave once the offspring emerge.

### 4.2. Inspiration Source

The SO model is based on the mating behavior of snakes, which has to occur in low temperatures and with available food, or else the snakes would search only for food or consume existing food. This brings about two processes: exploitation and exploration. The exploration involves environmental factors, i.e., cold temperature and food. The exploitation factors involve different cases, which is a combination of the temperature and availability of food to obtain a more global efficiency towards the probability of mating. This probability also involves the fight mode before mating with a female, which may produce new snakes in the successful case.

### 4.3. Algorithm and Mathematical Model

The SO model is illustrated in Figure 4 and explained in detail below [29].

#### 4.3.1. Initialization

The SO optimization algorithm process begins by generating random populations with uniform distributions, similar to other metaheuristic models. This Initial population could be processed by utilizing the following equation:

$$X_i = X_{min} + r \times (X_{max} - X_{min}) \tag{11}$$

In Equation (11), $r$ is a random number between 0 and 1, and $X_{min}$ and $X_{max}$ reflect the lower and upper bounds of the problem.

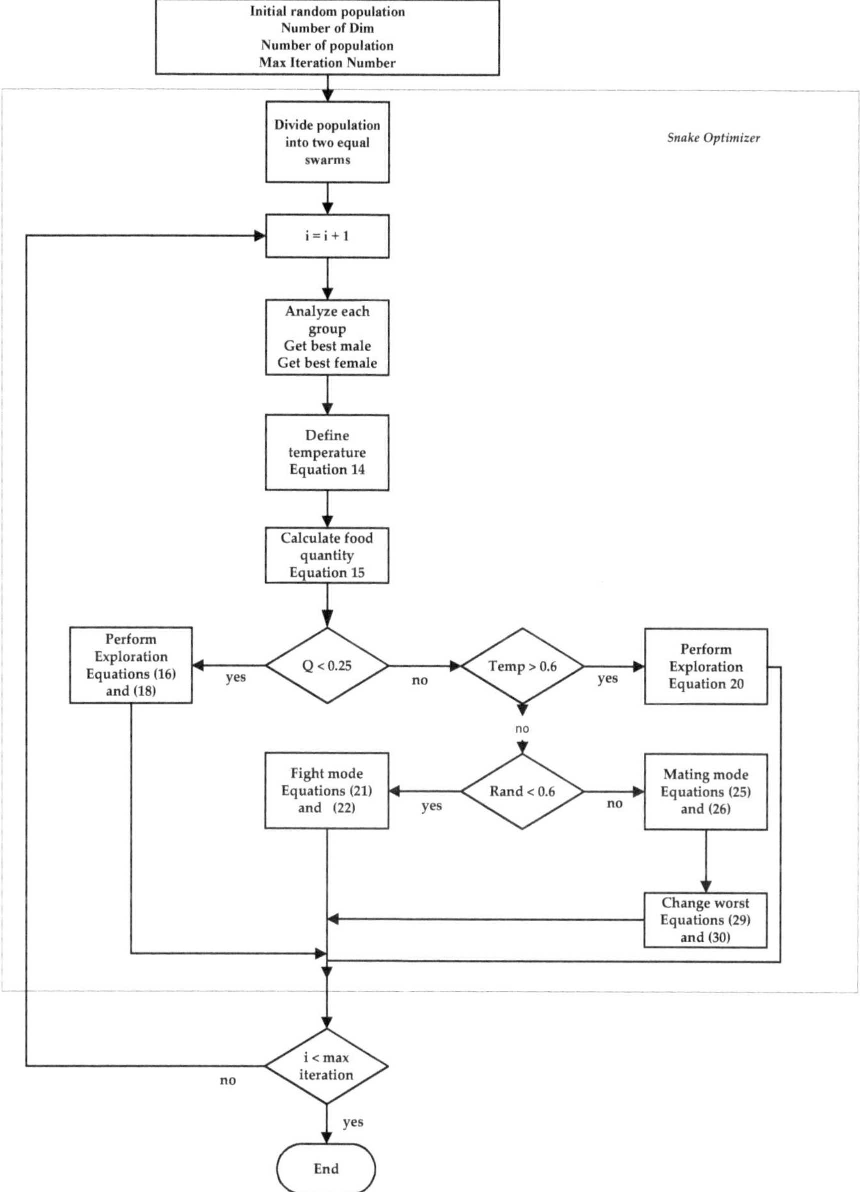


**Figure 4.** Snake optimization model flow.

### 4.3.2. Division of the Swarm into Two Groups: Females and Males

The female group and the male group make up 50% of the population each, and the division is obtained by applying Equations (12) and (13)

$$N_m \approx \frac{N}{2} \tag{12}$$

$$N_f = N - N_m \tag{13}$$

In Equations (12) and (13), $N_f$ represents the females while $N_m$ represents the males, and $N$ is the total number of females and males.

### 4.3.3. Food Quality and Temperature and Each Group Valuation

The best male ($f_{best,m}$) and best female ($f_{best,f}$) and also the food position ($f_{food}$) is evaluated within each group. Equation (14) defines temperature *Temp*:

$$Temp = exp\left[\frac{-t}{T}\right] \tag{14}$$

where $T$ represents the maximum number of iterations and $t$ represents the current iteration of the algorithm. Food quantity is calculated by utilizing Equation (15), where $c_1$ is a constant equal to 0.5:

$$Q = c_1 \times exp\left[\frac{t - T}{T}\right] \tag{15}$$

### 4.3.4. Exploration Phase (No Food)

There is a threshold at which the snake is required to search for food and any random position is selected, updating their position. This only occurs if Q < threshold (threshold = 0.25). The formulation of the exploration phase within the model is as follows:

$$X_{i,m}\,(t+1) = X_{rand,m}(t) \pm c_2 \times A_m \times ((X_{max} - X_{min}) \times rand + X_{min}) \tag{16}$$

In the above equation, *rand* is a random number between 0 and 1, $X_{i,m}$ is the $i$th male position, $X_{rand,m}$ is the position of a random male, and $A_m$ is the capability of the male to source food and is calculated as:

$$A_m = exp\left[\frac{-f_{rand,m}}{f_{i,m}}\right] \tag{17}$$

Just to note that in Equation (17), $f_{rand,m}$ is the fitness of $X_{rand,f}$, and $f_{i,f}$ is the fitness of the $i$th individual within the male group.

$$X_{if} = X_{rand,f}(t+1) \pm c_2 \times A_f \times ((X_{max} - X_{min}) \times rand + X_{min}) \tag{18}$$

$X_{i,f}$ above represents the $i$th female place and $X_{rand,f}$ is the random female's location. The capability of the female to source food, $A_f$, is shown below:

$$A_f = exp\left[\frac{-f_{rand,f}}{f_{i,f}}\right] \tag{19}$$

Similar to the male individual, $f_{randf}$ refers to the fitness of $X_{rand,f}$ and the fitness of the $i$th individual in the female group is $f_{i,f}$.

### 4.3.5. Exploitation Phase (Existing Food)

If Q > threshold and the temperature > threshold (0.6) (hot).
The formulae above show when snakes would move to the food only.

$$X_{i,j}(t+1) = X_{food} \pm c_3 \times Temp \times rand \times (X_{food} - X_{i,j}(t)) \tag{20}$$

In Equation (20), $C_3$ is a constant with a value of 2, $X_{i,j}$ is the individual male and female position, and $X_{food}$ is the position of the best male and female.

When the temperature is less than 0.6%, which is the threshold, the snake will be in mating or fight mode.

$$X_{i,m}(t+1) = X_{i,m}(t) + c_3 \times FM \times rand \times \left[Q \times X_{best,f} - X_{i,m}(t)\right] \tag{21}$$

FM refers to the male's fighting ability within the fight mode in Equation (21), $X_{i,m}$ represents the $i$th position for the male, and $X_{best,f}$ represents the best female in the group.

$$X_{i,f}(t+1) = X_{i,f}(t+1) + c_3 \times FF \times rand \times [Q \times X_{best,m} - X_{i,F}(t+1)] \tag{22}$$

*FF* refers to the fighting capability of the female, $X_{i,f}$ represents the *i*th position of the female, and $X_{best,m}$ represents the best male individual in the male group.

*FF* and *FM* can be equated as shown below:

$$FM = exp\left[\frac{-f_{best,f}}{f_i}\right] \tag{23}$$

$$FF = exp\left[\frac{-f_{best,m}}{f_i}\right] \tag{24}$$

In Equations (23) and (24), $f_{best,f}$ and $f_{best,m}$ are the fitness of the best female individual in the female group and the best male individual in the male group, respectively, and $fi$ is the fitness agent. If the fitness levels are reached, the expected next level would be the mating mode, which is given below:

$$X_{i,m}(t+1) = X_{i,m}(t) + c_3 \times M_m \times rand \times \left[Q \times X_{i,f}(t) - X_{i,m}(t)\right] \tag{25}$$

$$X_{i,f}(t+1) = X_{i,f}(t) + c_3 \times M_f \times rand \times \left[Q \times X_{i,m}(t) - X_{i,f}(t)\right] \tag{26}$$

Equations (25) and (26) contain the formulation of the *i*th female group and male group as $X_{i,f}$ and $X_{i,m}$, respectively. $M_m$ and $M_f$ represents the mating capability of the male and female, respectively, as shown below:

$$M_m = exp\left[\frac{-f_{i,f}}{f_{i,m}}\right] \tag{27}$$

$$M_f = exp\left[\frac{-f_{i,m}}{f_{i,f}}\right] \tag{28}$$

This brings the algorithm to the decision that if the egg hatches, it will replace the worst male or female.

$$X_{worst,m} = X_{min} + rand \times (X_{max} - X_{min}) \tag{29}$$

$$X_{worst,f} = X_{min} + rand \times (X_{max} - X_{min}) \tag{30}$$

From Equations (29) and (30), the replacement of the $X_{worst,m}$ and $X_{worst,f}$, i.e., the worst female probability and the worst male probability, is one of the unique ways that the SO optimizes performance. The operator $\pm$ known as the flag direction operator and also called the diversity factor provides the possibility to decrease or increase the solution of positions. This provides many opportunities to change the direction of agents which, in turn, can search the space in all probable directions. Within any metaheuristic algorithm, it is essential to achieve randomization, and using the SO to perform this search is most beneficial. This operator is implemented in many metaheuristic algorithms; one that stands out is from the hunger games search (HGS). The SO algorithm continues for several iterations until the criterion has been met.

## 5. Results and Analysis

The present work is dedicated to research, modeling, and designing a novel controller for a diesel–wind isolated system. This model is capable of meeting the electrical energy requirement of a small community or society. The present isolated system consists of a diesel engine generator with a capacity of 150 kW and a wind turbine system with a capacity of 150 kW. The diesel system has a speed governor mechanism capable of increasing or lowering the power generation to match the standard frequency requirement. Due to the wind system and due to high variability of load, there may be high-frequency power deviations; hence, to provide quality power to the customers, it is essential to minimize these deviations in the shortest amount of time, which could be possible by adding the controllers for diesel and wind separately, and minimize the error which is the difference between actual and desired output. The idea behind the present study is to explore a well-known and simpler control structure, i.e., PID, for research and investigations. However, the PID design is enhanced via a new SO algorithm. The full details of this algorithm are given in Section 4.

The aim of SO is to minimize the error definition and, for the present study, three different error definitions were used, which are the integral of time multiplied absolute error (*ITAE*), the integral of time multiplied square error (*ITSE*), and integral of absolute error (*IAE*) as shown in Equations (31)–(33). *IAE* is known to yield a slow response without knowledge of time. *ITSE* can provide a fast dynamic response, but it results in notable sustained oscillations owing to squaring the error and tolerating the error at the start of the response. By adding a time weight to *IAE*, *ITAE* is obtained, enabling a more desirable response with a shorter settling time and mild or no overshoot [30–32]. As such, *ITAE* was preferred in this research as an objective function during the optimization by SO.

$$ITAE = \int_0^{ts} t\, |\Delta F|\, dt \tag{31}$$

$$IAE = \int_0^{ts} |\Delta F|\, dt \tag{32}$$

$$ITSE = \int_0^{ts} |\Delta F|^2\, t\, dt \tag{33}$$

During the course of optimization, the PID controller gains were not optimized considering only one profile of load disturbance. Instead, they were procured under random changes in the load demand to accommodate the new condition. The random load input was limited to the range of $-0.4$ p.u. to 1.6 p.u., and it can be either in an increasing trend or decreasing trend, changing in steps with a specific duration of time.

The convergence profile of SO during the minimization of the *ITAE* value is displayed in Figure 5, when the number of snakes and maximum iteration number was set to 50 and 100, respectively. As can be seen, after a sharp decrease in the early stages of the optimization, *ITAE* was reduced to 396.3 from 436.2 with no undesirable oscillation. Therefore, it can be inferred that SO can exhibit promising convergence characteristics. It may also be noted from Figure 5 that the algorithm stagnated and achieved the final solution after 50–60 iterations. No further improvement in the value of *ITAE* was gained afterward. This justifies our selection of the maximum iteration as 100 for the present study.

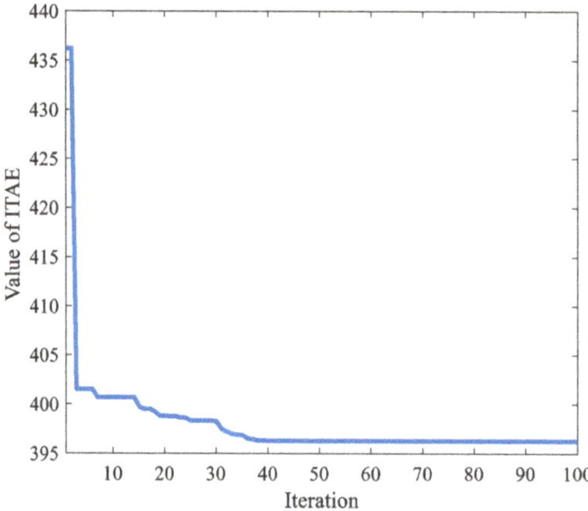

**Figure 5.** SO-based convergence profile for *ITAE* minimization.

To provide better insight into the SO-based optimization procedure, the algorithm balance between exploration and exploitation properties is analyzed in Figure 6. The red line shows the global exploration capacity, while the blue curve stands for the local exploitation ability. It is apparent from Figure 6 that both exploration and exploitation existed in the algorithm, and a balanced transition between them can solve the global optimum solution effectively. After emphasizing a solid exploration in the early period of the optimization, SO started to promote exploitation as the algorithm iterates. On average, the portion of the local exploitation search of SO was larger than that of the global exploration search.

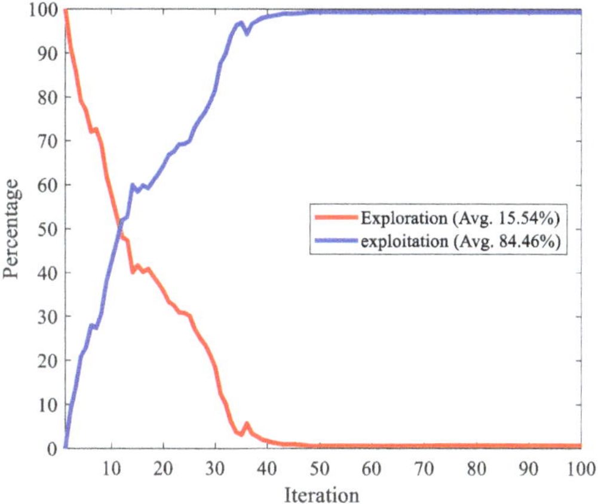


**Figure 6.** Balance analysis of SO.

SO was run 50 times, and the best solution among the 50 runs was taken as the final solution. The time required by SO to complete one run was measured as 25 min on a Windows 10 computer with an AMD Ryzen 9 5900HX CPU and 32 GB of memory. It should be noted that an essential portion of this time consumption was allocated by objective function calculation that covers calling the Simulink model from the m-file script, simulating the whole power system to obtain the dynamic responses, and returning the calculated *ITAE* value back to the m-file script. The m-file script contains the source codes of the SO and it is linked to the Simulink model of the studied power system.

The statistical data showing the minimum, maximum, average, and standard deviation of the best solutions achieved over multiple independent runs with SO-PID is presented in Table 1 to testify to the robustness of the used algorithm.

**Table 1.** The statistical analysis of the numerical results for SO-PID controller.

| Statistical Measure | Minimum | Maximum | Average | Standard Deviation |
|---|---|---|---|---|
| SO-PID | 396.3405 | 396.7219 | 396.5133 | 0.1435 |

As per the results in Table 1, it is clear that the difference between the minimum and maximum values of the *ITAE* objective function was very small, thus, the standard deviation was calculated as close to zero. This confirms that the SO performs stably and that its optimal performance for the considered problem is robust.

The PID controller gains procured through SO for the smallest value of *ITAE* are given in Table 2. The SO-optimized PID controllers are available for diesel and wind systems, and the output signal from SO-PID was applied to the diesel and the wind system simultaneously so that the frequency and power deviations can be suppressed within a few seconds only, and the operation of this isolated system can be stabilized for diverse working conditions. At first, the output and performance of the SO-PID were obtained by applying 1.0 p.u. step load alteration in a diesel–wind isolated system, and the simulation was run for 50 s. The SO-PID performance was compared with ZNM-PID [1] and with noticeable results obtained via Ganguly using QOHS-PID [26] for the same system. The calculated gains of PID via SO and through ZNM and QOHS are given in Table 2. Table 2 also shows the value of *ITAE*, *IAE*, and *ITSE* for all three techniques. It can be seen that the *ITAE* achieved through ZNM-PID was 4249, *IAE* was 89.91, and *ITSE* was 2590. The obtained values of *ITAE*, *IAE*, and *ITSE* were very high and cannot be appreciated for diesel–wind isolated systems. Furthermore, the *ITAE* obtained via QOHS-PID was 59.54, *IAE* was 7.257, and *ITSE* was 13.97; hence, there was a remarkable reduction in these values for the same model under the same disturbance. Furthermore, the SO-PID reduced the error value, i.e., *ITAE* to 37.1 from 59.54, *IAE* to 4.706 from 7.257, and *ITSE* to 5.762 from 13.97; hence, SO-PID outperformed all the other methods in terms of *ITAE*, *IAE*, and *ITSE*. It can also be seen that there was no need to calculate values of Kp and Kd as these gains are set to be zero in comparison to gains obtained via ZNM-PID and QOHS-PID, and still, the performance of the proposed technique was superior. From the results of Figure 7a,b, it can be seen that SO-PID offered reduced overshoot and better settling time. The system results returned to the reference value straight after the disturbance in 10 s compared to responses obtained via ZNM-PID and QOHS-PID.

**Table 2.** Comparison of optimized controller gains.

| Design | Kp (DEG) | Ki (DEG) | Kd (DEG) | Kp (WTG) | Ki (WTG) | Kd (WTG) | ITAE | IAE | ITSE |
|---|---|---|---|---|---|---|---|---|---|
| ZNM-PID [1] | 0.096 | 0.036 | 0.062 | 0.12 | 0.057 | 0.062 | 4249 | 89.91 | 2590 |
| QOHS-PID [26] | 0.9124 | 0.9976 | 0.0349 | 0.9996 | 0.0011 | 0.6519 | 59.54 | 7.257 | 13.97 |
| SO-PID [Proposed] | 2 | 1.99387 | 1.99453 | 0 | 2 | 0 | 37.1 | 4.706 | 5.762 |

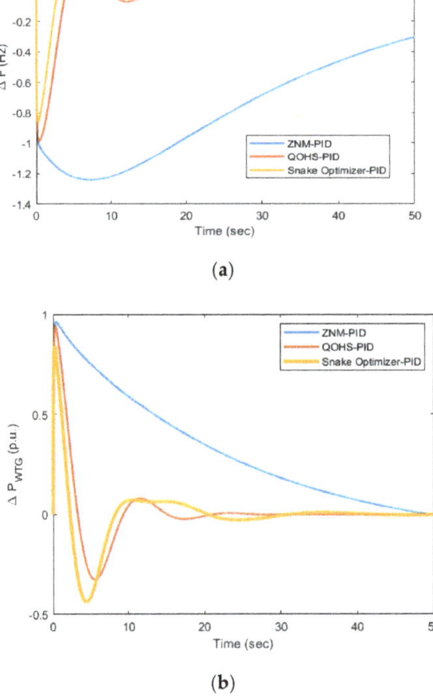


**Figure 7.** (a,b) Output of diesel–wind isolated system for 1.0 p.u. step load alteration.

In the next step of the research and investigations, a 0.01 p.u. step change in the wind speed from the standard speed and the results of SO-PID were compared with those using ZNM-PID and QOHS-PID. The gains of the controllers were kept the same for the three techniques and the values of *ITAE*, *IAE*, and *ITSE* were obtained and are shown in Table 3. From results of Table 3, it was calculated that the *ITAE* obtained via ZNM-PID was 12.51 and was further reduced via QOHS-PID to 0.4405 and remarkably reduced to 0.1873 through SO-PID. The same trend was also seen for *IAE* and *ITSE* and the value of these errors were reduced to 0.01979 and 0.000156 in comparison to that obtained via QOHS-PID and ZNM-PID and hence SO performed better for this case as well. The graphical results are shown in Figure 8a,b for these techniques for a 0.01 p.u. step change in the wind speed. From the graphs in Figure 8a,b, it is critically seen that $\Delta P_{WTG}$ (p.u.) responses showed higher oscillations and the system was not able to settle to a zero value within 50 s through ZNM-PID and QOHS-PID. There was a major steady-state error seen in the responses with a major overshoot. However, SO-PID was able to achieve minimum overshoot, fewer oscillations, and zero steady-state error in comparison to ZNM-PID and QOHS-PID. For $\Delta F$, which is in Hz, the outcome of SO-PID was much better in comparison to ZNM-PID and QOHS-PID.

**Table 3.** Proposed PID output matched for 0.01 p.u. step change in wind speed.

| Design | *ITAE* | *IAE* | *ITSE* |
|---|---|---|---|
| ZNM-PID [1] | 12.51 | 0.3219 | 0.04365 |
| QOHS-PID [26] | 0.4405 | 0.02259 | 0.000176 |
| **SO-PID [Proposed]** | **0.1873** | **0.01979** | **0.000156** |

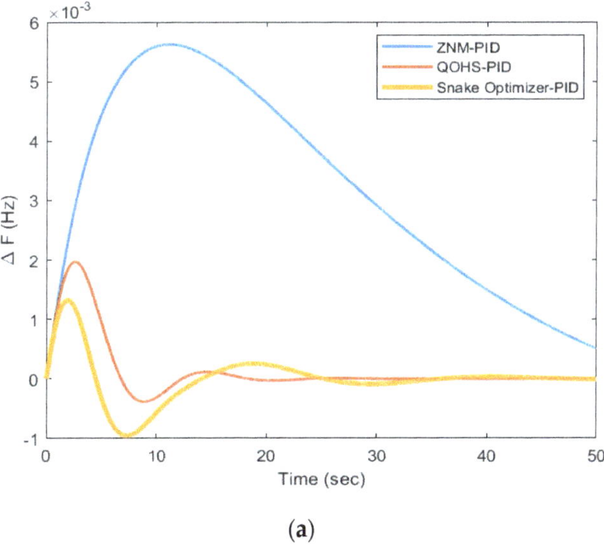


(**a**)

**Figure 8.** *Cont.*

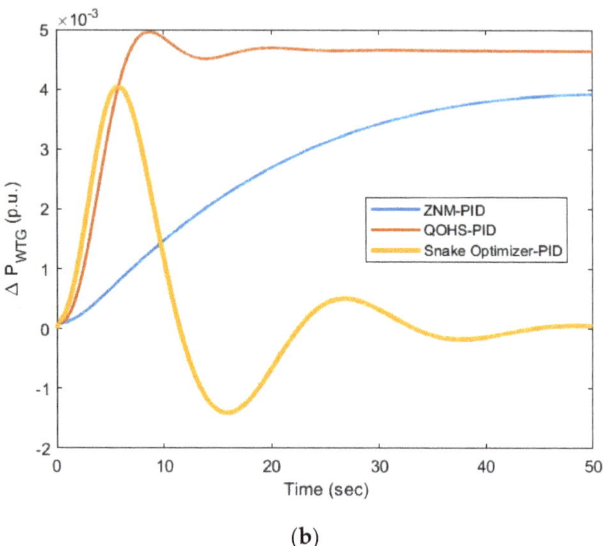

**(b)**

**Figure 8.** (**a**,**b**) Output of diesel–wind isolated system for 0.01 p.u. step change in wind speed.

Further, the investigations were also performed for load changes at various instances over the period of simulation. Figure 9a shows the load changes at 0, 20, and 40 s of time in a diesel–wind isolated system with diverse magnitudes, and the results of the frequency and power deviations were obtained and compared for SO-PID, QOHS-PID, and ZNM-PID. The graphical outcomes for this load change are shown in Figure 9b,c, and it can be seen that SO-PID was able to efficiently suppress the frequency and power deviations compared to ZNM-PID and QOHS-PID. The investigations were further extended for random load change in the diesel–wind isolated system with diverse magnitudes. This random load change is shown in Figure 10a, and the diesel–wind responses are given in Figure 10b,c. It was observed that ZNM-PID was unsuccessful in tracking the continuous change in load demand for frequency and power deviations. Compared to ZNM-PID, QOHS-PID was better, and SO-PID was the best at minimizing frequency and power deviations with a smaller overshoot and with better settling time. Hence, the research was extended to see the impact of energy storage devices on a diesel–wind isolated system with SO-PID. From the literature, ESD has been shown to have the capability to improve the dynamic performance. UCs and RFBs recently came out as powerful ESDs, and therefore, the impact of UC and RFB with SO needs to be explored for the diesel–wind isolated system in order to obtain a better dynamic performance for various working conditions.

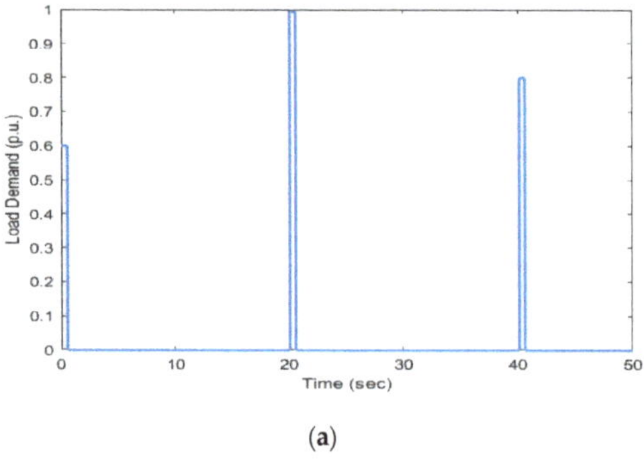

**(a)**

**Figure 9.** *Cont.*

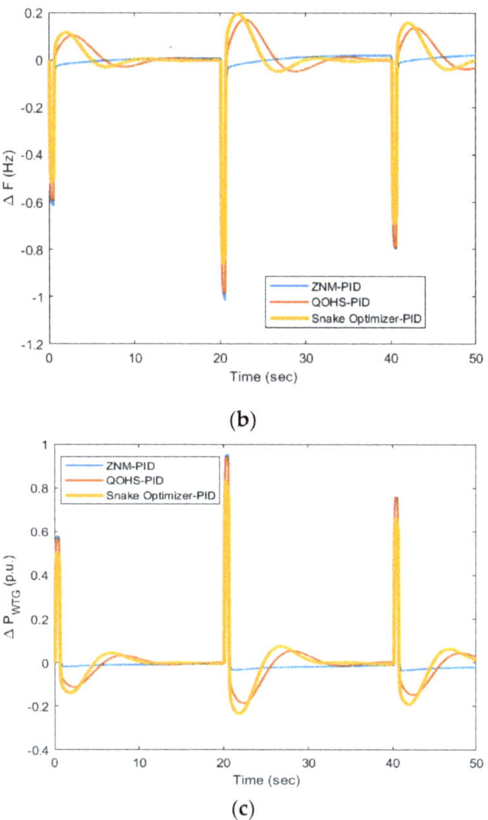

(b)

(c)

**Figure 9.** (**a**–**c**) Output of diesel–wind isolated system for load alterations at different instant of time.

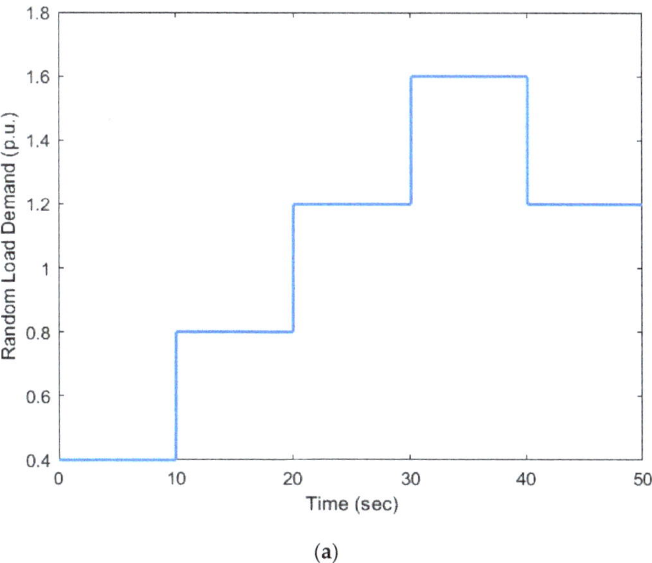

(a)

**Figure 10.** *Cont.*

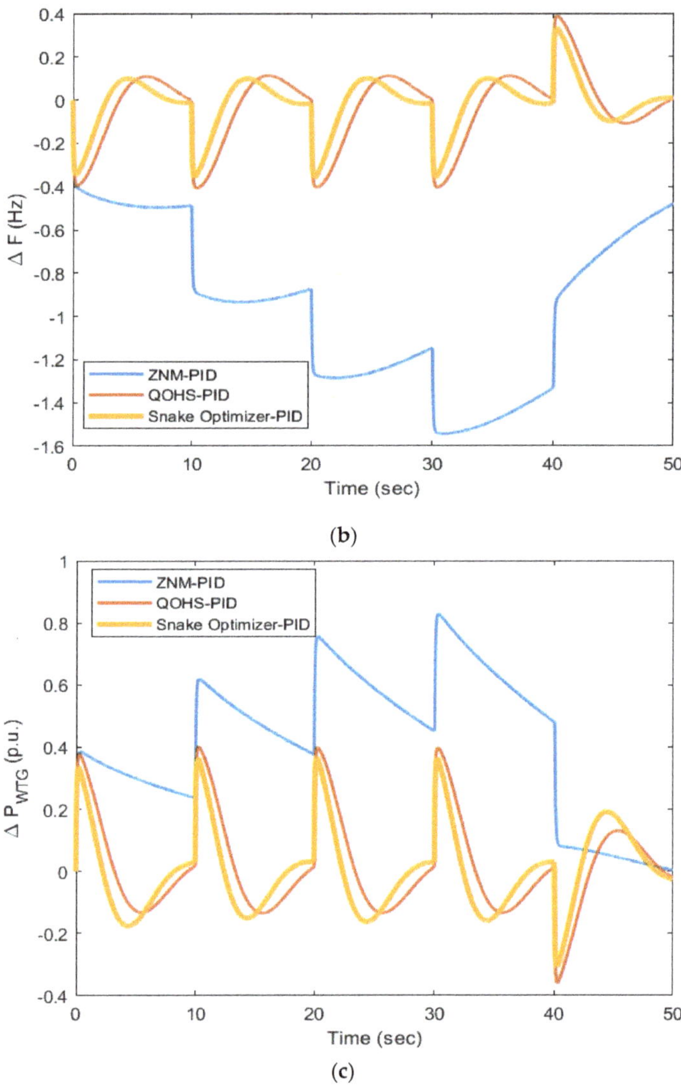

**Figure 10.** (**a–c**) Output of diesel–wind isolated system for random load alterations.

Due to load disturbance, the frequency and power output from wind will be highly disturbed, and the impact of these ESDs needs to be studied and investigated for various cases of load changes. The investigations started by applying a 1.0 p.u. step load alteration in a diesel–wind isolated system and compared with the results of SO-PID after integrating a UC or RFB into a diesel–wind isolated system. The values of *ITAE*, *IAE*, and *ITSE* are given in Table 4 and it was clearly visible that *ITAE* with SO-PID was 37.1 and was reduced to 32.46 with the integration of an ultracapacitor and further reduced to 5.336 after linking an RFB into the diesel–wind isolated system under similar disturbances. The trend was same for *IAE* and *ITSE*, and it can be seen that *IAE* became 1.411 and *ITSE* was reduced to 0.453 from 4.706 and 5.762; hence, it is evident that the UC and RFB with SO were able to improve the performance of the diesel–wind isolated system. Figure 11a,b shows the dynamic performance for this case and it was evident that the RFB outperformed the UC by managing to achieve a minimum first peak and enhanced settling of system responses. The trend of settling was very fast in comparison to the UC and with the results of SO-PID only.

**Table 4.** SO-PID output matched by adding an RFB or UC. The results were obtained for a 1.0 p. u. step load alteration.

| Design | *ITAE* | *IAE* | *ITSE* |
|---|---|---|---|
| SO-PID | 37.1 | 4.706 | 5.762 |
| SO-PID + Ultracapacitor | 32.46 | 3.411 | 2.458 |
| **SO-PID + Redox Flow Battery** | **5.336** | **1.411** | **0.453** |

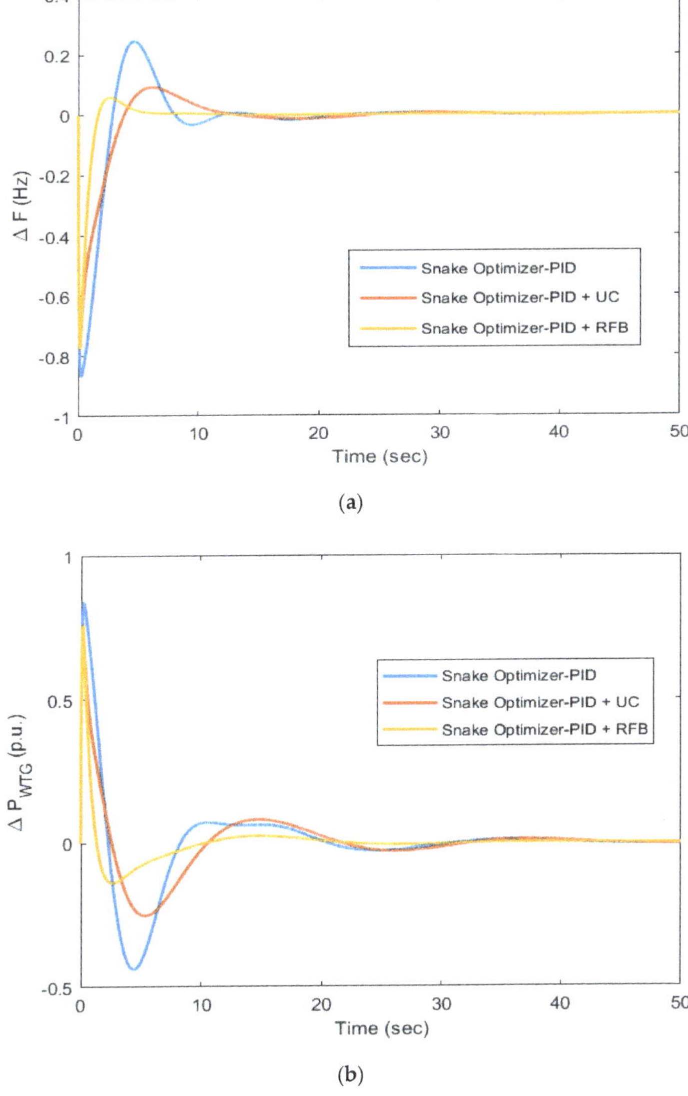


**Figure 11.** (**a**,**b**) Comparative output of PID with a UC or RFB for diesel–wind isolated system for a 1.0 p. u. step load alteration.

In the next step, the load was varied at 0, 20, and 40 s and the frequency and power deviation responses are shown in Figure 12a,b. These results showed the positive impact of the UC and RFB for diesel–wind isolated systems and that they were able to significantly improve the system responses. Here, SO-PID with RFB was also much better at reducing the frequency and power deviations in comparison to SO-PID with UC and with SO-PID only. The RFB and UC removed all oscillations from the system responses and improved the settling time for load changes at different times. Lastly, the impact of these ESDs were also checked for random load patterns applied to the diesel–wind isolated system and the results in Figure 13a,b clearly showed that if there is a continuous load change in the diesel–wind isolated system, the UC or RFB, with SO-PID can minimize unnecessary frequency and power excursions and maintained the system results at the reference value and can track the load demand more effectively. Still, the RFB and SO-PID combination was able to reduce the continuous oscillations and improve the dynamic output of the diesel–wind system for diverse cases.

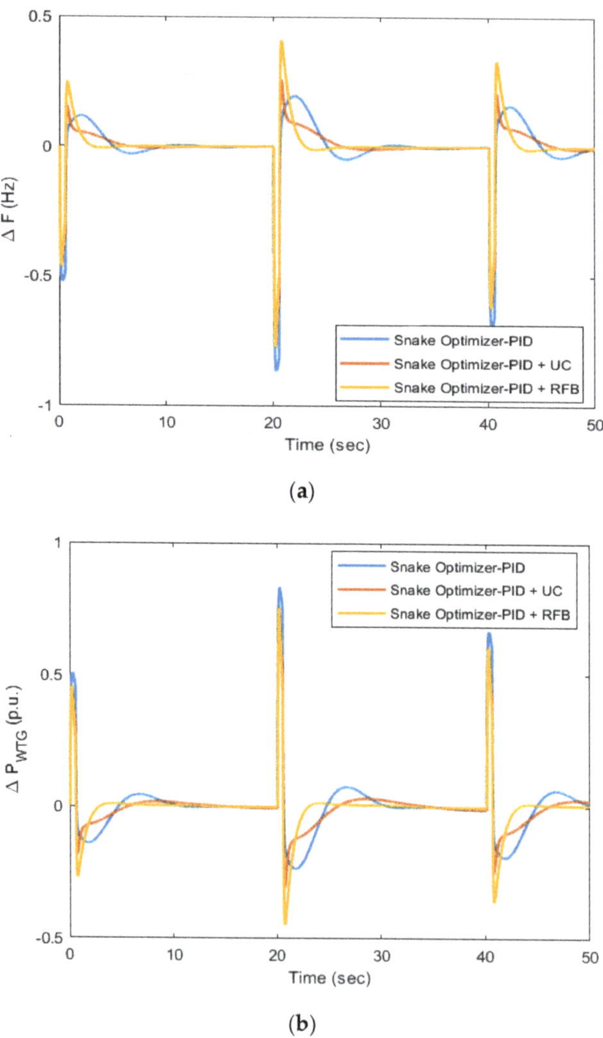


**Figure 12.** (**a**,**b**) The comparative output of PID with a UC or RFB for a diesel–wind isolated system for load alterations at different instant of time.

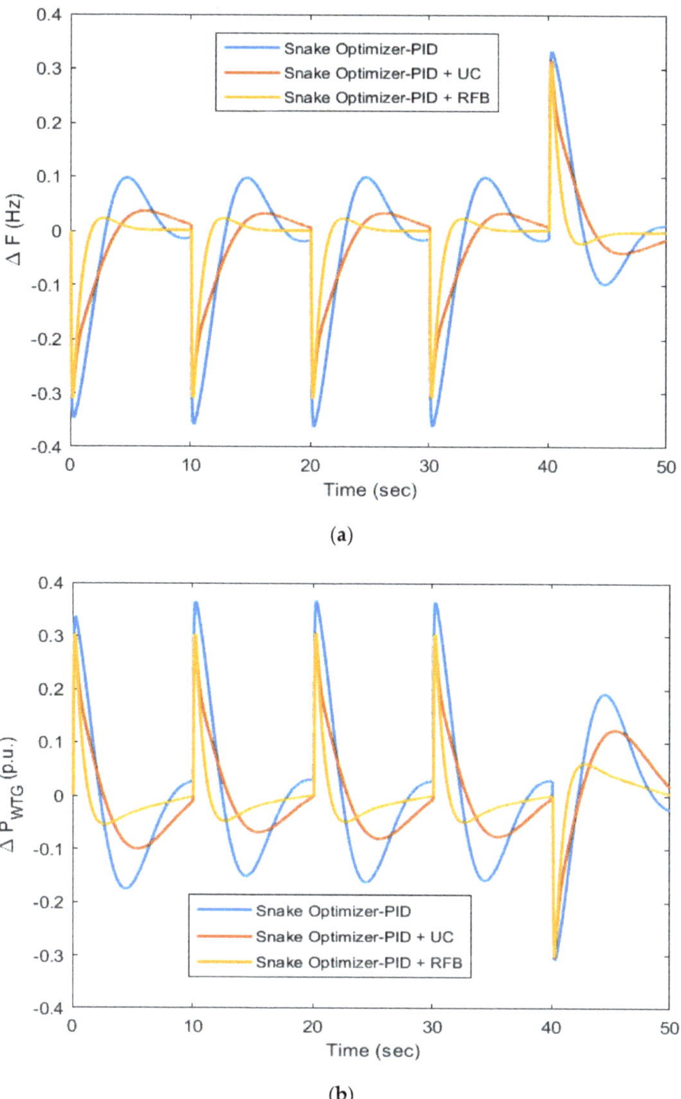



**Figure 13.** (**a**,**b**) The comparative output of PID with a UC or RFB for the diesel–wind isolated system for random load alterations.

## 6. Conclusions

The present work was focused on designing a novel control strategy to minimize frequency and power deviations for a diesel–wind isolated system generating diesel and wind power with enough capability to cater to the energy requirements of a small community or society. The SO-PID was proposed and investigated for various possible cases for a diesel–wind isolated system and the output of this strategy was compared with those of ZNM-PID and QOHS-PID. The following conclusions were drawn from the present research work:

- The SO-PID gave the minimum frequency and wind power deviations for a step change in load, step change in wind speed, load changes at different instances, and for a random load pattern compared to ZNM-PID and QOHS-PID under similar working conditions.

- The output of SO-PID was compared via calculating the gains of PID and through error values such as *ITAE*, *IAE*, and *ITSE*.
- The *ITAE* achieved through ZNM-PID was 4249, *IAE* was 89.91, and *ITSE* was 2590. Further, the *ITAE* obtained via QOHS-PID was 59.54, *IAE* was 7.257, and *ITSE* was 13.97; hence, it can be clearly seen that a remarkable reduction in these values was obtained for the same model with the same disturbance.
- The SO-PID reduced the error value, i.e., *ITAE* to 37.1 from 59.54, *IAE* to 4.706 from 7.257, and *ITSE* to 5.762 from 13.97 and hence, SO-PID outperformed all the other methods in terms of *ITAE*, *IAE*, and *ITSE*. The graphical results supported the numerical results for the various considered cases.
- The research was extended to see the impact of a UC or RFB with SO-PID for a diesel–wind isolated system. The *ITAE* obtained with SO-PID was 37.1 and was reduced to 32.46 with the integration of a UC and further reduced to 5.336 after linking an RFB into the diesel–wind isolated system under similar disturbances.
- The trend was the same for *IAE* and *ITSE* and *IAE* became 1.411 and *ITSE* was reduced to 0.453 from 4.706 and 5.762; hence, it was evident that a UC and RFB with SO were able to improve the dynamic performance of the diesel–wind isolated system.
- The RFB and SO-PID combination effectively suppressed frequency and power deviations of the diesel–wind isolated system for a step change in load, load changes at different instances, and random load patterns when compared to SO-PID with UC and with SO-PID only.

**Author Contributions:** All authors planned the study and contributed to the idea; introduction, V.R.; software, P.N.B., G.S. and E.Ç.; analysis, V.R., G.S., P.N.B. and E.Ç.; conclusion, V.R. and G.S.; writing—original draft preparation, V.R.; writing—review and editing, V.R.; supervision, G.S. and P.N.B. All authors have read and agreed to the published version of the manuscript.

**Funding:** This research received no external funding.

**Data Availability Statement:** Not applicable.

**Conflicts of Interest:** The authors declare no conflict of interest.

## Abbreviations

| | |
|---|---|
| *FF* | Fossil Fuel |
| UC | Ultra Capacitor |
| RES | Renewable Energy Storage |
| ESD | Energy Storage Device |
| RFB | Redox Flow Battery |
| HAE | Hydrogen Generative Aqua Electrolyzer |
| FC | Fuel Cell |
| FL | Fuzzy Logic |
| SO | Snake Optimization |
| DEG | Diesel Engine Generator |
| *WTG* | Wind Turbine Generator |
| ZNM | Ziegler–Nichols Method |
| DPM | D Partition Method |
| *ITAE* | Integral Time Absolute Error |
| *IAE* | Integral Absolute Error |
| ISE | Integral Squared Error |
| *Kp* | Proportional Gain |
| *Ki* | Integral Gain |
| *Kd* | Derivative Gain |
| PID | Proportional Integral Derivative |
| QOHS | Quasi-Opposition Harmony Search |

## Appendix A

**Table A1.** Nominal system parameters of the diesel–wind model [3,26].

| Model | Parameters | | | | | | |
|---|---|---|---|---|---|---|---|
| Diesel Unit | $K_D = 0.3333$ | $R_D = 3.0\ \text{Hz/pu}$ | $T_{D1} = 1.00\ \text{s}$ | $T_{D2} = 2.00\ \text{s}$ | $T_{D3} = 0.025\ \text{s}$ | $K_P = 120$ | $T_P = 14.4\ \text{s}$ |
| Wind Unit | $K_{P1} = 1.25$ | $K_{P2} = 1.00$ | $K_{P3} = 1.40$ | $K_{TP} = 0.0033$ | $K_{IG} = 0.9969$ | $T_W = 13.25\ \text{s}$ | $K_{PC} = 0.080$ |
| | $T_{P1} = 0.60\ \text{s}$ | $T_{P2} = 0.041\ \text{s}$ | $T_{P3} = 1.0\ \text{s}$ | | | | |
| RFB | $K_O = 0.45\ \text{p.u. MW/Hz}$ | $T_d = 0\ \text{s}$ | $K_{rb} = 1$ | $T_{rb} = 6.7\ \text{s}$ | $\Delta P_{rb}^{max} = +0.004$ p.u. MW | $\Delta P_{rb}^{min} = -0.004$ p.u. MW | |

## References

1. Mallesham, G.; Mishra, S.; Jha, A.N. Ziegler-Nichols based controller parameters tuning for load frequency control in a microgrid. In Proceedings of the 2011 International Conference on Energy, Automation and Signal (ICEAS), Bhubaneswar, India, 28–30 December 2011; pp. 1–8.
2. Veronica, J.; Kumar, N.S.; Longatt, F.G. Design of Load Frequency Control for a Microgrid Using D-partition Method. *Int. J. Emerg. Electr. Power Syst.* **2020**, *21*, 1–11. [CrossRef]
3. Dekaraja, B.; Chandra Saikia, L. Impact of RFB and PLL Dynamic on Combined ALFC-AVR Regulation of Multiarea Multisource System Under Deregulated Environment with AC/Accurate HVDC link. *IETE J. Res.* **2022**, 1–25. [CrossRef]
4. Bevrani, H.; Ghosh, A.; Ledwich, G. Renewable energy sources and frequency regulation: Survey and new perspectives. *IET Renew. Power Gener.* **2010**, *4*, 438. [CrossRef]
5. Eduardo, M.G.; Manuel, A.M. Evaluating operational risk in a power system with a large amount of wind power. *Electr. Power Syst. Res.* **2009**, *79*, 734–739.
6. Senjyu, T.; Nakaji, T.; Uezato, K.; Funabashi, T. A hybrid power system using alternative energy facilities in isolated island. *IEEE Trans. Energy Conver.* **2005**, *20*, 406–414. [CrossRef]
7. Rekioua, D.; Bensmail, S.; Bettar, N. Development of hybrid photovoltaic-fuel cell system for stand-alone application. *Int. J. Hydrogen Energy* **2014**, *39*, 1604–1611. [CrossRef]
8. Mokrani, Z.; Rekioua, D.; Rekioua, T. Modeling, control and power management of hybrid photovoltaic fuel cells with battery bank supplying electric vehicle. *Int. J. Hydrogen Energy* **2014**, *39*, 15178–15187. [CrossRef]
9. Mezzai, N.; Rekioua, D.; Rekioua, T.; Mohammedi, A.; Idjdarane, K.; Bacha, S. Modeling of hybrid photovoltaic/wind/fuel cells power system. *Int. J. Hydrogen Energy* **2014**, *39*, 15158–15168. [CrossRef]
10. Aissou, S.; Rekioua, D.; Mezzai, N.; Rekioua, T.; Bacha, S. Modeling and control of hybrid photovoltaic wind power system with battery storage. *Energy Convers. Manag.* **2015**, *89*, 615–625. [CrossRef]
11. Tamalouzt, S.; Benyahia, N.; Rekioua, T.; Rekioua, D.; Abdessemed, R. Performances analysis of WT-DFIG with PV and fuel cell hybrid power sources system associated with hydrogen storage hybrid energy system. *Int. J. Hydrogen Energy* **2016**, *41*, 21006–21021. [CrossRef]
12. Wang, L.; Lee, D.; Lee, W.; Chen, Z. Analysis of a novel autonomous marine hybrid power generation/energy storage system with a high-voltage direct current link. *J. Power Sources* **2008**, *185*, 1284–1292. [CrossRef]
13. Ngamroo, I. Robust frequency control of wind-diesel hybrid power system using superconducting magnetic energy storage. *Int. J. Emerg. Electr. Power Syst.* **2009**, 10–18. [CrossRef]
14. Francis, R.; Chidambaram, I.A. Optimized PI+ load frequency controller using BWNN approach for an interconnected reheat power system with RFB and hydrogen electrolyser units. *Int. J. Elect. Power Energy Syst.* **2015**, *67*, 381–392. [CrossRef]
15. Sahu, R.K.; Gorripotu, T.S.; Panda, S. A hybrid DE-PS algorithm for load frequency control under deregulated power system with UPFC and RFB. *Ain Shams Eng. J.* **2015**, *6*, 893–911. [CrossRef]
16. Sharma, G.; Narayanan, K.; Adefarati, T.; Sharma, S. Frequency regularization of a linked wind–diesel system using dual structure fuzzy with ultra-capacitor. *Prot. Control Mod. Power Syst.* **2022**, *7*, 12. [CrossRef]
17. Ibraheem Kumar, P.; Kothari, D.P. Recent philosophies of automatic generation control strategies in power systems. *IEEE Trans. Power Syst.* **2005**, *20*, 346–357. [CrossRef]
18. Yang, H.; Wei, Z.; Chengzhi, L. Optimal design and techno-economic analysis of a hybrid solar-wind power generation system. *Appl. Energy* **2009**, *86*, 163–169. [CrossRef]
19. Ray, P.; Mohanty, S.; Kishor, N. Proportional-integral controller based small signal analysis of hybrid distributed generation systems. *Energy Convers. Manag.* **2011**, *52*, 1943–1954. [CrossRef]
20. Abou El Ela, A.A.; El-Sehiemy, R.; Shaheen, A.M.; Diab, A.E.G. Optimal Design of PID Controller Based Sampe-Jaya Algorithm for Load Frequency Control of Linear and Nonlinear Multi-Area Thermal Power Systems. *Int. J. Eng. Res. Afr.* **2020**, *50*, 79–93. [CrossRef]
21. Das, D.C.; Roy, A.; Sinha, N. PSO based frequency controller for wind-solar-diesel hybrid energy generation/energy storage system. *Proc. ICEAS* **2011**, 458–463. [CrossRef]

22. Zhou, T.; Sun, W. Optimization of battery super-capacitor hybrid energy storage station in wind/solar generation system. *IEEE Tran. Sustain. Energy* **2014**, *5*, 408–415. [CrossRef]
23. Maleki, A.; Askarzadeh, A. Artificial bee swarm optimization for optimum sizing of a stand-alone PV/WT/FC hybrid system considering LPSP concept. *Sol. Energy* **2014**, *107*, 227–235. [CrossRef]
24. Das, D.C.; Roy, A.; Sinha, N. GA based frequency controller for solar thermal diesel–wind hybrid energy generation/energy storage system. *Electr. Power Energy Syst.* **2012**, *43*, 262–279. [CrossRef]
25. Pan, I.; Das, S. Fractional order fuzzy control of hybrid power system with renewable generation using chaotic PSO. *ISA Trans.* **2016**, *62*, 19–29. [CrossRef] [PubMed]
26. Gangulya, S.; Shivab, C.K.; Mukherjeec, V. Frequency stabilization of isolated and grid connected hybrid power system models. *J. Energy Storage* **2018**, *19*, 145–159. [CrossRef]
27. El-Sehiemy, R.; Shaheen, A.; Ginidi, A.; Al-Gahtani, S.F. Proportional-Integral-Derivative Controller Based-Artificial Rabbits Algorithm for Load Frequency Control in Multi-Area Power Systems. *Fractal Fract.* **2023**, *7*, 97. [CrossRef]
28. Abou El-Ela, A.A.; El-Sehiemy, R.A.; Shaheen, A.M.; Diab, A.E.G. Enhanced coyote optimizer-based cascaded load frequency controllers in multi-area power systems with renewable. *Neural Comput. Appl.* **2021**, *33*, 8459–8477. [CrossRef]
29. Hashim, F.A.; Hussien, A.G. Snake Optimizer: A novel meta-heuristic optimization algorithm. *Knowl.-Based Syst.* **2022**, *242*, 108320. [CrossRef]
30. Çelik, E.; Öztürk, N. Novel fuzzy 1PD-TI controller for AGC of interconnected electric power systems with renewable power generation and energy storage devices. *Eng. Sci. Technol.* **2022**, *35*, 101166. [CrossRef]
31. Çelik, E.; Öztürk, N.; Houssein, E.H. Improved load frequency control of interconnected power systems using energy storage devices and a new cost function. *Neural Comput. Appl.* **2023**, *35*, 681–697. [CrossRef]
32. Çelik, E.; Öztürk, N.; Houssein, E.H. Influence of energy storage device on load frequency control of an interconnected dual-area thermal and solar photovoltaic power system. *Neural Comput. Appl.* **2022**, *34*, 20083–20099. [CrossRef]

*Review*

# Real-Time Grid Monitoring and Protection: A Comprehensive Survey on the Advantages of Phasor Measurement Units

Chinmayee Biswal [1], Binod Kumar Sahu [1], Manohar Mishra [2,*] and Pravat Kumar Rout [2]

[1]  Department of Electrical Engineering, Siksha 'O' Anusandhan (Deemed to Be University), Bhubaneswar 751030, India; chinmayeebiswal0895@gmail.com (C.B.); binoditer@gmail.com (B.K.S.)
[2]  Department of Electrical and Electronics Engineering, Siksha 'O' Anusandhan (Deemed to Be University), Bhubaneswar 751030, India; pkrout_india@yahoo.com
*  Correspondence: manoharmishra@soa.ac.in

**Abstract:** The emerging smart-grid and microgrid concept implementation into the conventional power system brings complexity due to the incorporation of various renewable energy sources and non-linear inverter-based devices. The occurrence of frequent power outages may have a significant negative impact on a nation's economic, societal, and fiscal standing. As a result, it is essential to employ sophisticated monitoring and measuring technology. Implementing phasor measurement units (PMUs) in modern power systems brings about substantial improvement and beneficial solutions, mainly to protection issues and challenges. PMU-assisted state estimation, phase angle monitoring, power oscillation monitoring, voltage stability monitoring, fault detection, and cyberattack identification are a few prominent applications. Although substantial research has been carried out on the aspects of PMU applications to power system protection, it can be evolved from its current infancy stage and become an open domain of research to achieve further improvements and novel approaches. The three principal objectives are emphasized in this review. The first objective is to present all the methods on the synchro-phasor-based PMU application to estimate the power system states and dynamic phenomena in frequent time intervals to observe centrally, which helps to make appropriate decisions for better protection. The second is to discuss and analyze the post-disturbance scenarios adopted through better protection schemes based on accurate and synchronized measurements through GPS synchronization. Thirdly, this review summarizes current research on PMU applications for power system protection, showcasing innovative breakthroughs, addressing existing challenges, and highlighting areas for future research to enhance system resilience against catastrophic events.

**Keywords:** wide-area measurement systems (WAMS); state estimation (SE); synchro-phasor technology; fault detection; phasor measurement units (PMUs); fault localization; power swing; backup protection; renewable energy sources (RESs)

**Citation:** Biswal, C.; Sahu, B.K.; Mishra, M.; Rout, P.K. Real-Time Grid Monitoring and Protection: A Comprehensive Survey on the Advantages of Phasor Measurement Units. *Energies* **2023**, *16*, 4054. https://doi.org/10.3390/en16104054

Academic Editors: Saeed Sepasi and Quynh Thi Tu Tran

Received: 3 April 2023
Revised: 4 May 2023
Accepted: 10 May 2023
Published: 12 May 2023

## 1. Introduction

In the conventional power grid, the energy management system (EMS) and Supervisory Control and Data Acquisition (SCADA) system can afford the steady state view of the power system with high data flow latency [1]. The critical issue in adopting a better protection scheme with these setups is the inability to measure the phase angles of the bus voltage and current in real-time in a synchronized way [2]. The instant dynamic behavior assessment with faster monitoring and measurement is essential for a reliable, protective, and secure operation. The transformation to the smart grid concept of the modern power system needs faster monitoring, more accurate, and synchronized phasor measurement of electrical parameters to achieve the above objective with the use of a Global Positioning System (GPS) to power system operation [3]. Among many emerging devices to facilitate better system monitoring, measurement, and signal communication, PMUs are very attractive in comparison due to the ability to provide the phasor signals in microseconds with accuracy and a faster way through the GPS [4–6]. The application of the PMUs in the system network facilitates tracking grid dynamics in real-time, obtaining data for wide-area monitoring, assessing the state estimation, predicting system failure, helping to plan energy delivery proactively, and providing improved protection and control [7–9]. Due to this perspective of the PMUs, research of the application of the PMUs in power systems is up-and-coming with many research gaps to handle the solution issues and challenges, particularly to power system protection. The objective of this review includes WAMS-based power network protection, PMU measurement for enhancing power grid monitoring and protection, fault monitoring, localization and detection, and PMU-based state and security estimation.

The conventional measurements were asynchronously reported in 2–5 s intervals [1]. This time interval limited the capabilities of monitoring and measurement operation for the quick and correct decision to provide a secure protective operation in real-time. The introduction of PMU enhances the power system monitoring and subsequent control and protection aspects in many ways, according to the intent of the application [10]. The measuring features such as the fast reporting rate (50–100 measurements/s), synchronization of measurements through GPS signals, measurements of frequency and frequency rate of change, and the provision of phase measurements (instead of only magnitude measurements) are the significant factors to put together the PMUs as the critical enablers for real-time measurement and protection. Firstly, even a wide power network can be fully or partially observable by PMUs, which helps monitor power systems dynamically to take corrective solutions for steady and stable operation even under abnormal conditions [11]. Secondly, the integration of PMUs is optimally restructured along with the current SCADA system to establish the wide-area monitoring and protection system (WAMPS) [8]. This will provide real-time visualization of the system variation and improve the capabilities of WAMPS for dynamic state estimator, phase angle monitoring, oscillation detection and monitoring, and voltage stability [11,12]. This, in turn, considerably improves many crucial operational benefits such as enhancing reliability by reducing the number of blackouts and power cuts, cost-effective power system operation, and extensively allowing the integration of RESs in the smart power system [9]. Thirdly, the uncertainties associated with renewable energy sources (RESs), electrical energy storage (EES), electric vehicle charging (EVC), and changes in load scenarios make control and protection more challenging [11]. Wide-area monitoring with PMUs can overcome new challenges in the operation, planning, protection, and control of distribution and transmission grids to enhance flexibility for the system operation [9]. In addition, the new fast communication technologies support the WAMPS's fast access to power grid information. The PMU application to the power transmission and distribution sector is indispensable and needs further research to enhance the protection to make the system secure and reliable.

The installations of PMUs in intelligent grid systems reach for distributed and hierarchical controlled WAMS that facilitate various monitoring, protection, and control applications [13]. The conventional protection schemes restrict the distance relays to operate for faults occurring between the location of the concerned relay and the related reach point [14]. It remains unaffected in the cases of all the faults outside this domain, leading to unreliable and insecure protection for the system. The WAMS technology based on the PMUs' synchronized phasor measurement brings a better solution to the above issue [9]. Secondly, a wide-area differential protection (WADP) can be established with the PMUs' measurement to suit the shipboard applications' needs. Thirdly, a secured, coordinated, and synchronized backup protection can be established to function as an intelligent processing system. In a nutshell, WAMS with the PMUs' measurement provide the grid dynamic state in terms of angular and voltage stability and transfer capacity at different instances on various transmission and distribution systems to control and regulate the power flow satisfying the constraints related to the grid parameters, integrating various protection schemes and identifying corrective actions with an accurate decision [12]. Lastly, the security of both the transmission and distribution sector is reasonably necessary with the provision of the WAMS to cope with the deregulated and restructured power system. Overall, WAMS facilitate a developed intelligent grid of adaptive islanding, a secure and self-healing system.

Even though very few review articles have been published recently, as shown in Table 1, a proper, concise review confined to PMU application to grid protection aspects is yet to present for the researchers and power engineers. Few review research articles mention the broad scope of PMUs and WAMS with their possible applications. This article extensively discusses and analyzes PMUs' utilization in the power system protection domain at the transmission and distribution levels. The significant contributions of this study can be enumerated as follows:

- It emphasizes various applications of PMUs to provide better protection and secure microgrid systems. The investigations were conducted on protection aspects only.
- The ideas and newness in the proposed PMU-based backup protection schemes by various researchers are systematically presented along with their pros and fault detection times.
- This review provides an overview of the standards for installing, testing, and designing PMUs in power system protection.
- The research gaps are extensively summarized to provide the futuristic scope of the PMU-based protection scheme design and formulation.

**Table 1.** Recent literature review papers on PMU-based protection.

| Ref. No. | Title | Highlights | Shortcomings |
|---|---|---|---|
| [1] | A Comprehensive Survey on Phasor Measurement Unit Applications in Distribution Systems | • Surveys the primary challenges of PMU applications. <br> • Provides possible future applications for distribution networks. <br> • It encompasses using synchro-phasor-based intelligent electronic devices in distribution systems. | • Does not provide any comparative analysis of the PMU-based protection scheme. <br> • Does not discuss the challenges and problems that PMUs may face concerning their role in protection systems. |
| [10] | D-PMU based applications for emerging active distribution systems: A review | • Surveys the implementation of D-PMUs in the distribution network. <br> • Reviews the application, monitoring, and control of D-PMUs. | • Does not explain the operation and principles of PMUs. <br> • Does not provide any insight into the standard of PMU application. |
| [13] | Synchro-phasor measurement applications and optimal PMU placement:A review | • Surveys a comprehensive overview of synchro-phasor technology, including its principles, operation, and optimal placement techniques. <br> • Covers various applications in the transmission and distribution system. | • Does not offer any information or understanding regarding the standard utilized in PMUs. <br> • Does not offer any details or explanation regarding communication systems based on synchro-phasors. |
| [2] | A Survey on the Micro-Phasor Measurement Unit in Distribution Networks | • Provides the advantages of using μPMUs in distribution networks, including their ability to improve reliability and robustness. <br> • Covers environmental sustainability and social and economic benefits. | • Does not provide detailed information related to the wide-area backup protection. <br> • No specific data has been mentioned regarding the PMU standards. |

The remaining part of the manuscript is organized as follows: Section 2 addresses the various aspects of PMUs and WAMS, looking at their architecture, design, operation, and implementation. Section 3 depicts different issues and challenges related to PMU application, particularly for protection aspects. Section 4 summarizes various PMU-based protection methods, including backup protection. Section 5 presents several applications of PMUs. Sections 6 and 7 enumerate the standards and the future scope, respectively.

Finally, Section 8 ends with the concluding remarks from the work performed on the PMU application for protection to date.

## 2. PMU and WAMS for Power System Protection

This section presents the basic principle of PMUs and WAMS technology operation and how it improves power system monitoring and protection. Moreover, researchers have conducted studies on several monitoring and protection techniques that employ PMUs, as demonstrated by the comparative publication statistics illustrated in Figure 1, on an annual basis. A distribution pie chart of articles published in various journals covering PMU and protection applications is available, as shown in Figure 2.

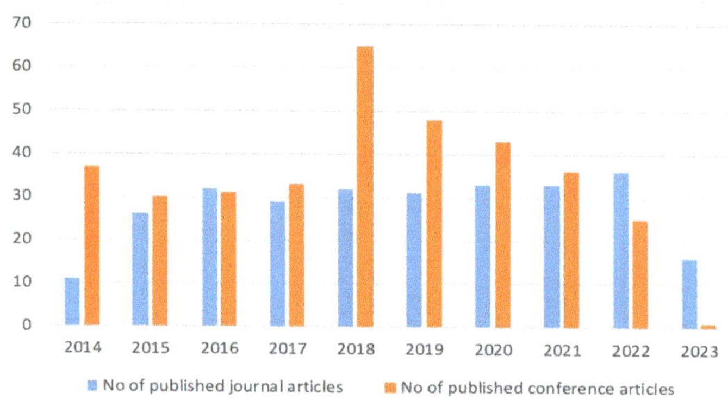


**Figure 1.** Annual publication data on PMU-based protection.

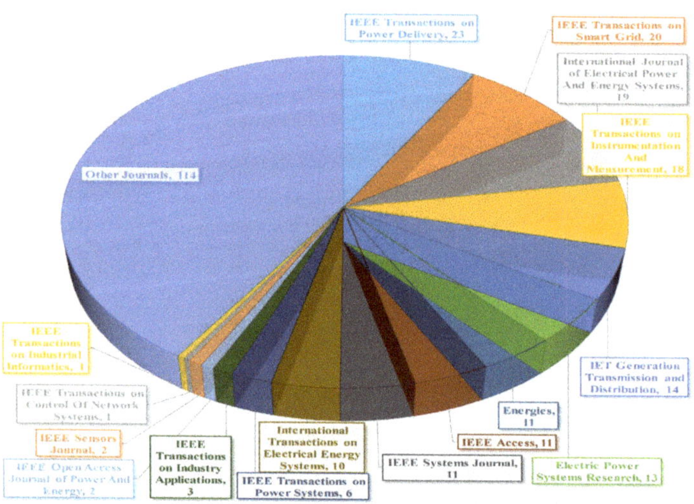


**Figure 2.** PMU-based protection schemes in various academic publications.

### 2.1. Phasor

A phasor is a mathematical representation of a sinusoidal wave quantity in which the complex number's modulus represents the cosine/sine wave's amplitude, and the angle represents the cosine/sine wave's phase angle. Using phasors simplifies the analysis of circuits and systems that involve sinusoidal wave quantities, making it easier to perform calculations and visualize the relationships between different signals [15,16].

A sinusoidal signal is generally represented as follows:

$$x(t) = X_m \cos(\omega t + \varphi). \tag{1}$$

The phasor representation of this expression is represented by the following:

$$X = \frac{X_m}{\sqrt{2}} e^{j\varphi} = \frac{X_m}{\sqrt{2}} (\cos \varphi + j \sin \varphi). \tag{2}$$

Here, $X_m$ and $\varphi$ are denoted as the peak value of the sinusoid and the signal's phase angle, respectively. The phase angle measurement is regarded as positive when measured in an anti-clockwise direction from the positive real axis. It relies on the frequency inherent in the phasor representation. For different signals to be represented in the same phasor diagram, they must have the same frequency. Still, input signals may not remain stationary in real-time conditions, as shown in Figure 3. A finite data window is used to consider the input signals to address this. PMUs require at least one period of the input signal's fundamental frequency in the data window [8]. However, when non-harmonic and harmonic components are present, the frequency is not necessarily the same as the fundamental frequency. Thus, PMUs use frequency tracking algorithms to separate the fundamental frequency components and estimate the fundamental frequency period before measuring the phasor.

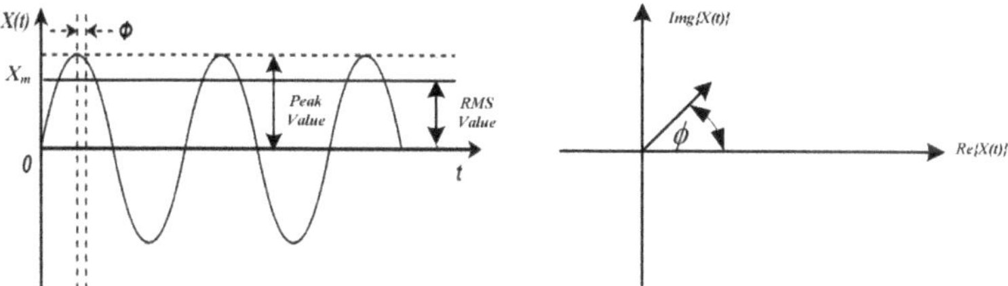


**Figure 3.** Synchro-phasor representation.

## 2.2. Synchro-Phasor and PMUs

A PMU is a tool used to calculate the phase angle and magnitude of a grid-wide electrical quantity (such as voltage or current) using a standard time reference. The time synchronization is given by a GPS and according to some protocols such as the IEEE 1588 Precision Time Protocol [17]. This enables synchronized real-time measurements of several distant places on the grid and helps to provide a wide-area monitoring and signal retrieval system. PMUs can quickly take samples from a waveform and rebuild the phasor quantity, which consists of measurements of both the angle and magnitude. These resulting measurements are known as the synchro-phasor data. These time-synchronized observations are crucial to analyze the power system from a control, protection, and energy management point of view to smooth the whole system's operation.

Figure 4 depicts the operational block diagram of a PMU. It processes information quickly and accurately and estimates the phasors and timestamps of the incoming signals. The anti-aliasing filter eliminates signal elements whose frequency is equal to/higher than half the Nyquist rate. An analog-to-digital converter (A/D converter) converts the input voltage and current phasors from analog to digital AC signals. A phase-locked oscillator (PLO) transforms the GPS one pulse/sec signal into a series of high-frequency pulses (HFP). The phasor microprocessor performs the phasor computations, and the resulting phasors are merged to provide positive sequence readings. The information from the GPS and clock is eventually used to tag the phasors. At last, by using a modem, the ultimate phasor value is sent to a data center [13].

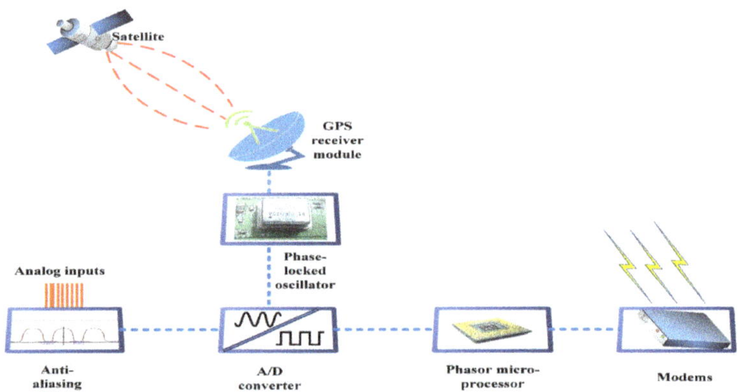


**Figure 4.** The principle of operation of PMUs as suggested initially at Virginia Tech., USA (Blacksburg, VA, USA).

### 2.3. Synchro-Phasor-Based Communication

Communication features are required for applications that demand phasor data in isolated locations. In each communication activity, two data flow components are crucial [18,19]. The first factor is channel capacity (the rate of data change that could be maintained on the current dataset), calculated in kilobits per second or megabits per second. The second factor is latency, which is the interval between the creation of data and its availability for use by the intended purpose [20]. There are two basic categories of communication networks, including wired (e.g., telephonic line, overhead line, coaxial lines, power line communication (PLC), optical fiber) and wireless (e.g., Wi-Fi, radio frequency, satellite, microwaves). Wired networks may be divided into star and mesh networks [21,22]. Figure 5 represents an overall view of the synchro-phasor-based communication network.

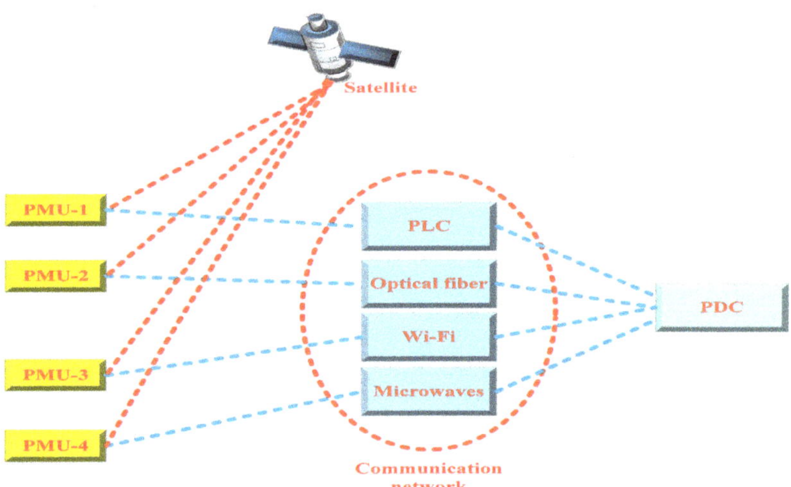


**Figure 5.** Synchro-phasor-based communication network.

A star network is one in which all communication nodes are connected to the central node. However, additional communication cables and towers are required to establish such a network. Furthermore, real-world power systems, such as NASPInet, have already adopted star networks [23]. On the other hand, a mesh network is a network whose communication topology resembles a network of power lines. It can utilize power line communication (PLC) technology [24] and has proven to be effective in the past [25,26]. Depending on the physical communication medium, the PMU data transfer modes may be categorized [27]. Leased line connections (LLC) were one of the earliest forms of communication for these uses. Switched telephone circuits (STC) are also viable if transfer latency is not critical. Additionally, traditional electric utility communication techniques, including PLC and microwave links, have been widely adopted and are still prevalent in many current applications [20].

Despite the challenges such as noise, signal loss, and attenuation, PLC remains a feasible option for PMU communications at medium and low voltage distribution levels due to its low cost, easy accessibility, and widespread deployment. However, the latest fifth-generation (5G) mobile services are considered the most promising alternative for PMU communications in the future. With a transfer rate of 20 gigabits per second (Gbps), 5G is almost 100 times faster than 4G, enabling advanced wireless broadband, extensive machine-type communications, and ultra-reliable low-latency communications [28]. Although this technology has not been deployed in functioning power systems, its security must be ensured for future applications. Lastly, fiber-optic networks are the preferred medium today because they have unmatched channel capacity, rapid data transmission speeds, and resistance to electromagnetic interference.

### 2.4. Synchro-Phasor-Based Wide-Area Measuring System (WAMS)

Power systems are evolving due to the increased renewable energy production, open access to transmission networks, and deregulation of the electrical industry. The functioning of electrical power systems is under strain due to this reality, which increases the need to enhance the control and protection systems provided by WAMS based on PMU [29]. The critical elements of WAMS are phasor data concentrators (PDCs), communication networks, and PMUs, as shown in Figure 6.

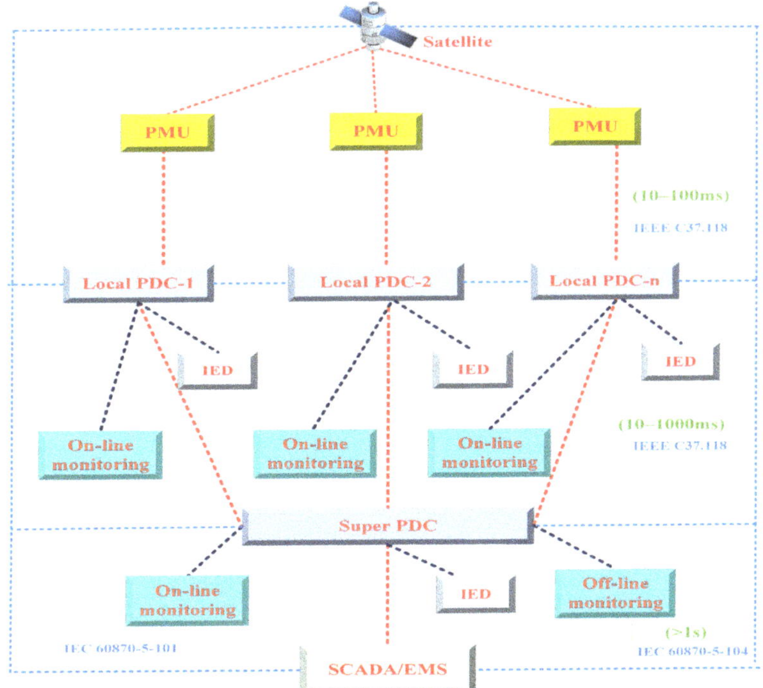


**Figure 6.** Hierarchical architecture of WAMS.

In the hierarchically constructed WAMS, the PDC arranges the phasors by the GPS timestamps using the measured output data from PMUs [30]. The PMUs collect and transmit critical information about the power system, such as voltage, current, frequency, and rate of change of frequency (ROCOF), using the IEEE C37.118.2 or IEC 61850-90-5 standards. These data are then received by nearby PDCs (within 10 ms–100 ms), which synchronize and distribute them to super PDCs. The PDCs are responsible for monitoring, storing, and analyzing the received data and detecting and responding to faults in the system [31]. Additionally, it may utilize intelligent electronic devices (IEDs), such as protective relays, switches, and capacitor banks, to quickly respond to disturbances and ensure the stability and security of the power system at both the distribution and transmission levels. However, transferring all the data from the synchro-phasor network to the SCADA or EMS is not required, as receiving higher-level online alerts is sufficient. According to a suggestion in [1], the future of synchro-phasor networks involves using the IEC 61850-90-5 standard for efficient communication between all network components, replacing the current IEEE C37.118.2 standard. Furthermore, rapid relay communication can be achieved using the IEC 61850 standard. This approach allows for additional data review at a later time while ensuring quick and reliable communication between all components of the synchro-phasor network.

### 2.5. Synchro-Phasor-Based Load Shedding

Currently, power systems are operated closer to their stability limits, mainly due to the deregulation of the electricity markets and the rise in electrical energy demand. However, due to significant disturbances, such as tripped generators or transmission line faults [32], the following active and reactive power unbalance can ultimately cause substantial simultaneous voltage and frequency variations along with instability. As a result, there is a greater chance that the entire system could collapse [33]. Load shedding is a technique that can prevent system collapse by shedding some of the load demand when there is an imbalance between the power generation and load demand, as shown in Figure 7. The technique can be represented as follows [34]:

$$Min \sum_{j=m} c_j x_j \tag{3}$$

where $c_j$ represents the weight of the load; $x_j = 1$ represents that load $j$ needs to be shed.

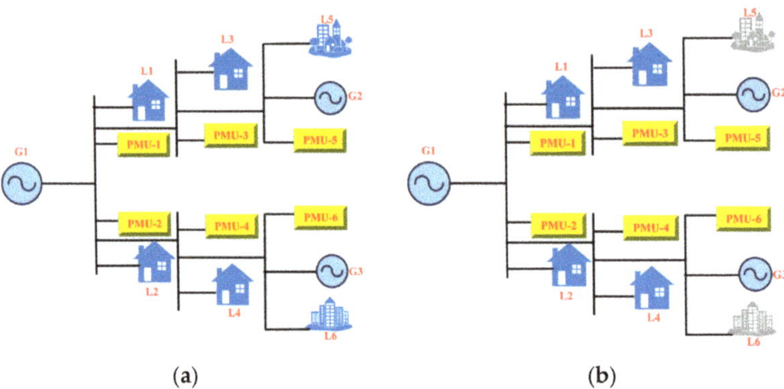

(a)        (b)

**Figure 7.** A simplified load-shedding scenario (**a**) before and (**b**) after.

The technique is classified into two main categories based on the objectives and measurements; these include under-frequency load shedding (UFLS) and under-voltage load shedding (UVLS) [35]. In the past, various under-frequency load shedding (UFLS) methods have been developed and can be broadly categorized into classic, adaptive, semi-adaptive, and algorithmic approaches [36,37]. In conventional UFLS methods, relays use locally measured frequency as input and continuously compare it against predetermined thresholds. The key benefits of these systems are simplicity, ease of operation, and reliability. However, they can also have drawbacks, such as treating steep and gradual frequency gradients equally, potentially resulting in excess or insufficient load shedding [38]. The enormous scope is suitable for the use of PMUs to develop a new methodology for optimal load shedding to prevent voltage and frequency instability phenomena with proper power regulation by adopting the concept of WAMS.

PMUs are a potentially effective solution for improving UFLS [39–41], as they can provide highly accurate frequency and ROCOF measurements with synchronized time information [42]. However, the reliability of PMU-based measurements may be affected by time-varying conditions and transient events since they rely on a static signal model [43,44]. Therefore, the control scheme should consider the potential measurement uncertainty [45, 46]. A new approach was proposed in reference [47] to address the above challenges. This approach utilizes synchro-phasor reading obtained from PMUs to calculate the absolute frequency and ROCOF. The novel approach employs two sets of ROCOF criteria to calculate the load that must be shed and initiates the load shedding. This approach results in a more sensitive system that can handle quick dynamics in a modern power network (MPN). However, adopting PMUs improves the estimate and monitoring rate accuracy, offering a more effective solution for UFLS in MPN. In addition, in [48], a method is suggested to address measurement errors, communication failure, and time delays. The above factors are incorporated into the problem formulation to make the results more accurate and reflective of real-world scenarios. Finally, [49] proposes an approach that considers the connections across load buses and is robust to PMU failure and communication delay, further improving the load-shedding solution's effectiveness. Lastly, PMU is also very helpful in creating credible and practical tools for UVLS issues that may successfully minimize voltage collapse [50–52]. In addition, references [53,54] discuss the problem of voltage collapse in power systems due to high reactive power mismatch. However, the above methods are unsuitable for dynamic real-time conditions and an open research field for power engineers.

## 3. Issues and Challenges of PMU-Based Protection

Recently, many suggestions have been proposed to adopt a secure, reliable, and self-protective capability for the intelligent transmission and distribution sector. However, integrating WAMS and PMUs with the traditional protection scheme and topology raises many inherent critical issues and challenges in real-time, including the following:

- The architecture of wide-area protection systems (WAPS): The architecture of WAPS needs to be designed to provide adaptive relaying for primary protection and long-term voltage instability at the local stage of operation. The frequency instability, angular instability, and control coordination need to be considered for small-signal stability at the global stage of operation. A proper architecture can enhance the system security, improve the speed of information exchange between components, and provide high system reliability [55].
- Dynamic and nonlinear loading: The dynamics of loads and electric vehicle charging in the distribution sector can cause frequency instability. However, they are not substantially considered in frequency instability protection measures based on WAMS. The protection scheme must focus on the dynamic and nonlinear loading characteristics through WAPS [56].
- Overreaching distance relay zones: Overreaching distance relay zones under stressed abnormal conditions can lead to cascaded tripping and blackouts. WAMS-based anticipatory adaptive relaying schemes may be a better solution for handling these situations in the future [57].
- Operational uncertainties of FACT devices: The series of connected FACT devices have a critical impact on the distance protection of the transmission lines due to their

associated operational uncertainties. WAMS-based flexible protection schemes can be designed to handle the uncertainties related to distance protection and acquire the control parameters of FACTS devices in a brief period [4].

- Optimal PMU placement: Without a comprehensive plan for wide-area protection, there is a lack of optimal PMU placement that can provide complete observability of the power system. Proper PMU placement is essential to detect and prevent different types of instabilities and uncertainties in the system [57].
- Time delays: Time delays during information exchange and transfer in PMUs can impact the accuracy and effectiveness of WAPS. Proper measures must be taken to handle time delays accurately and correctly through real-time adaptive identification techniques [57].
- GPS spoof attacks: GPS spoof attacks can influence PMU timestamps by injecting counterfeit GPS signals into the antenna of the PMU's time reference receiver. This can violate the PMU's maximum phase error and inaccurate signaling for protection and control. Mitigation strategies must be developed to prevent GPS spoof attacks [58,59].
- Calibration of PMUs: Characterizing PMUs according to prescribed measurement standards is essential to ensure the reliability and consistency of the operation and control of WAPS. The calibration of PMUs under steady-state conditions traceable to prescribed standards is necessary [60].
- False data injection attacks (FDIA): Cybersecurity is a significant concern for intelligent power grids, and FDIA can cause operational, physical, and economic damage. Continuous cyber-physical system monitoring is essential to achieve secure and reliable operations under cyber threats. The WAPS should be designed to prevent and detect FDIA research gaps, aiming for a better solution to these factors [61].
- Role of PMUs in inverter-based resources: PMUs can be crucial in monitoring and controlling power systems that incorporate inverter-based resources, especially relevant in regions with high resource penetration. However, this deployment also presents challenges, such as the high cost of PMUs, the need for precise time synchronization, and a lack of standardization that may create interoperability issues. Nevertheless, PMUs also present opportunities, including real-time monitoring and control that can improve power system stability and reliability and more precise information to enhance inverter-based resource performance [62].

## 4. PMU-Based Protection Methods

Synchro-phasor measurements have revolutionized the field of power system protection by providing highly accurate and time-synchronized phasor data. This data detects and responds to various power system events, including faults, out-of-step conditions, and other abnormal operating conditions. To address these abnormal conditions, the synchro-phasor has provided several solutions to the above protection problems in power systems. They have enabled the development of sophisticated protection schemes such as adaptive out-of-step protection (OSP), differential protection, and backup protection, which rely on accurate and time-synchronized phasor data to detect and respond to power system events [3].

Differential protection is a commonly used protection scheme in power systems. It protects power transformers, generators, and other critical equipment from damage due to internal faults. It relies on measuring the difference in current flowing into and out of a protected zone. When a fault occurs within the zone, the current balance is disrupted, and the differential relay detects the fault and initiates a trip signal to disconnect the equipment. Adaptive OSP protects power system generators from damage due to the loss of synchronism, which can occur when two generators are connected to the same power system with unsynchronized rotational speeds. Lastly, backup protection is used as a secondary protection scheme to protect against equipment failures or the malfunction of primary protection systems. The schemes rely on synchro-phasor measurements to detect and isolate faults quickly and accurately, ensuring the power system remains stable and reliable.

### 4.1. Backup Protection

With the emergence of interlinked power systems, systemic disruptions (e.g., transmission line faults, blackout events) inside the power system may create an inexorably problematic situation for the power industry. To prevent the deterioration of the power grid, protection and control measures must be taken to recover the system to normal operating conditions. Generally, primary and backup protection is used to protect the transmission lines [63,64]. Conventional backup protection is mainly performed based on independent judgment following local data. Furthermore, conventional backup protection cannot adjust to varying loading scenarios and fault impedance [65]. As a result, symmetrical disturbances cannot be detected from other loading conditions, including load invasion, power swing, and generator breakdowns [66].

Additionally, backup protection may operate incorrectly, leading to cascading failures or blackout events [66,67]. The implementation of phasor measurement units (PMUs) in power systems has recently been widely acknowledged. The voltage and current synchro-phasor are delivered by PMUs using GPS. Researchers are proposing wide-area backup protection (WABP) schemes that utilize data collected from various points in the grid to detect transmission line faults. There are three significant categories documented in WABP research [68]. The first involves using electrical quantity parameters, the second involves WABP based on the protection relay device's switching status information, and the third category combines electrical quantity and circuit breaker/protective relay operating information.

Figure 8 depicts a schematic diagram of WABP consisting of a protection device (PD) that detects the network's electrical properties and transfers the data to the regional network's system protection center (SPC). After that, SPC examines the data and runs the protective algorithm to decide whether or not a defect has occurred. Once a defect is discovered, the SPC sends a trip order to the appropriate circuit breaker, isolating the faulty area of the network. Table 2 depicts the backup protection approaches recommended by many authors in recent years, and a detailed examination of these techniques is covered in the following sections.

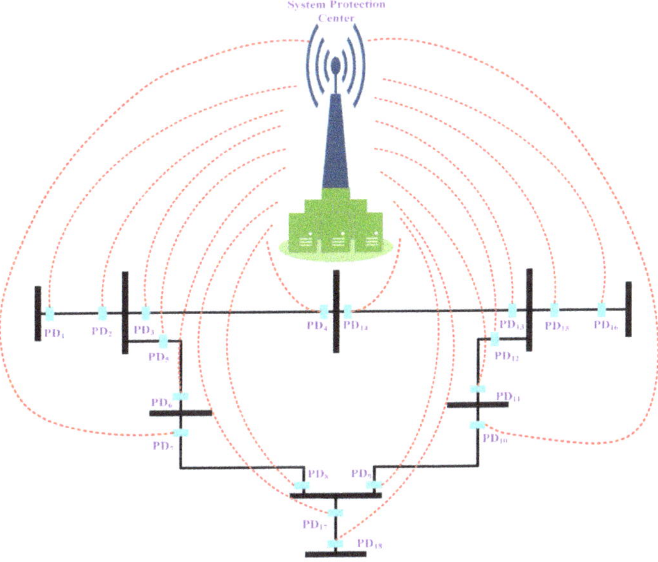


**Figure 8.** Architecture of WABP.

### 4.1.1. WABP Scheme Based on Electrical Quantity Parameters

This scheme consists of WABP based on information on electrical parameters, such as the evaluation of the fault voltage [69], the comparison of the sequence current phase and amplitude, the comparison of the voltage and current composite directional component [70–72], and the cumulative impedance comparison algorithm. As mentioned above, the schemes can realize WABP by gathering wide-area data without modifying the typical backup protection architecture. However, by comparing the variation of electrical quantity between the calculated and estimated values to a certain threshold, the defective line was successfully located, as shown in Figure 9.

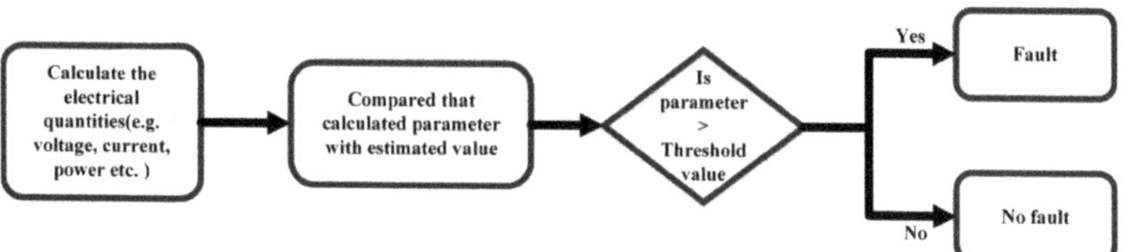


**Figure 9.** Block diagram of WABP scheme based on electrical quantity.

Identifying a faulty transmission line was achieved by analyzing variations in the positive-sequence current angles across all transmission lines. To determine the bus closest to the defective line, [72] analyzed the positive-sequence voltage (PSV) magnitudes of the buses and calculated the absolute difference in the positive-sequence current (PSC) angles at both ends of each line connected to the identified bus. However, this method has limitations, particularly for high impedance faults where the measurement of the bus voltage magnitude may be low, leading to insignificant differences in current angles. Several alternative techniques have been proposed to address this limitation. For example, a unique WABP algorithm was proposed in [70] to identify faulty branches based on the steady-state fault component.

In contrast, another special WABP algorithm based on fault component voltage distribution was suggested in [69]. Later, ref. [73] proposed a WABP scheme based on PSV readings to determine the location of the damaged line. At the same time, the Koopman analysis was used in another study to generate WABP and detect the faulty line based on the PMU data, but it has a high computational burden [74]. Further, reference [75] introduced a novel WABP algorithm that does not consider system variables such as transmission line impedance, but only uses the phasor data of voltages and currents, which was proven effective. In [76], a decentralized method was presented to detect and clear transmission line faults in power systems. This approach utilizes the gain in momentum data of all generators and PSV magnitudes to identify the susceptible protection zone and the nearest bus to the fault.

Moreover, the techniques described in references [72–76] do not address two crucial factors, wide-area fault detection and location for cross-country and evolving faults. In this direction, a novel WABP scheme was proposed in [68] to handle the associated issues. The scheme requires only a limited number of synchronized PMUs, making it a cost-effective solution for power systems. The main drawback of all these schemes is that they need a lot of observations and relevant data, such as transmission line voltage and current. As a result, when examining the real-time implementation in a large region, such schemes could be both costly and challenging.

### 4.1.2. WABP Scheme Based on Switching Status of Protection Relay

This scheme consists of WABP based on switching information and updates, such as WABP-based genetic and fault tolerance algorithm [77]. Several techniques are suggested in [78–81]; these backup protection principle schemes are framed by observing and analyzing the switching status of the protection device, as demonstrated in Figure 10.

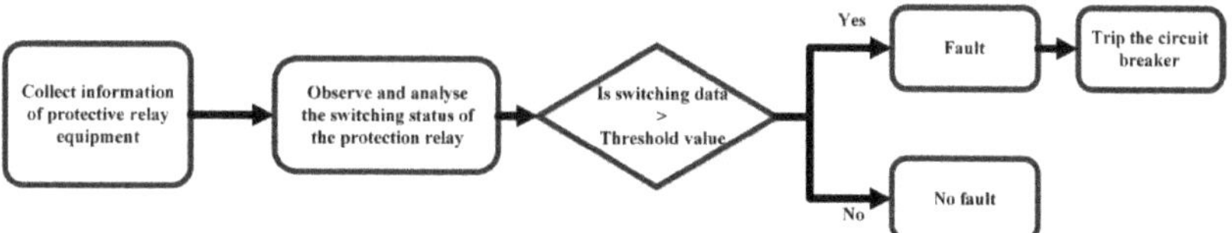


**Figure 10.** Block diagram of WABP scheme based on the status of circuit breaker.

In [78], the authors used broad-area data, such as the statuses and operations of conventional protection relays, to detect faults through the collaboration between intelligent agents. Another backup protection scheme, outlined in [79], relies on the status of primary and auxiliary protective relays and circuit breakers to simplify the settings and avert cascading trips. In [80], the authors suggested a comprehensive backup protection scheme that considers the contribution of all established distance relays across the network, utilizing a protection fitness function. While [81] presented a backup protection approach that relies on the status of the distance relay's second and third zones, its key restrictions include higher end-to-end latencies and data congestion at the phasor data concentrator (PDC). The main drawback of these schemes is that they frequently need the circuit breaker's status, which could make them impracticable and expensive.

### 4.1.3. WABP Scheme Based on Electrical Quantity and Switching Status of Protection Relay

In this study, novel methods for the transmission line backup protection are provided by employing synchro-phasor measurements. These schemes consist of WABP based on data on electrical parameters and the operational status of protective devices. The above methods are suggested to be implemented into two phases at the predefined buses. The first stage involves identifying the fault and strained circumstances. However, in the second phase, the defective line is cut off from the electrical supply, as shown in Figure 11.

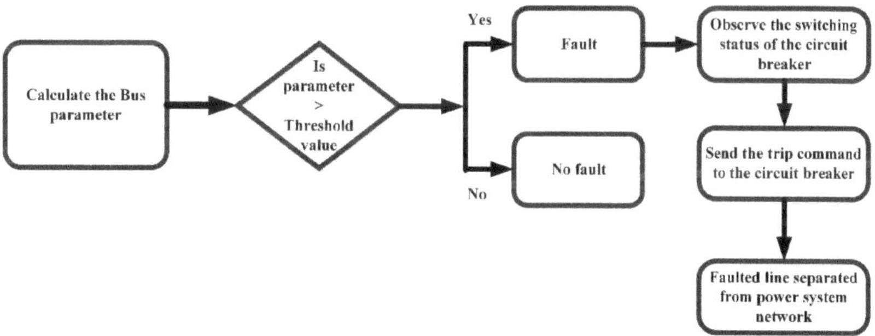


**Figure 11.** Block diagram of WABP scheme based on the bus data and circuit breaker status.

In [82], a synchro-phasor state estimator's residual vector is utilized to detect faults and improve the security of the present remote backup protection scheme. In contrast, ref. [83] proposed a WABP technique for transmission lines that operate based on synchronized data and the actions of third-zone distance relays. The proposed algorithm offers a significant advantage over other wide-area methods, requiring fewer synchronized devices. Therefore, the proposed algorithm represents a valuable contribution to the backup protection of power transmission lines, as it can achieve accurate results using fewer synchronized devices. Furthermore, ref. [84] proposed a novel WABP strategy based on device detection to address the device failure issue. This strategy is valuable to the power system protection field, as it can efficiently and reliably locate faults and provide backup protection.

In summary, the above method proposes innovative approaches to enhance the security and reliability of power systems through backup protection. However, utilizing wide-area information, synchronized data, and device detection techniques are significant advancements that can significantly improve the accuracy and effectiveness of backup protection strategies. These contributions are crucial in ensuring the stability and efficiency of power systems, which are essential for the modern world's functioning. The major drawback is that the strategy's effectiveness may be limited to communication failures or delays.

**Table 2.** Backup protection scheme.

| Ref. No. | Contribution | Detection Time |
|---|---|---|
| [85] | • It applies to various types of faults.<br>• It provides high security and dependability in identifying the faulted line.<br>• The computational burden is less. | 65–95 ms |
| [86] | • It increases the accuracy of fault identification.<br>• It can be implemented with a smaller number of PMUs.<br>• The implementation costs less. | 403 ms |
| [87] | • It can quickly identify the faulted line and prevent damage to the power system.<br>• It can appropriately maintain the dependability–security of the power system network during system transient. | 400 ms |
| [88] | • It is easier to implement and maintain.<br>• It reduces the computational burden.<br>• It reduces the risk of maloperation. | 740 ms |
| [89] | • This proposed strategy effectively identifies the faulted line, which can prevent further damage to the power system.<br>• It reduces the number of PMUs needed, saving costs and resources while maintaining the same level of protection. | - |
| [90] | • It can detect the fault in a different location with a high accuracy.<br>• It can detect all high-resistance asymmetrical faults and is resistant to measurement errors, improving the reliability of the protection system. | 50 ms |
| [91] | • This suggested method can identify the faulted line correctly.<br>• It can easily classify single-phase-to-ground faults, which can help improve the overall system stability.<br>• It is applicable for modern power system networks with reduced system inertia. | 20 ms |
| [92] | • This method can detect high-impedance faults.<br>• It has less computational burden.<br>• It can prevent unnecessary power interruptions. | 0.48 ms |
| [93] | • This scheme is highly effective during power oscillation, generator outage, load shedding, and voltage fluctuation.<br>• It can detect different types of fault locations.<br>• It can operate in a wide range of scenarios. | 300 ms |

**Table 2.** *Cont.*

| Ref. No. | Contribution | Detection Time |
|---|---|---|
| [94] | • This method requires significantly less time training the classification models than other classifiers, reducing the overall capital cost.<br>• It can operate quickly, and has accurate fault detection and classification to protect the power system from the cascade impact of the maloperation of relays. | 140 ms |
| [95] | • This proposed technique limits the computational requirements.<br>• It requires a smaller number of PMUs.<br>• It reduces the computational burden. | 12 ms |
| [96] | • This technique is reliable for conventional protection systems.<br>• It can help to prevent power outages and ensure the stability and efficiency of the power system. | 81 ms |
| [97] | • This approach provides accurate fault location in the presence of measurement and parameter errors.<br>• It is a versatile and reliable tool for system operators and engineers. | 50 ms |

## 5. Synchro-Phasor-Based Application in Transmission and Distribution System

The real-time management of power systems necessitates using fast and synchronized data collection and processing techniques. PMUs, with their synchronized timing and high-frequency sampling capabilities, play a crucial role in analyzing the complicated dynamics of the power system. These measurements also assist in detecting system abnormalities and implementing prompt corrective actions. Despite these benefits, minor variations in voltage phasors at different buses in the distribution system must also be accounted for, as they are crucial in achieving better control and protection. Extensive research has been conducted in this field. The following sections outline the applications of synchro-phasor measurements in transmission and distribution systems, as shown in Figure 12.

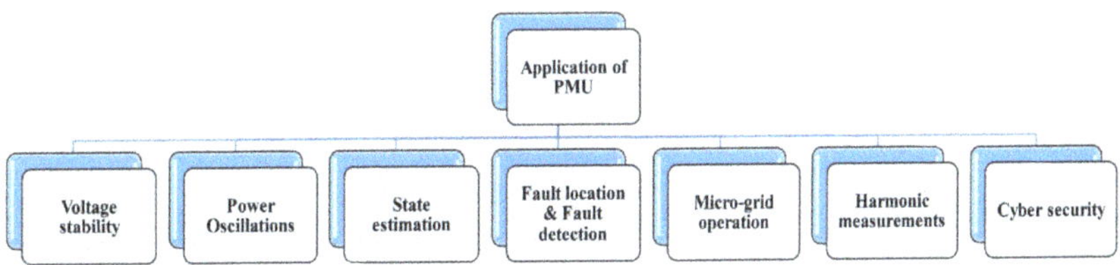


**Figure 12.** Applications of PMU.

*5.1. Voltage Stability*

Voltage stability refers to the capability of a system to sustain its voltage levels after a disturbance occurs. This concept is typically divided into two categories, short-term voltage stability and long-term voltage stability, which are differentiated based on the type and duration of the disturbance [98]. The system voltage stability margin (VSM) is expressed as follows:

$$\text{Percentage of VSM} = \frac{\lambda_M - \lambda_C}{\lambda_C} \times 100\% \tag{4}$$

where $\lambda_M$ represents the maximum loading point and $\lambda_C$ represents the current operating point.

Traditionally, only SCADA and EMS data were used to monitor the P–V and Q–V curves, but a model-free technique was developed in [13] to measure the voltage magnitude. In [99], a new strategy was designed to improve the accuracy and speed of measuring the voltage magnitude, which is essential in addressing voltage instability and preventing blackouts. However, this approach involves using WAMS and optimizing parallel processes with guidelines to minimize calculation times. Later, an innovative approach was proposed to estimate the voltage stability margin in a power grid to enhance operators' awareness [100]. This method utilizes PMUs and incorporates a preprocessing technique for PMU measurements to eliminate any inconsistency or uncertainty due to random load disturbances. This approach promises to provide a reliable and accurate estimation of the voltage stability margin, which is crucial for ensuring a power system's safety and stable operation. In [101], a new technique for monitoring voltage stability in a transmission system was introduced. This method was developed by combining the impedance matching method and the previous local identification of voltage emergency technique (LIVET). The proposed method can detect voltage instability soon after it occurs by monitoring the changes in the active power transfer. A novel multi-bus voltage stability index was introduced in [102]. This index can calculate the stability margin and the load-shedding percentage. Several voltage stability evaluations were conducted under various operating conditions [103–109]. Table 3 depicts some recent research on voltage stability using PMU data.

**Table 3.** Application of PMU on voltage stability.

| Ref. No. | Contribution |
| --- | --- |
| [103] | This approach can reduce the number of contingency scenarios and produce accurate results with noisy input. |
| [104] | This method provides faster predictions of fluctuations, giving the transmission system operator ample time to implement corrective measures. |
| [105] | This suggested approach demonstrates improved real-time applicability for accelerating high-dimensional PMU tests and higher tolerance to instabilities brought on by high solar integration. |
| [106] | This approach can accurately identify the critical bus and quantify stress in the system in various test scenarios, including noisy measurements, load increases in different directions, and line and generator outages. |
| [107] | This approach is robust to reactive power limitations of generators. |
| [108] | This proposed technique has a high accuracy and fast execution speed. |
| [109] | Compared to the Lyapunov exponent technique, this suggested method has a faster detection time and is more successful in identifying short-term voltage instability. |

### 5.2. Power Oscillations

Distance relay failure contributes significantly to most cascaded blackouts worldwide [27]. Severe disruptions such as faults and load failure may cause the system to experience power swings (PSs) or power oscillations (POs) [110]. The swing equation can be expressed as follows [111]:

$$\dot{\delta} = \omega_n(\omega_r - 1) \tag{5}$$

$$2H\dot{\omega}_r = P_m - P_e - D(\omega_r - \omega_n) \tag{6}$$

where $\dot{\delta}$ represents power angle; $\omega_n$ and $\omega_r$ shows the actual and angular velocity, respectively; $P_m$ and $P_e$ are the mechanical and electrical power; $H$ is the inertia constant; and $D$ is the damping coefficient.

Depending on the severe disturbances, POs can either be stable (synchronous generators maintain synchronism) or unstable (synchronous generators lose synchronism) [112]. During stable power oscillation (SPO), blocking the performance of the distance relay is desirable. On the other hand, alternative events are offered during unstable power oscillation (UPO). Depending on the location, the distance relay could block or trip in the first event. In the second event, the distance relay could be plugged (close to the electrical center) and proceed by forcing separation at a different system site [113]. However, UPOs can cause significant changes in the voltage and current, which can be hazardous for the operation of the power grid. In addition to that, UPOs can also cause the failure of specific protective systems, such as distance and under-impedance relays, leading to widespread blackouts and cascading outages. The above issues highlight the importance of detecting and mitigating UPOs in power systems to ensure the stability and reliability of the power grid [114].

To maintain the stability of the power grid, an out-of-step (OOS) prevention mechanism is provided, which includes pole slip protection (PSP) of synchronous generators, exceptional security, additional control, power swing blocking (PSB) of distance protection, and OOS tripping in transmission networks [114]. The above mechanism works to detect and prevent OOS events, ensuring the stability and reliability of the power grid. However, conventional blind-based schemes, such as pole slip protection (PSP) for synchronous generators, face limitations in differentiating between stable and unstable swings and require a detailed stability study for their implementation [115]. Other techniques, such as the rate of change of impedance and resistance [115], changing the zone shape of distance relays [116], and the swing-center voltage (SCV) method [117], were also reported, but have limitations such as requiring extensive stability studies or being applicable only for two-source equivalent systems.

Various techniques were proposed in the power system industry to prevent out-of-step (OOS) events and detect POs to ensure stable operation and prevent blackouts [116]. These include the frequency of voltage value-based approaches [116], pattern recognition-based strategies [117–119], admittance trajectory-based methods [120], the relative speed of a hypothetical identical machine [121], PMU information-based methods [122,123], and predictive analysis-based methods [124]. These methods aim to distinguish stable power swings from unstable ones, increase the protection system's reliability and security, and reduce false tripping and communication failure. However, each method has its limitations and challenges, such as the requirement for a vast amount of training data, improvement in filter design, and the necessity for PMU information. Table 4 tabulates some recent research on power oscillation using PMU data.

**Table 4.** Application of PMU on power oscillation.

| Ref. No. | Contribution |
| --- | --- |
| [111] | This proposed approach has several advantages, including the prevention of zone-one trip of distance relay, maintaining the dependability of the overall protection system, the ability to predict the transient stability status of the system, and ease of integration with existing numerical relays. |
| [125] | This method is computationally efficient and provides improved reliability as it is immune to external faults and switching events and remains unaffected by noisy signals. |
| [126] | This approach can identify three-phase faults while detecting and discriminating between stable, unstable, and multi-mode power swings. |
| [127] | This approach can identify an OOS state more accurately and reliably, with a smaller probability of false alarms or missed detection. |
| [128] | This proposed algorithm provides faster tripping (up to 200 ms) than traditional impedance-based OOS methods, making it more reliable. |
| [129] | This proposed approach has improved performance compared to existing methods under various conditions, including current transformer (CT) saturation, single-pole operation, and load switching. |
| [130] | This method can detect HIFs within a brief time (less than 4 ms), regardless of the fault's characteristics, such as resistance, inception angle, or location. |
| [131] | This proposed model outperforms existing deep learning and matrix completion-based methods regarding prediction accuracy, making it a more reliable choice for filling in missing PMU data during power oscillations. |

### 5.3. State Estimation

Power system state estimation (PSSE), a crucial real-time application, typically happens every 30 s. An appropriate SE provides the necessary input data so that operators can learn about the security and dependability of the system and, at the same time, achieve the capability to reduce capital expenditures. The SE involves calculating the amplitude of the bus voltage and phase angles by utilizing redundant active and reactive power injection data [132]. However, the estimation can be performed using several SE techniques. All approaches often use the same measurement items, the foundation for establishing the algorithm's goal function. The SE measuring model that is most frequently employed is the following:

$$y = h(x) + e \tag{7}$$

where $y$ is the measurement vector entered into the SE method, $x$ represents the selected state variables, and $h(x)$ and $e$ represent the measurement function and error vectors, respectively. Since poor SE performance can have severe financial and practical repercussions, much research has already been conducted to enhance SE performance, and numerous strategies have been suggested in this article.

Conventionally, the data measurement is asynchronous, resulting in static state estimation (SSE). The following have been explored through SSE: (i) the enhancement that was obtained when PMUs were introduced and multiplied; (ii) the ideal placement and positioning of PMUs to provide high reliability, the identification of incorrect data, and the detection of system element issues; (iii) the inclusion of PMUs in the estimation technique, such as by immediately replacing the actual values inside the state variables; (iv) calculating the weights attached to these PMUs; and (v) the advantages of PMUs for enhancing data security [132]. However, synchro-phasor technology allows for dynamic state estimation (DSE), providing real-time information about the voltage magnitude and phase angle at each node in the power system. This approach provides a more accurate and dynamic understanding of the system's state, enabling better decision making and control [133].

Several studies have shown the benefits of using PMU data with an extended Kalman filter (EKF) for online state and parameter estimation. References [134,135] demonstrated the potential applications of this approach. References [136,137] used the EKF to estimate a single-machine infinite bus system's generator states, unknown inputs, and dynamic states. References [138,139] proposed using an ensemble Kalman filter (EnKF) for simultaneous state and parameter estimation and the use of an extended particle filter (EPF) for dynamic state estimation, respectively. The main disadvantage of these methods is the lower sampling rate. A comparative study based on dynamic state estimation was suggested to address this above issue in [140].

Later, an SE method for the distribution system was deployed in reference [141]. This method offers several advantages, including low latency and high state estimation frame rates. In [132], a novel hybrid power system static state estimation (SE) method was introduced, which utilizes PMU observations as a multivariate time series and incorporates existing time and cross-correlation through VAR models. Additionally, the process provides better confidence intervals, and has the ability to anticipate the power state and detect sudden biases, resulting in more robust state estimation. Despite the attempts to capture the dynamic conditions of the power system through various methods, the challenge of decentralized processing and decoupling with power plant controls persists. To overcome this, ref. [142] proposed an EKF technique that features decentralized processing and decoupling with power plant controls. Even with noise, the framework accurately estimates the states, outputs, and unknown inputs. It was also demonstrated that the above method was resistant to generator component errors and efficient in various situations. In [143], a new method that combined SCADA and PMU measurements was presented. It shows a significantly reduced computing time compared to existing approaches, especially in large-scale networks with multiple faulty SCADA measurements in PMU observable areas. Reference [144] proposed an improved dynamic state estimation scheme. The scheme's performance was evaluated by considering the impact of probabilistic communication interruptions and delays, ensuring robustness and reliability in real-world conditions. The centralized approach integrates and aggregates information from multiple sources, providing a more comprehensive and accurate state estimate reference. Table 5 presents some recent research on SE using PMU data.

**Table 5.** Application of PMU on state estimation.

| Ref. No. | Contribution |
|---|---|
| [145] | This suggested approach performs better in terms of efficiency, dependability, and robustness over a range of operational circumstances and calibration properties of the measuring instruments. It may be utilized for modern distributed network design and management. |
| [146] | This approach is valuable for locating the cyber threat and filtering false data streams. |
| [147] | The equivalent circuit formulation framework provides a novel approach to state estimation that is both efficient and accurate, with the potential for future developments and applications in the power system monitoring and control area. |
| [148] | This proposed hybrid SE approach reduces the computational complexities. It provides a more efficient and accurate way to estimate the state of the power system, leading to better monitoring and control. |
| [149] | This technique enhances state estimation accuracy with the ability to handle disturbances and uncertainties in the active distribution grids and microgrids. |
| [150] | This proposed method shows a higher accuracy and reliability than traditional methods, with the ability to quickly and accurately determine the exact location of faults in both meshed and radial topologies. |
| [151] | This suggested approach has substantial advantages, specifically when operating at off-nominal frequencies, and may be employed in various monitoring systems. |
| [152] | The least-absolute-value estimator is more computationally efficient, leads to improved accuracy, optimizes the objective function, handles gross errors quickly, and eliminates the need to deal with multiple insufficient data. |
| [153] | The low computational cost of the estimation stage makes this algorithm even more attractive, as it can be run in real-time without sacrificing accuracy. |
| [154] | This proposed solution for solving the SE problem for three-phase systems is a computationally efficient and straightforward alternative to the conventional approach. |
| [155] | These proposed next-generation distribution systems incorporate earthing resistances as state variables and field measurements, improving observability, accuracy, and reliability. The PMU infrastructure provides high-speed visibility and helps prevent equipment and personal losses due to temporary overvoltages. |
| [156] | The generalized algorithm is linear, efficient, and robust. This proposed method is comparable in accuracy and computational burden to the conventional weighted least-squares estimator and outperforms it in the presence of gross measurement errors. |
| [157] | To maintain grid operation, this suggested Convolutional Neural Network (CNN)-based filter can operate as an extra layer of security by removing erroneous data before a state estimate is carried out. This would allow system operators to make better decisions. |

*5.4. Fault Location and Fault Detection*

Fault location is the process of determining the location of a fault in an electrical power grid. PMU-based approaches were proposed to prevent widespread power outages or other disruptions. These PMUs provide real-time voltage, current, and frequency readings, allowing the electrical power system to be fully observable at all times, thereby assisting system analysts in identifying faults quickly and avoiding blackouts. The faulty bus model can be expressed as follows:

$$\text{Percentage of positive sequence voltage } (PSVn) = \left| \frac{PSV_{nbf} - PSV_{naf}}{PSV_{nbf}} \right| \times 100 \quad (8)$$

where $n$ = bus number; $PSV_{nbf}$ = positive sequence voltage (before fault); and $PSV_{naf}$ = positive sequence voltage (after fault).

However, to further improve the efficiency of the protection system, it is recommended to use the WAMS, which provide a more effective backup protection scheme.

The WAMS feature PMUs equipped with intelligent electronic devices and synchronized data from both ends of the transmission lines [158,159]. It is essential to accurately identify faulty transmission lines and perform a dynamic security analysis of the power system to enhance protection through PMUs [160]. Implementing the differential protection philosophy based on Kirchhoff's laws is also crucial to demonstrate the efficiency of PMUs in protecting transmission lines [161]. The study conducted in [70] was one of the first studies to develop a differential protection scheme for a 33 kV transmission line and substation, using numerical relays and incorporating GPS synchronization to streamline the sample data. Moreover, the recursive algorithm for updating phasors was found to increase the estimation errors over time due to the continuous nature of electrical power systems (EPS). To effectively monitor and protect EPS, it is crucial to have an accurate synchronization of phasors and a precise measurement of the phase angles. The phase angles of the voltages and currents play a crucial role in defining the location of the faults in the transmission line when using a minimum positive-sequence voltage as a backup for traditional protection systems. This information can then be used to classify faulty buses. In addition, the use of Wavelet and Fourier transform for fault classification in transmission lines was discussed in reference [162]. Reference [163] proposed a method that detects and classifies transmission line faults using only PMU data from the generator bus. The methodology employed PMU readings from a single bus voltage for the whole grid, decreasing the price and complexity of the surveillance system. Recently, there has been an increase in the use of artificial intelligence (AI) or machine learning (ML) techniques for fault diagnosis, including detection, identification, and localization of faults in power transmission networks [164]. Table 6 tabulates recent research on fault location and detection using PMU data.

**Table 6.** Application of PMU on fault location and fault detection.

| Ref. No. | Contribution |
|---|---|
| [164] | This designed intelligent algorithm is explicitly data-driven and successfully treats the nonlinear time-varying behavior of the system. |
| [165] | This proposed fault identification method achieves fast and accurate results with 100% accuracy in less than 20 ms. It applies to conventional and contemporary networks, regardless of DG types and sizes. |
| [166] | This proposed method addresses the issue of high labeling cost and utilizes unlabeled data for classification with a high accuracy. Compared with previous work, a high impedance fault (HIF) location method is developed to locate faults with a small estimation error using μPMUs. |
| [167] | This proposed scheme is practical, robust, and reliable under different operating conditions, fault types, distances, and resistance. |
| [168] | This algorithm accurately classifies faulted buses and identifies the faulty line, making it suitable for application in practical power systems. The proposed algorithm offers improved fault classification accuracy, robustness, faster response, and lower computational complexity, making it valuable to PMU-based wide-area measurements. |
| [169] | This proposed method exhibits considerably lower computational complexity. |
| [170] | This method can detect the fault type and send the reclosing command around six cycles after the secondary arc extinction in case of a single-phase fault occurrence. |
| [171] | This proposed graph-based faulted line identification algorithm using μPMU data in distribution systems offers efficient and accurate faulted subgraph identification and faulted-line location, with potential for future improvement through robust SE and topology identification methods. |
| [172] | It can capture similar information on fault events across different parts of the distribution networks, which can improve fault detection accuracy. |
| [173] | This framework addresses storage issues and reduces data processing time by employing a limited data window, making it more efficient and practical for real-time applications. |

### 5.5. Microgrid Operation

Microgrids can operate in two different modes, islanded mode and grid-connected mode. The microgrid is connected to the main power grid in the grid-connected mode and can import or export electricity as needed. This mode allows the microgrid to take advantage of the benefits of the main grid, such as the backup power and access to additional energy resources. In GC mode, the microgrid must maintain a stable frequency and voltage to ensure the reliable operation of connected loads and to avoid any negative impact on the main power grid [174]. Using PMUs, the microgrid can monitor the power system parameters and quickly detect any changes or disturbances in the grid.

Islanding is a situation that occurs when a portion of an electrical power grid becomes disconnected from the primary grid, forming an isolated island. This can occur due to several factors, including equipment failures, natural disasters, or cyberattacks. Islanding can pose a significant threat to the stability of the power grid, leading to unanticipated power outages and other system disruptions. It is crucial to employ effective islanding detection methods to mitigate this problem. PMUs were identified as a potential solution for islanding detection. PMUs are specialized hardware devices that track and gauge the electrical power grid, providing instantaneous voltage, current, and frequency measurements. By utilizing these measurements, PMUs can detect when a portion of the grid becomes disconnected from the primary grid and initiate an alert or another appropriate response.

To transfer data between PMUs in smart grids, different communication technologies such as wire lines, fiber-optic cables, 4G/5G networks, and power line communication are utilized [175]. PMUs provide time-synchronized signals from various locations in the microgrids (MG), which are particularly crucial when there is an increased installation of DGs [176]. PMU-based detection is a remote technique that offers fast, reliable, and precise islanding detection in different operating conditions. Even if a circuit breaker (CB) open is detected, the DG can still operate in islanded mode if there is a sufficient generation-load balance [177]. However, relying solely on local measurements reduces the ability to control DGs, which limits flexibility in active distribution system management. Therefore, gathering system component parameters to achieve resilient and stochastic energy management in MGs is crucial, which can be accomplished through PMU measurements [178].

Synchronized measurements obtained from multiple PMU sites are instrumental in accurately detecting islanding events in MGs [179,180]. By analyzing power system variables such as voltage phasor, current phasor, frequency, and ROCOF, operators can make decisions and prevent threats to the system. However, the reporting rate of PMUs can be extremely high, with 60 to 120 frames per second for various variables, resulting in massive amounts of data that need to be processed [181]. Identifying an actual islanding event from embedded dynamics in the MGs is a daunting task that requires sophisticated techniques such as multivariate statistical methods, including independent component analysis, principal component analysis (PCA), and partial least squares to compress the data. PCA is a technique that reduces correlation among the observed variables without significantly losing data by reducing the input data dimension [182]. The PCA model is updated as new data arrives to achieve a more accurate detection since the power network data is subjected to change over time. PCA has been utilized in various studies for islanding detection [183–186]. Table 7 tabulates recent research on islanding detection using PMU data.

**Table 7.** Application of PMU on islanding detection.

| Ref. No. | Contribution |
| --- | --- |
| [187] | These test results demonstrate the effectiveness of the machine learning strategy in harnessing synchro-phasor data for power system applications with reasonable accuracy in classifying scenarios and detecting islanding events. |
| [188] | This method provides efficient islanding detection without a non-detecting zone (NDZ) and power quality issues. This reduces the cost of maintaining physical devices and provides more reliability in terms of functioning. |

**Table 7.** *Cont.*

| Ref. No. | Contribution |
| --- | --- |
| [189] | This proposed scheme can detect islanding cases with a frequency deviation of 0.01 Hz in just 0.25 s. The scheme is unaffected by synchronization errors in μPMU data and by measurement noise up to a signal-to-noise ratio (SNR) value of 35 dB. |
| [178] | This proposed method has a high detection speed (30 ms) and is resistant to false alarms so that it can distinguish between islanding and non-islanding events with high accuracy and precision. |
| [190] | This algorithm is programmed within the μPMU, which reduces cyberattacks. It is robust and accurate, as it is simulated for various conditions such as islanding and faults. |
| [191] | This method can detect islanding within two cycles of system frequency under the worst-case scenario of perfect power balance, ensuring rapid response and preventing damage to the power system. It is highly reliable and accurate, resulting in a zero non-detection zone in a zero-power imbalance. |
| [192] | This proposed approach shows that the phase angle difference islanding detection method is practical and affordable and overcomes the issues affecting the frequency change rate on electrical grids with lower inertia. |
| [193] | This proposed method develops a novel differential protection strategy for microgrids that use inexpensive PMUs or μPMUs. The method can distinguish between internal, external, and severe no-fault circumstances. |
| [194] | This method is dependable and flexible for different microgrid networks without requiring system model data or depending on particular disruption characteristics because it uses a regressive vector model, an indicator function to separate events from inaccurate measurements. |

### 5.6. Harmonic Measurements

Monitoring and controlling harmonic distortions are critical in power system management, particularly in power distribution systems. With the increasing power of electronic devices, nonlinear loads, and inverter-based energy resources, the importance of this task has increased in recent years. Such equipment can introduce harmonic distortions, which may compromise the dependability and security of power distribution systems by causing conductor overheating or interfering with power protection systems. Hence, utilities must monitor and control harmonic levels in power distribution systems to prevent pollution and ensure secure and reliable operation. However, PMUs offer high-speed and accurate harmonic measurements that enable power system operators to identify and analyze harmonic distortion problems in real-time. By measuring the amplitude and frequency of harmonics, PMUs can provide valuable information to power system operators to identify the source of harmonic distortions and take corrective actions. For every harmonic injection, the total harmonic vector error ($TVE$) for the $n$th harmonic order is expressed as follows:

$$TVE = \frac{|V(n)_{measured} - V(n)_{ideal}|}{|V(n)_{ideal}|} \tag{9}$$

A novel algorithm for calculating harmonic phasors using PMUs based on the principle of fundamental phasor calculation was proposed in [195]. By accurately calculating harmonic phasors, this algorithm enables the efficient utilization of PMU calculation resources and enhances the overall functionality of PMUs in power system monitoring and control. In [196], two estimation techniques were developed from recent PMU-based applications by examining their estimation results using a real-world measurement dataset. While a novel harmonic phasor estimator for P-class PMUs was proposed in [197], it was found that the technique suffered from the inter harmonic injection. A phasor signal processing model was proposed to overcome this issue, which involves adding a preprocessing filtering step before PMUs [198]. This step incorporates estimating signal parameters using either the estimation of signal parameters via rotational invariance techniques (ESPRIT) or the matrix pencil (MP) algorithm for signal processing. In addition, a design for a PMU using a PXI modulator and GPS receivers for accurate harmonic synchro-phasor measurement in distribution networks was presented in [199]. The presented PMU prototype and measurement procedures can help improve the performance and efficiency of distribution grids, which are essential for the reliable and cost-effective integration of renewable energy sources and other distributed energy resources. Table 8 depicts recent research on harmonic measurement using PMU data.

**Table 8.** Application of PMU on harmonic measurement.

| Ref. No. | Contribution |
| --- | --- |
| [198] | This algorithm is robust against measurement noise (it can work with a 65 dB SNR). The proposed algorithm has less computational complexity than the standard matrix pencil method, making it suitable for deployment in embedded platforms. |
| [200] | This method can operate with a higher accuracy and faster speed than traditional methods. This method can also reduce the number of required PMU placements, saving costs and improving the overall system efficiency. |
| [201] | This method has been experimentally validated on an inductive voltage transformer, demonstrating its practical applicability. The black-box approach of this method allows it to be applied to other types of voltage transducers, which increases its versatility. |
| [202] | This proposed scheme is robust to harmonic and fundamental current measurement errors, which increases the system's reliability. |
| [203] | This method extracts sparsity patterns of the harmonic state variables, which enhances its efficiency and accuracy. It also can detect the number and locations of the harmonic sources on the network. |
| [204] | This method further improves clustering accuracy by incorporating measurements from fundamental and harmonic phasors, especially for unbalanced event types. |

*5.7. Cybersecurity*

In recent years, PMUs have been applied to the field of cybersecurity, as they have the potential to detect and respond to cyberattacks on power systems. In December 2015, over 200,000 customers were impacted by cyberattacks that targeted three electric distribution companies in Ukraine [205]. The cyberattack detection model is expressed as follows:

$$\text{Percentage of true positive rate} = \frac{True\,positive}{P} \times 100 \tag{10}$$

$$\text{Percentage of false positive rate} = \frac{False\,positive}{N} \times 100 \tag{11}$$

where $P$ = the positive number of attack PMUs; $N$ = the number of safe PMUs.

Several approaches have investigated the use of PMUs in cybersecurity, focusing on detecting irregularities and suspicious behavior in real-time. An approach for detecting the manipulation of PMU data involves utilizing transmission line parameters, including shunt admittance and series impedance, as they remain constant and can help identify potential attacks and recover the correct data [206,207]. However, this approach requires the placement of PMUs on every bus on the grid, which can be costly. Time synchronization attacks are also typical attacks that can affect the accuracy of voltage and current phase angle measurements, resulting in adverse impacts on power system operations such as fault localization, voltage stability detection, and line parameter identification [208]. Detecting such attacks can be achieved using various methods, including cross-layer defense mechanisms [209], data-driven models [210], and phasor measurement analysis [211]. In addition to attack detection, several methods exist for recovering accurate data once a cyberattack has been detected. For instance, a new framework was proposed for restoring missing PMU measurements. However, its efficacy remains untested during transient power system operation [212]. Table 9 tabulates recent research on cybersecurity using PMU data.

**Table 9.** Application of PMU on cybersecurity.

| Ref. No. | Contribution |
| --- | --- |
| [213] | These techniques worked well for spotting and recognizing cyberattacks on micro-PMUs. This offers empirical proof that upholds the recommended approaches' effectiveness and raises its credibility. |
| [214] | This method demonstrated that data-driven methods successfully and accurately detect attacks on solar PV systems using the PMU data. |
| [215] | This detection technique can spot attacks using grid power flow equations. |
| [216] | Using the extended Kalman filter (EKF) over the traditional EKF constitutes a substantial advance since this technique offers a more durable and reliable solution for sampling errors, topological mistakes, and cyber threats. |
| [217] | This proposed method is efficient during false data detection. |
| [218] | This proposed technique includes a delayed alarm triggering mechanism to ensure reliable PMU-based data manipulation attack detection. This suggested technique enhances the system's noise immunity. |

## 6. Standards

The IEEE Standards for PMUs provide guidelines for designing, installing, and testing PMUs used in power systems. The standards specify communication protocols, performance requirements, and testing procedures for synchro-phasors, including PMUs, which are essential for the accurate and reliable measurement of power system dynamics. The standards also provide recommendations for hardware, software, and communication protocols for PMUs and power system stabilizers (PSSs) and define communication protocols for PMUs used in distributed energy resource (DER) systems. The IEEE standards ensure that the PMUs and synchro-phasors from different manufacturers can communicate and work together effectively, providing a common framework for designing, installing, and testing. Table 10 depicts the advantages and limitations of the following IEEE standards:

- IEEE Std 1344-1995: This was the first synchro-phasor standard formulated for measurement and communication specifications, emphasizing achieving better interoperability among PMUs [219].
- IEEE C37.118-2005: This standard was developed to address the limitations of IEEE Std 1344-1995, specifically regarding the performance of PMUs at off-nominal frequencies. This standard restricts the frequency deviation of off-nominal frequency inputs within a range from the nominal frequency [220].
- IEEE Std C37.118.1-2011: This standard identifies the measurement of electrical parameters such as synchro-phasor, frequency, and ROCOF under steady-state and dynamic conditions. It also includes compliance requirements and evaluation methods and defines P and M classes as two performance classes for PMUs [221].
- IEEE Std C37.118.1a-2014: This standard is a revision to IEEE Std C37.118.1-2011 and aims to eliminate specific limitations, mainly frequency, and ROCOF measurements. It also corrects latency and measurement discrepancies and refines ramp tests to guarantee repeatable results while evaluating PMUs with anti-aliasing filters [222].
- IEEE C37.244-2013: This standard defines the functional, performance, and communication needs of power dispatching centers (PDCs) to provide better system monitoring, protection, and control. It includes an information annex on report rate conversion and filtering issues and outlines various tests and test methodologies to ensure protocol support, cybersecurity, and communication media compatibility [223].

**Table 10.** Advantages and limitations of IEEE standard.

| Standards | Contribution | Limitation |
|---|---|---|
| IEEE Std 1344-1995 | • It provides measurement and communication requirements for PMUs. | • The off-nominal frequencies are not similar.<br>• It does not specify the response of PMUs correctly during system transients. |
| IEEE Std C37.118-2005 | • It restricts frequency deviation to ±5 Hz.<br>• It provides better accuracy in data reporting. | • It does not specify PMUs' response during system transients.<br>• It failed to operate accurately during power system dynamic performance. |
| IEEE Std C37.118.1-2011 | • It establishes two performance classes.<br>• It provides compliance requirements and evaluation methods to improve PMUs' performance under both conditions. | • It does not respond appropriately during power system transient. |
| IEEE C37.244-2013 | • It covers an information annex on report rate conversion and filtering issues.<br>• It summarizes various tests and test methodologies to ensure compatibility with protocol support, cybersecurity, and communication media. | • It does not provide specific guidance on protecting PMUs from cybersecurity threats. |
| IEEE Std C37.118.1a-2014 | • It provides better protection and control under a steady state and dynamic conditions. | • They do not provide specific guidelines for the communication protocols used by PMUs to transmit data to the protection and control system. |

## 7. Future Scope

- Integrating PMUs with other grid monitoring and control technologies, such as SCADA systems, could enhance the capabilities of PMUs in detecting and responding to cyberattacks and other threats to the power grid. Future work could explore the integration of PMUs with AI and ML algorithms, enabling automatic and real-time detection and response to cyber threats.
- The research can be conducted on developing more secure PMUs resistant to cyberattacks and hacking, ensuring the reliability and stability of the electrical power grid. This could involve using advanced encryption techniques and secure communication protocols to protect the data transmitted by PMUs and develop secure firmware and software resistant to malware and other security threats.
- Harmonic measurement can be used to monitor and analyze the performance of renewable energy systems, such as wind turbines and solar photovoltaic systems, which are known to produce significant harmonic distortion. By detecting and addressing these issues, the performance of these systems can be improved, leading to better energy production and efficiency.
- To boost the overall effectiveness and dependability of the power system, islanding detection technology might be used with control strategies such as load shedding and islanding control. This may lessen the effects of islanding occurrences and increase the power system's resistance to faults and disturbances.
- The islanding approach may be further examined from the cybersecurity perspective, particularly regarding possible attacks and weaknesses. Through the identification of potential threats and the development of mitigation techniques, the security of the power system may be improved.
- Developing an AI model for identifying fault detection and replacing lost or defective information with correct information is a complex task that requires expertise in data science and ML.

- One of the primary goals of system state estimation is to accurately determine the state variables of the power system, such as voltage and phase angle. In the future, researchers can work toward developing more accurate and reliable PMUs that can provide more precise measurements and improve the accuracy of system state estimation.

## 8. Conclusions

PMUs are advanced instruments designed to measure voltage and current phasors in electric power systems accurately. Due to their high precision and accuracy, they have become widely used in transmission and distribution networks for various purposes, including power system protection. With the increasing use of AI and ML techniques in power systems, PMUs are expected to become increasingly essential. This paper explains the operating principle and various applications of PMUs in terms of their capability to establish a better protective system using other established technologies. It also highlights the numerous benefits of PMU data in areas such as backup protection, voltage monitoring, power swing detection, state estimation, harmonic measurements, and cybersecurity. The thorough discussions and analyses reflect that using PMUs improves network reliability, resiliency, cost savings, and environmental safety, and provides a better protection system.

**Author Contributions:** Conceptualization, B.K.S., M.M., and P.K.R.; methodology, B.K.S. and P.K.R.; software, C.B.; validation, C.B., M.M., and P.K.R.; formal analysis, B.K.S.; investigation, C.B.; resources, P.K.R.; data curation, C.B.; writing—original draft preparation, C.B., P.K.R., and B.K.S.; writing—review and editing, C.B., B.K.S., and P.K.R.; visualization, M.M.; supervision, B.K.S. and P.K.R. All authors have read and agreed to the published version of the manuscript.

**Funding:** This research received no external funding.

**Data Availability Statement:** Not applicable.

**Conflicts of Interest:** The authors declare no conflict of interest.

## References

1. Hojabri, M.; Dersch, U.; Papaemmanouil, A.; Bosshart, P. A comprehensive survey on phasor measurement unit applications in distribution systems. *Energies* **2019**, *12*, 4552. [CrossRef]
2. Dusabimana, E.; Yoon, S.G. A survey on the micro-phasor measurement unit in distribution networks. *Electronics* **2020**, *9*, 305. [CrossRef]
3. Phadke, A.G.; Bi, T. Phasor measurement units, WAMS, and their applications in protection and control of power systems. *J. Mod. Power Syst. Clean Energy* **2018**, *6*, 619–629. [CrossRef]
4. Arefin, A.A.; Baba, M.; Singh, N.S.S.; Nor, N.B.M.; Sheikh, M.A.; Kannan, R.; Abro, G.E.M.; Mathur, N. Review of the Techniques of the Data Analytics and Islanding Detection of Distribution Systems Using Phasor Measurement Unit Data. *Electronics* **2022**, *11*, 2967. [CrossRef]
5. Singh, B.; Sharma, N.K.; Tiwari, A.N.; Verma, K.S.; Singh, S.N. Applications of phasor measurement units (PMUs) in electric power system networks incorporated with FACTS controllers. *Int. J. Eng. Sci. Technol.* **2011**, *3*, 2967. [CrossRef]
6. Hagan, T.; Senaratne, D.; Meier, R.; Cotilla-Sanchez, E.; Kim, J. Implementing Power System Protection Algorithms in a Digital Hardware-in-the-Loop Substation. *IEEE Open Access J. Power Energy* **2022**, *10*, 270–282. [CrossRef]
7. Bertsch, J.; Carnal, C.; Karlson, D.; McDaniel, J.; Vu, K. Wide-area protection and power system utilization. *Proc. IEEE* **2005**, *93*, 997–1003. [CrossRef]
8. Cruz, M.A.; Rocha, H.R.; Paiva, M.H.; Segatto, M.E.; Camby, E.; Caporossi, G. An algorithm for cost optimization of PMU and communication infrastructure in WAMS. *Int. J. Electr. Power Energy Syst.* **2019**, *106*, 96–104. [CrossRef]
9. De La Ree, J.; Centeno, V.; Thorp, J.S.; Phadke, A.G. Synchronized phasor measurement applications in power systems. *IEEE Trans. Smart Grid* **2010**, *1*, 20–27. [CrossRef]
10. Liu, Y.; Wu, L.; Li, J. D-PMU based applications for emerging active distribution systems: A review. *Electr. Power Syst. Res.* **2020**, *179*, 106063. [CrossRef]
11. Seferi, Y.; Cetina RG, Q.; Blair, S.M. Review of PMU algorithms suitable for real-time operation with digital sampled value data. In Proceedings of the 2021 IEEE 11th International Workshop on Applied Measurements for Power Systems (AMPS), Virtual, 29 September–1 October 2021; IEEE: Piscataway, NJ, USA, 2021; pp. 1–6.
12. Asprou, M. Synchronized measurement technology: A blessing for power systems [Trends in Future I&M]. *IEEE Instrum. Meas. Mag.* **2017**, *20*, 25–42.
13. Joshi, P.M.; Verma, H.K. Synchrophasor measurement applications and optimal PMU placement: A review. *Electr. Power Syst. Res.* **2021**, *199*, 107428. [CrossRef]

14. Khandare, B.B.; Deshmukh, B.T. A literature review on wide area protection technique using PMU. In Proceedings of the 2017 International Conference on Energy, Communication, Data Analytics and Soft Computing (ICECDS), Chennai, India, 1–2 August 2017; IEEE: Piscataway, NJ, USA, 2017; pp. 1449–1454.
15. Usman, M.U.; Faruque, M.O. Applications of synchrophasor technologies in power systems. *J. Mod. Power Syst. Clean Energy* **2019**, *7*, 211–226. [CrossRef]
16. Martin, K.E. Synchrophasor measurements under the IEEE standard C37. 118.1-2011 with amendment C37. 118.1 a. *IEEE Trans. Power Deliv.* **2015**, *30*, 1514–1522. [CrossRef]
17. Carta, A.; Locci, N.; Muscas, C.; Pinna, F.; Sulis, S. GPS and IEEE 1588 synchronization for the measurement of synchrophasors in electric power systems. *Comput. Stand. Interfaces* **2011**, *33*, 176–181. [CrossRef]
18. Khatib, A.R.A. Internet-Based Wide Area Measurement Applications in Deregulated Power Systems. Doctoral Dissertation, Virginia Polytechnic Institute and State University, Blacksburg, VA, USA, 2002.
19. Snyder, A.F.; Ivanescu, D.; HadjSaid, N.; Georges, D.; Margotin, T. Delayed-input wide-area stability control with synchronized phasor measurements and linear matrix inequalities. In Proceedings of the 2000 Power Engineering Society Summer Meeting (Cat. No. 00CH37134), Como, Italy, 27 July 2000; IEEE: Piscataway, NJ, USA, 2000; Volume 2, pp. 1009–1014.
20. Phadke, A.G.; Thorp, J.S. *Synchronized Phasor Measurements and Their Applications*; Springer: New York, NY, USA, 2008; Volume 1, p. 81.
21. Ye, F.; Bose, A. Multiple communication topologies for pmu-based applications: Introduction, analysis and simulation. *IEEE Trans. Smart Grid* **2020**, *11*, 5051–5061. [CrossRef]
22. Li, G.W.; Ju, W.Y.; Shi, D.Y. Functional vulnerability assessment of SCADA network. In Proceedings of the 2012 Asia-Pacific Power and Energy Engineering Conference, Shanghai, China, 27–29 March 2012; IEEE: Piscataway, NJ, USA, 2012; pp. 1–4.
23. Myrda, P.T.; Koellner, K. Naspinet-the internet for synchrophasors. In Proceedings of the 2010 43rd Hawaii International Conference on System Sciences, Honolulu, HI, USA, 5–8 January 2010; IEEE: Piscataway, NJ, USA, 2010; pp. 1–6.
24. Galli, S.; Scaglione, A.; Wang, Z. For the grid and through the grid: The role of power line communications in the smart grid. *Proc. IEEE* **2011**, *99*, 998–1027. [CrossRef]
25. Aquilue, R.; Gutierrez, I.; Pijoan, J.L.; Sanchez, G. High-voltage multicarrier spread-spectrum system field test. *IEEE Trans. Power Deliv.* **2009**, *24*, 1112–1121. [CrossRef]
26. Pighi, R.; Raheli, R. On multicarrier signal transmission for high-voltage power lines. In Proceedings of the International Symposium on Power Line Communications and Its Applications, Vancouver, BC, Canada, 6–8 April 2005; IEEE: Piscataway, NJ, USA, 2005; pp. 32–36.
27. Horowitz, S.H.; Phadke, A.G.; Henville, C.F. *Power System Relaying*; John Wiley Sons: Hoboken, NJ, USA, 2022.
28. What Is 5G: Everything You Need to Know about 5G: 5G FAQ: Qualcomm. Wireless Technology & Innovation. (n.d.). Retrieved 22 February 2023. Available online: https://www.qualcomm.com/5g/what-is-5g#Whatis5G? (accessed on 25 March 2023).
29. Bobba, R.B.; Dagle, J.; Heine, E.; Khurana, H.; Sanders, W.H.; Sauer, P.; Yardley, T. Enhancing grid measurements: Wide area measurement systems, NASPInet, and security. *IEEE Power Energy Mag.* **2011**, *10*, 67–73. [CrossRef]
30. Fesharaki, F.H.; Hooshmand, R.A.; Khodabakhshian, A. Simultaneous optimal design of measurement and communication infrastructures in hierarchical structured WAMS. *IEEE Trans. Smart Grid* **2013**, *5*, 312–319. [CrossRef]
31. Borlase, S. (Ed.) . *Smart Grids: Infrastructure, Technology, and Solutions*; CRC Press: Boca Raton, FL, USA, 2017.
32. Abdelwahid, S.; Babiker, A.; Eltom, A.; Kobet, G. Hardware implementation of an automatic adaptive centralized underfrequency load shedding scheme. *IEEE Trans. Power Deliv.* **2014**, *29*, 2664–2673. [CrossRef]
33. Tang, J.; Liu, J.; Ponci, F.; Monti, A. Adaptive load shedding based on combined frequency and voltage stability assessment using synchrophasor measurements. *IEEE Trans. Power Syst.* **2013**, *28*, 2035–2047. [CrossRef]
34. Xu, J.; Xie, B.; Liao, S.; Yuan, Z.; Ke, D.; Sun, Y.; Li, X.; Peng, X. Load shedding and restoration for intentional island with renewable distributed generation. *J. Mod. Power Syst. Clean Energy* **2021**, *9*, 612–624. [CrossRef]
35. Monti, A.; Muscas, C.; Ponci, F. *Phasor Measurement Units and Wide Area Monitoring Systems*; Academic Press: Cambridge, MA, USA, 2016.
36. Amraee, T.; Darebaghi, M.G.; Soroudi, A.; Keane, A. Probabilistic under frequency load shedding considering RoCoF relays of distributed generators. *IEEE Trans. Power Syst.* **2017**, *33*, 3587–3598. [CrossRef]
37. HaesAlhelou, H.; Hamedani-Golshan, M.E.; Njenda, T.C.; Siano, P. Wide-area measurement system-based optimal multi-stage under-frequency load-shedding in interconnected smart power systems using evolutionary computing techniques. *Appl. Sci.* **2019**, *9*, 508.
38. Derviškadić, A.; Zuo, Y.; Frigo, G.; Paolone, M. Under frequency load shedding based on PMU estimates of frequency and ROCOF. In Proceedings of the 2018 IEEE PES Innovative Smart Grid Technologies Conference Europe (ISGT-Europe), Sarajevo, Bosnia and Herzegovina, 15 August 2018; IEEE: Piscataway, NJ, USA, 2018; pp. 1–6.
39. Karimi, M.; Wall, P.; Mokhlis, H.; Terzija, V. A new centralized adaptive underfrequency load shedding controller for microgrids based on a distribution state estimator. *IEEE Trans. Power Deliv.* **2016**, *32*, 370–380. [CrossRef]
40. Rudez, U.; Mihalic, R. WAMS-based underfrequency load shedding with short-term frequency prediction. *IEEE Trans. Power Deliv.* **2015**, *31*, 1912–1920. [CrossRef]

41. Jegarluei, M.R.; Cortés, J.S.; Azizi, S.; Terzija, V. Wide-area event identification in power systems: A review of the state-of-the-art. In Proceedings of the 2022 International Conference on Smart Grid Synchronized Measurements and Analytics (SGSMA), Split, Croatia, 24–26 May 2022; IEEE: Piscataway, NJ, USA, 2022; pp. 1–7.
42. *IEEE Std C37 118.1-2011*; IEEE Standard for Synchrophasor Measurements for Power Systems–Amendment 1: Modification of Selected Performance Requirements. IEEE Std C37. 118.1 a-2014 (Amendment to IEEE Std C37. 118.1-2011). IEEE: Piscataway, NJ, USA, 2014.
43. Phadke, A.G.; Kasztenny, B. Synchronized phasor and frequency measurement under transient conditions. *IEEE Trans. Power Deliv.* **2008**, *24*, 89–95. [CrossRef]
44. Frigo, G.; Colangelo, D.; Derviškadić, A.; Pignati, M.; Narduzzi, C.; Paolone, M. Definition of accurate reference synchrophasors for static and dynamic characterization of PMUs. *IEEE Trans. Instrum. Meas.* **2017**, *66*, 2233–2246. [CrossRef]
45. Roscoe, A.J.; Dyśko, A.; Marshall, B.; Lee, M.; Kirkham, H.; Rietveld, G. The case for redefinition of frequency and ROCOF to account for AC power system phase steps. In Proceedings of the 2017 IEEE International Workshop on Applied Measurements for Power Systems (AMPS), Liverpool, UK, 20–22 September 2017; IEEE: Piscataway, NJ, USA, 2017; pp. 1–6.
46. Frigo, G.; Derviškadić, A.; Zuo, Y.; Paolone, M. PMU-based ROCOF measurements: Uncertainty limits and metrological significance in power system applications. *IEEE Trans. Instrum. Meas.* **2019**, *68*, 3810–3822. [CrossRef]
47. Zuo, Y.; Frigo, G.; Derviškadić, A.; Paolone, M. Impact of synchrophasor estimation algorithms in ROCOF-based under-frequency load-shedding. *IEEE Trans. Power Syst.* **2019**, *35*, 1305–1316. [CrossRef]
48. Golpira, H.; Bevrani, H.; Messina, A.R.; Francois, B. A data-driven under frequency load shedding scheme in power systems. *IEEE Trans. Power Syst.* **2022**, *38*, 1138–1150. [CrossRef]
49. Bekhradian, R.; Sanaye-Pasand, M.; Mahari, A. Adaptive Wide-Area Load Shedding Scheme Based on the Sink and Source Concept to Preserve Power System Stability. *IEEE Syst. J.* **2022**, *17*, 503–512. [CrossRef]
50. Glavic, M.; Van Cutsem, T. Wide-area detection of voltage instability from synchronized phasor measurements. Part I: Principle. *IEEE Trans. Power Syst.* **2009**, *24*, 1408–1416. [CrossRef]
51. Gu, W.; Wan, Q. Linearized voltage stability index for wide-area voltage monitoring and control. *Int. J. Electr. Power Energy Syst.* **2010**, *32*, 333–336. [CrossRef]
52. Mahari, A.; Seyedi, H. A wide area synchrophasor-based load shedding scheme to prevent voltage collapse. *Int. J. Electr. Power Energy Syst.* **2016**, *78*, 248–257. [CrossRef]
53. Modarresi, J.; Gholipour, E.; Khodabakhshian, A. A new undervoltage load shedding method to reduce active power curtailment. *Int. Trans. Electr. Energy Syst.* **2017**, *27*, e2291. [CrossRef]
54. Adewole, A.C.; Tzoneva, R.; Apostolov, A. Adaptive under-voltage load shedding scheme for large interconnected smart grids based on wide area synchrophasor measurements. *IET Gener. Transm. Distrib.* **2016**, *10*, 1957–1968. [CrossRef]
55. Phadke, A.G.; Wall, P.; Ding, L.; Terzija, V. Improving the performance of power system protection using wide area monitoring systems. *J. Mod. Power Syst. Clean Energy* **2016**, *4*, 319–331. [CrossRef]
56. Liu, W.; Lin, Z.; Wen, F.; Ledwich, G. A wide area monitoring system based load restoration method. *IEEE Trans. Power Syst.* **2013**, *28*, 2025–2034. [CrossRef]
57. Itiki, R.; Libonati, F.; Burgués, H.; Martini, M.; Essakiappan, S.; Manjrekar, M.; Bai, L.; Di Santo, S.G. A proposed wide-area stabilization system through a large-scale fleet of electric vehicles for grid. *Int. J. Electr. Power Energy Syst.* **2022**, *141*, 108164. [CrossRef]
58. Shepard, D.P.; Humphreys, T.E.; Fansler, A.A. Evaluation of the vulnerability of phasor measurement units to GPS spoofing attacks. *Int. J. Crit. Infrastruct. Prot.* **2012**, *5*, 146–153. [CrossRef]
59. Fan, Y.; Zhang, Z.; Trinkle, M.; Dimitrovski, A.D.; Song, J.B.; Li, H. A cross-layer defense mechanism against GPS spoofing attacks on PMUs in smart grids. *IEEE Trans. Smart Grid* **2014**, *6*, 2659–2668. [CrossRef]
60. Hu, Y.; Novosel, D. Progresses in PMU testing and calibration. In Proceedings of the 2008 Third International Conference on Electric Utility Deregulation and Restructuring and Power Technologies, Nanjing, China, 6–9 April 2008; IEEE: Piscataway, NJ, USA, 2008; pp. 150–155.
61. Ahmed, A.; Krishnan, V.V.G.; Foroutan, S.A.; Touhiduzzaman; Rublein, C.L.; Srivastava, A.K.; Wu, Y.; Hahn, A.; Suresh, S. Cyber physical security analytics for anomalies in transmission protection systems. *IEEE Trans. Ind. Appl.* **2019**, *55*, 6313–6323. [CrossRef]
62. Chamorro, H.R.; Gomez-Diaz, E.O.; Paternina, M.R.A.; Andrade, M.A.; Barocio, E.; Rueda, J.L.; Gonzalez-Longatt, F.; Sood, V.K. Power system coherency recognition and islanding: Practical limits and future perspectives. *IET Energy Syst. Integr.* **2023**, *5*, 1–14. [CrossRef]
63. Phadke, A.G.; Thorp, J.S. *Computer Relaying for Power Systems*; John Wiley Sons: Hoboken, NJ, USA, 2009.
64. Anderson, P.M.; Henville, C.F.; Rifaat, R.; Johnson, B.; Meliopoulos, S. *Power System Protection*; John Wiley Sons: Hoboken, NJ, USA, 2022.
65. Adamiak, M.; Apostolov, A.; Begovic, M.; Henville, C.; Martin, K.; Michel, G.; Phadke, A.; Thorp, J. Wide area protection—Technology and infrastructures. *IEEE Trans. Power Deliv.* **2006**, *21*, 601–609. [CrossRef]
66. Horowitz, S.H.; Phadke, A.G. Third zone revisited. *IEEE Trans. Power Deliv.* **2005**, *21*, 23–29. [CrossRef]
67. Tan, J.C.; Crossley, P.A.; McLaren, P.G.; Gale, P.F.; Hall, I.; Farrell, J. Application of a wide area backup protection expert system to prevent cascading outages. *IEEE Trans. Power Deliv.* **2002**, *17*, 375–380. [CrossRef]

68. Saber, A.; Emam, A.; Elghazaly, H. Wide-area backup protection scheme for transmission lines considering cross-country and evolving faults. *IEEE Syst. J.* **2018**, *13*, 813–822. [CrossRef]
69. He, Z.; Zhang, Z.; Chen, W.; Malik, O.P.; Yin, X. Wide-area backup protection algorithm based on fault component voltage distribution. *IEEE Trans. Power Deliv.* **2011**, *26*, 2752–2760. [CrossRef]
70. Ma, J.; Li, J.; Thorp, J.S.; Arana, A.J.; Yang, Q.; Phadke, A.G. A fault steady state component-based wide area backup protection algorithm. *IEEE Trans. Smart Grid* **2011**, *2*, 468–475. [CrossRef]
71. Nayak, P.K.; Pradhan, A.K.; Bajpai, P. Wide-area measurement-based backup protection for power network with series compensation. *IEEE Trans. Power Deliv.* **2014**, *29*, 1970–1977. [CrossRef]
72. Eissa, M.M.; Masoud, M.E.; Elanwar, M.M.M. A novel back up wide area protection technique for power transmission grids using phasor measurement unit. *IEEE Trans. Power Deliv.* **2009**, *25*, 270–278. [CrossRef]
73. Zare, J.; Aminifar, F.; Sanaye-Pasand, M. Communication-constrained regionalization of power systems for synchrophasor-based wide-area backup protection scheme. *IEEE Trans. Smart Grid* **2015**, *6*, 1530–1538. [CrossRef]
74. Dubey, R.; Samantaray, S.R.; Panigrahi, B.K.; Venkoparao, V.G. Koopman analysis based wide-area back-up protection and faulted line identification for series-compensated power network. *IEEE Syst. J.* **2016**, *12*, 2634–2644. [CrossRef]
75. Mirhosseini, S.S.; Akhbari, M. Wide area backup protection algorithm for transmission lines based on fault component complex power. *Int. J. Electr. Power Energy Syst.* **2016**, *83*, 1–6. [CrossRef]
76. Jena, M.K.; Samantaray, S.R.; Panigrahi, B.K. A new decentralized approach to wide-area back-up protection of transmission lines. *IEEE Syst. J.* **2017**, *12*, 3161–3168. [CrossRef]
77. Ma, J.; Liu, C.; Thorp, J.S. A wide-area backup protection algorithm based on distance protection fitting factor. *IEEE Trans. Power Deliv.* **2015**, *31*, 2196–2205. [CrossRef]
78. Tong, X.; Wang, X.; Wang, R.; Huang, F.; Dong, X.; Hopkinson, K.M.; Song, G. The study of a regional decentralized peer-to-peer negotiation-based wide-area backup protection multi-agent system. *IEEE Trans. Smart Grid* **2013**, *4*, 1197–1206. [CrossRef]
79. Li, Z.; Yin, X.; Zhang, Z.; He, Z. Wide-area protection fault identification algorithm based on multi-information fusion. *IEEE Trans. Power Deliv.* **2013**, *28*, 1348–1355.
80. Yu, F.; Booth, C.; Dysko, A.; Hong, Q. Wide-area backup protection and protection performance analysis scheme using PMU data. *Int. J. Electr. Power Energy Syst.* **2019**, *110*, 630–641. [CrossRef]
81. Chen, M.; Wang, H.; Shen, S.; He, B. Research on a distance relay-based wide-area backup protection algorithm for transmission lines. *IEEE Trans. Power Deliv.* **2016**, *32*, 97–105. [CrossRef]
82. Navalkar, P.V.; Soman, S.A. Secure remote backup protection of transmission lines using synchrophasors. *IEEE Trans. Power Deliv.* **2010**, *26*, 87–96. [CrossRef]
83. Sharafi, A.; Sanaye-Pasand, M.; Aminifar, F. Transmission system wide-area back-up protection using current phasor measurements. *Int. J. Electr. Power Energy Syst.* **2017**, *92*, 93–103. [CrossRef]
84. Li, W.; Tan, Y.; Li, Y.; Cao, Y.; Chen, C.; Zhang, M. A new differential backup protection strategy for smart distribution networks: A fast and reliable approach. *IEEE Access* **2019**, *7*, 38135–38145. [CrossRef]
85. Zare, J.; Aminifar, F.; Sanaye-Pasand, M. Synchrophasor-based wide-area backup protection scheme with data requirement analysis. *IEEE Trans. Power Deliv.* **2014**, *30*, 1410–1419. [CrossRef]
86. Neyestanaki, M.K.; Ranjbar, A.M. An adaptive PMU-based wide area backup protection scheme for power transmission lines. *IEEE Trans. Smart Grid* **2015**, *6*, 1550–1559. [CrossRef]
87. Nougain, V.; Jena, M.K.; Panigrahi, B.K. Decentralised wide-area back-up protection scheme based on the concept of centre of reactive power. *IET Gener. Transm. Distrib.* **2019**, *13*, 4551–4557. [CrossRef]
88. Samantaray, S.R.; Sharma, A. Enhancing performance of wide-area back-up protection scheme using PMU assisted dynamic state estimator. *IEEE Trans. Smart Grid* **2018**, *10*, 5066–5074.
89. Abd el-Ghany, H.A. Optimal PMU allocation for high-sensitivity wide-area backup protection scheme of transmission lines. *Electr. Power Syst. Res.* **2020**, *187*, 106485. [CrossRef]
90. Ahmadinia, M.; Sadeh, J. A modified wide-area backup protection scheme for shunt-compensated transmission lines. *Electr. Power Syst. Res.* **2020**, *183*, 106274. [CrossRef]
91. Azizi, S.; Liu, G.; Dobakhshari, A.S.; Terzija, V. Wide-area backup protection against asymmetrical faults using available phasor measurements. *IEEE Trans. Power Deliv.* **2019**, *35*, 2032–2039. [CrossRef]
92. Chavez, J.J.; Kumar, N.V.; Azizi, S.; Guardado, J.L.; Rueda, J.; Palensky, P.; Terzija, V.; Popov, M. PMU-voltage drop based fault locator for transmission backup protection. *Electr. Power Syst. Res.* **2021**, *196*, 107188. [CrossRef]
93. Samantaray, S.R. A differential voltage-based wide-area backup protection scheme for transmission network. *IEEE Syst. J.* **2021**, *16*, 520–530.
94. Harish, A.; Prince, A.; Jayan, M.V. Fault Detection and Classification for Wide Area Backup Protection of Power Transmission Lines Using Weighted Extreme Learning Machine. *IEEE Access* **2022**, *10*, 82407–82417. [CrossRef]
95. Jegarluei, M.R.; Dobakhshari, A.S.; Azizi, S. Reducing the computational complexity of wide-area backup protection in power systems. *IEEE Trans. Power Deliv.* **2022**, *37*, 2421–2424. [CrossRef]
96. Shazdeh, S.; Golpîra, H.; Bevrani, H. A PMU-based back-up protection scheme for fault detection considering uncertainties. *Int. J. Electr. Power Energy Syst.* **2023**, *145*, 108592. [CrossRef]

97. Jegarluei, M.R.; Aristidou, P.; Azizi, S. Wide-Area backup protection against asymmetrical faults in the presence of renewable energy sources. *Int. J. Electr. Power Energy Syst.* **2023**, *144*, 108528. [CrossRef]
98. Goh, H.H.; Chua, Q.S.; Lee, S.W.; Kok, B.C.; Goh, K.C.; Teo, K.T.K. Evaluation for voltage stability indices in power system using artificial neural network. *Procedia Eng.* **2015**, *118*, 1127–1136. [CrossRef]
99. Li, H.; Bose, A.; Venkatasubramanian, V.M. Wide-area voltage monitoring and optimization. *IEEE Trans. Smart Grid* **2015**, *7*, 785–793. [CrossRef]
100. Su, H.Y.; Liu, C.W. Estimating the voltage stability margin using PMU measurements. *IEEE Trans. Power Syst.* **2015**, *31*, 3221–3229. [CrossRef]
101. Vournas, C.D.; Lambrou, C.; Mandoulidis, P. Voltage stability monitoring from a transmission bus PMU. *IEEE Trans. Power Syst.* **2016**, *32*, 3266–3274. [CrossRef]
102. Kamel, M.; Karrar, A.A.; Eltom, A.H. Development and application of a new voltage stability index for on-line monitoring and shedding. *IEEE Trans. Power Syst.* **2017**, *33*, 1231–1241. [CrossRef]
103. Mandoulidis, P.; Vournas, C. A PMU-based real-time estimation of voltage stability and margin. *Electr. Power Syst. Res.* **2020**, *178*, 106008. [CrossRef]
104. Safavizadeh, A.; Kordi, M.; Eghtedarnia, F.; Torkzadeh, R.; Marzooghi, H. Framework for real-time short-term stability assessment of power systems using PMU measurements. *IET Gener. Transm. Distrib.* **2019**, *13*, 3433–3442. [CrossRef]
105. Gao, H.; Cai, G.; Yang, D.; Wang, L. Real-time long-term voltage stability assessment based on eGBDT for large-scale power system with high renewables penetration. *Electr. Power Syst. Res.* **2023**, *214*, 108915. [CrossRef]
106. Guddanti, K.P.; Matavalam AR, R.; Weng, Y. PMU-based distributed non-iterative algorithm for real-time voltage stability monitoring. *IEEE Trans. Smart Grid* **2020**, *11*, 5203–5215. [CrossRef]
107. Hur Rizvi, S.M.; Kundu, P.; Srivastava, A.K. Hybrid voltage stability and security assessment using synchrophasors with consideration of generator Q-limits. *IET Gener. Transm. Distrib.* **2020**, *14*, 4042–4051. [CrossRef]
108. Kumar, S.; Tyagi, B.; Kumar, V.; Chohan, S. PMU-based voltage stability measurement under contingency using ANN. *IEEE Trans. Instrum. Meas.* **2021**, *71*, 1–11. [CrossRef]
109. Yang, H.; Zhang, W.; Chen, J.; Wang, L. PMU-based voltage stability prediction using least square support vector machine with online learning. *Electr. Power Syst. Res.* **2018**, *160*, 234–242. [CrossRef]
110. Jena, M.K.; Panigrahi, B.K. Transient potential power based supervisory zone-1 operation during unstable power swing. *IEEE Syst. J.* **2018**, *13*, 1823–1830. [CrossRef]
111. Cai, G.; Wang, B.; Yang, D.; Sun, Z.; Wang, L. Inertia estimation based on observed electromechanical oscillation response for power systems. *IEEE Trans. Power Syst.* **2019**, *34*, 4291–4299. [CrossRef]
112. Machowski, J. Selectivity of power system protections at power swings in power system. *Acta Energetica* **2012**, *4*, 96–123.
113. Machowski, J.; Lubosny, Z.; Bialek, J.W.; Bumby, J.R. *Power System Dynamics: Stability and Control*; John Wiley Sons: Hoboken, NJ, USA, 2020.
114. Desai, J.P.; Makwana, V.H. Phasor measurement unit incorporated adaptive out-of-step protection of synchronous generator. *J. Mod. Power Syst. Clean Energy* **2020**, *9*, 1032–1042. [CrossRef]
115. Gao, Z.D.; Wang, G.B. A new power swing block in distance protection based on a microcomputer-principle and performance analysis. In Proceedings of the 1991 International Conference on Advances in Power System Control, Operation and Management, APSCOM-91, Hong Kong, China, 5–8 November 1991; IET: London, UK, 1991; pp. 843–847.
116. So, K.H.; Heo, J.Y.; Kim, C.H.; Aggarwal, R.K.; Song, K.B. Out-of-step detection algorithm using frequency deviation of voltage. *IET Gener. Transm. Distrib.* **2007**, *1*, 119–126. [CrossRef]
117. Sriram, C.; Kumar, D.R.; Raju, G.S. Blocking the distance relay operation in third zone during power swing using polynomial curve fitting method. In Proceedings of the 2014 International Conference on Smart Electric Grid (ISEG), Guntur, India, 19–20 September 2014; IEEE: Piscataway, NJ, USA, 2014; pp. 1–7.
118. Nayak, P.K.; Pradhan, A.K.; Bajpai, P. Secured zone 3 protection during stressed condition. *IEEE Trans. Power Deliv.* **2014**, *30*, 89–96. [CrossRef]
119. Chothani, N.G.; Bhalja, B.R.; Parikh, U.B. New support vector machine-based digital relaying scheme for discrimination between power swing and fault. *IET Gener. Transm. Distrib.* **2014**, *8*, 17–25. [CrossRef]
120. Jafari, R.; Moaddabi, N.; Eskandari-Nasab, M.; Gharehpetian, G.B.; Naderi, M.S. A novel power swing detection scheme independent of the rate of change of power system parameters. *IEEE Trans. Power Deliv.* **2014**, *29*, 1192–1202. [CrossRef]
121. Kang, D.; Gokaraju, R. A new method for blocking third-zone distance relays during stable power swings. *IEEE Trans. Power Deliv.* **2016**, *31*, 1836–1843. [CrossRef]
122. Kundu, P.; Pradhan, A.K. Wide area measurement based protection support during power swing. *Int. J. Electr. Power Energy Syst.* **2014**, *63*, 546–554. [CrossRef]
123. GhanizadehBolandi, T.; Haghifam, M.R.; Khederzadeh, M. Real-time monitoring of zone 3 vulnerable distance relays to prevent maloperation under load encroachment condition. *IET Gener. Transm. Distrib.* **2017**, *11*, 1878–1888. [CrossRef]
124. Lavand, S.A.; Soman, S.A. Predictive analytic to supervise zone 1 of distance relay using synchrophasors. *IEEE Trans. Power Deliv.* **2016**, *31*, 1844–1854. [CrossRef]
125. Patel, B.; Bera, P.; nee Dey, S.H. A novel method to distinguish internal and external faults during power swing. *IEEE Trans. Power Deliv.* **2020**, *36*, 2595–2605. [CrossRef]

126. Nazari, A.A.; Razavi, F.; Fakharian, A. A new power swing detection method in power systems with large-scale wind farms based on modified empirical-mode decomposition method. *IET Gener. Transm. Distrib.* **2022**, *17*, 1204–1215. [CrossRef]
127. Alnassar, Z.; Nagarajan, S.T. Analysis of Oscillations during Out-of-Step Condition in Power Systems. *Int. Trans. Electr. Energy Syst.* **2023**, *2023*, 4303491. [CrossRef]
128. Tealane, M.; Kilter, J.; Popov, M.; Bagleybter, O.; Klaar, D. Online detection of out-of-step condition using PMU-determined system impedances. *IEEE Access* **2022**, *10*, 14807–14818. [CrossRef]
129. Rao, J.T.; Bhalja, B.R.; Andreev, M.; Malik, O.P. Discrimination between in-zone and out-of-zone faults during power swing condition using synchrophasor data. *Int. J. Electr. Power Energy Syst.* **2023**, *146*, 108769.
130. El-Sayed LM, A.; Ibrahim, D.K.; Gilany, M.I. Enhancing distance relay performance using wide-area protection for detecting symmetrical/unsymmetrical faults during power swings. *Alex. Eng. J.* **2022**, *61*, 6869–6886. [CrossRef]
131. Cheng, Y.; Foggo, B.; Yamashita, K.; Yu, N. Missing Value Replacement for PMU Data via Deep Learning Model With Magnitude Trend Decoupling. *IEEE Access* **2023**, *11*, 27450–27461. [CrossRef]
132. Chakhchoukh, Y.; Vittal, V.; Heydt, G.T. PMU based state estimation by integrating correlation. *IEEE Trans. Power Syst.* **2013**, *29*, 617–626. [CrossRef]
133. Johnson, T.; Moger, T. Latest Trends in Electromechanical Dynamic State Estimation for Electric Power Grid. In Proceedings of the 2022 3rd International Conference for Emerging Technology (INCET), Belgaum, India, 27–29 May 2022; IEEE: Piscataway, NJ, USA, 2022; pp. 1–6.
134. Huang, Z.; Du, P.; Kosterev, D.; Yang, S. Generator dynamic model validation and parameter calibration using phasor measurements at the point of connection. *IEEE Trans. Power Syst.* **2013**, *28*, 1939–1949. [CrossRef]
135. Fan, L.; Wehbe, Y. Extended Kalman filtering based real-time dynamic state and parameter estimation using PMU data. *Electr. Power Syst. Res.* **2013**, *103*, 168–177. [CrossRef]
136. Ghahremani, E.; Kamwa, I. Dynamic state estimation in power system by applying the extended Kalman filter with unknown inputs to phasor measurements. *IEEE Trans. Power Syst.* **2011**, *26*, 2556–2566. [CrossRef]
137. Ghahremani, E.; Kamwa, I. Online state estimation of a synchronous generator using unscented Kalman filter from phasor measurements units. *IEEE Trans. Energy Convers.* **2011**, *26*, 1099–1108. [CrossRef]
138. Zhou, N.; Huang, Z.; Li, Y.; Welch, G. Local sequential ensemble Kalman filter for simultaneously tracking states and parameters. In Proceedings of the 2012 North American Power Symposium (NAPS), Champaign, IL, USA, 9–11 September 2012; IEEE: Piscataway, NJ, USA, 2012; pp. 1–6.
139. Zhou, N.; Meng, D.; Lu, S. Estimation of the dynamic states of synchronous machines using an extended particle filter. *IEEE Trans. Power Syst.* **2013**, *28*, 4152–4161. [CrossRef]
140. Zhou, N.; Meng, D.; Huang, Z.; Welch, G. Dynamic state estimation of a synchronous machine using PMU data: A comparative study. *IEEE Trans. Smart Grid* **2014**, *6*, 450–460. [CrossRef]
141. Pignati, M.; Popovic, M.; Barreto, S.; Cherkaoui, R.; Flores, G.D.; Le Boudec, J.Y.; Mohiuddin, M.; Paolone, M.; Romano, P.; Sarri, S.; et al. Real-time state estimation of the EPFL-campus medium-voltage grid by using PMUs. In Proceedings of the 2015 IEEE Power Energy Society Innovative Smart Grid Technologies Conference (ISGT), Washington, DC, USA, 18–20 February 2015; IEEE: Piscataway, NJ, USA, 2015; pp. 1–5.
142. Ghahremani, E.; Kamwa, I. Local and wide-area PMU-based decentralized dynamic state estimation in multi-machine power systems. *IEEE Trans. Power Syst.* **2015**, *31*, 547–562. [CrossRef]
143. Wu, T.; Chung, C.Y.; Kamwa, I. A fast state estimator for systems including limited number of PMUs. *IEEE Trans. Power Syst.* **2017**, *32*, 4329–4339. [CrossRef]
144. Paul, A.; Kamwa, I.; Joos, G. Centralized dynamic state estimation using a federation of extended Kalman filters with intermittent PMU data from generator terminals. *IEEE Trans. Power Syst.* **2018**, *33*, 6109–6119. [CrossRef]
145. Prasad, S.; Kumar, D.V. Trade-offs in PMU and IED deployment for active distribution state estimation using multi-objective evolutionary algorithm. *IEEE Trans. Instrum. Meas.* **2018**, *67*, 1298–1307. [CrossRef]
146. Shereen, E.; Delcourt, M.; Barreto, S.; Dán, G.; Le Boudec, J.Y.; Paolone, M. Feasibility of time-synchronization attacks against PMU-based state estimation. *IEEE Trans. Instrum. Meas.* **2019**, *69*, 3412–3427. [CrossRef]
147. Jovicic, A.; Jereminov, M.; Pileggi, L.; Hug, G. A linear formulation for power system state estimation including RTU and PMU measurements. In Proceedings of the 2019 IEEE PES Innovative Smart Grid Technologies Europe (ISGT-Europe), Bucharest, Romania, 29 September–2 October 2019; IEEE: Piscataway, NJ, USA, 2019; pp. 1–5.
148. Kabiri, M.; Amjady, N. A new hybrid state estimation considering different accuracy levels of PMU and SCADA measurements. *IEEE Trans. Instrum. Meas.* **2018**, *68*, 3078–3089. [CrossRef]
149. Lin, C.; Wu, W.; Guo, Y. Decentralized robust state estimation of active distribution grids incorporating microgrids based on PMU measurements. *IEEE Trans. Smart Grid* **2019**, *11*, 810–820. [CrossRef]
150. Gholami, M.; Abbaspour, A.; Moeini-Aghtaie, M.; Fotuhi-Firuzabad, M.; Lehtonen, M. Detecting the location of short-circuit faults in active distribution network using PMU-based state estimation. *IEEE Trans. Smart Grid* **2019**, *11*, 1396–1406. [CrossRef]
151. Muscas, C.; Pegoraro, P.A.; Sulis, S.; Pau, M.; Ponci, F.; Monti, A. New Kalman filter approach exploiting frequency knowledge for accurate PMU-based power system state estimation. *IEEE Trans. Instrum. Meas.* **2020**, *69*, 6713–6722. [CrossRef]
152. Dobakhshari, A.S.; Azizi, S.; Abdolmaleki, M.; Terzija, V. Linear LAV-based state estimation integrating hybrid SCADA/PMU measurements. *IET Gener. Transm. Distrib.* **2020**, *14*, 1583–1590. [CrossRef]

153. Jovicic, A.; Hug, G. Linear state estimation and bad data detection for power systems with RTU and PMU measurements. *IET Gener. Transm. Distrib.* **2020**, *14*, 5675–5684. [CrossRef]

154. Khalili, R.; Abur, A. PMU-based decoupled state estimation for unsymmetrical power systems. *IEEE Trans. Power Syst.* **2021**, *36*, 5359–5368. [CrossRef]

155. De Oliveira-De Jesus, P.M.; Rodriguez, N.A.; Celeita, D.F.; Ramos, G.A. PMU-based system state estimation for multigrounded distribution systems. *IEEE Trans. Power Syst.* **2020**, *36*, 1071–1081. [CrossRef]

156. Dobakhshari, A.S.; Abdolmaleki, M.; Terzija, V.; Azizi, S. Robust hybrid linear state estimator utilizing SCADA and PMU measurements. *IEEE Trans. Power Syst.* **2020**, *36*, 1264–1273. [CrossRef]

157. Basumallik, S.; Ma, R.; Eftekharnejad, S. Packet-data anomaly detection in PMU-based state estimator using convolutional neural network. *Int. J. Electr. Power Energy Syst.* **2019**, *107*, 690–702. [CrossRef]

158. Mekhamer, S.F.; Abdelaziz, A.Y.; Ezzat, M.; Abdel-Salam, T.S. Fault location in long transmission lines using synchronized phasor measurements from both ends. *Electr. Power Compon. Syst.* **2012**, *40*, 759–776. [CrossRef]

159. Abdelaziz, A.Y.; Mekhamer, S.F.; Ezzat, M. Fault location of uncompensated/series-compensated lines using two-end synchronized measurements. *Electr. Power Compon. Syst.* **2013**, *41*, 693–715. [CrossRef]

160. Mukherjee, R.; De, A. Real-time dynamic security analysis of power systems using strategic PMU measurements and decision tree classification. *Electr. Eng.* **2021**, *103*, 813–824. [CrossRef]

161. Hauer, J.F. Validation of phasor calculations in the macrodyne PMU for California-Oregon transmission project tests of March 1993. *IEEE Trans. Power Deliv.* **1996**, *11*, 1224–1231. [CrossRef]

162. Abdollahi, A.; Seyedtabaii, S. Comparison of fourierwavelet transform methods for transmission line fault classification. In Proceedings of the 2010 4th International Power Engineering and Optimization Conference (PEOCO), Shah Alam, Malaysia, 23–24 June 2010; IEEE: Piscataway, NJ, USA, 2010; pp. 579–584.

163. Gopakumar, P.; Reddy MJ, B.; Mohanta, D.K. Transmission line fault detection and localisation methodology using PMU measurements. *IET Gener. Transm. Distrib.* **2015**, *9*, 1033–1042. [CrossRef]

164. Bakdi, A.; Bounoua, W.; Guichi, A.; Mekhilef, S. Real-time fault detection in PV systems under MPPT using PMU and high-frequency multi-sensor data through online PCA-KDE-based multivariate KL divergence. *Int. J. Electr. Power Energy Syst.* **2021**, *125*, 106457. [CrossRef]

165. Bansal, Y.; Sodhi, R. PMUs Enabled Tellegen's Theorem-Based Fault Identification Method for Unbalanced Active Distribution Network Using RTDS. *IEEE Syst. J.* **2020**, *14*, 4567–4578. [CrossRef]

166. Cui, Q.; Weng, Y. Enhance high impedance fault detection and location accuracy via $\mu$-PMUs. *IEEE Trans. Smart Grid* **2019**, *11*, 797–809. [CrossRef]

167. Sharma, N.K.; Samantaray, S.R. Assessment of PMU-based wide-area angle criterion for fault detection in microgrid. *IET Gener. Transm. Distrib.* **2019**, *13*, 4301–4310. [CrossRef]

168. Khan, M.Q.; Ahmed, M.M.; Haidar, A.M. An accurate algorithm of PMU-based wide area measurements for fault detection using positive-sequence voltage and unwrapped dynamic angles. *Measurement* **2022**, *192*, 110906. [CrossRef]

169. Liu, G.; Li, X.; Wang, C.; Chen, Z.; Chen, R.; Qiu, R.C. Hessian Locally Linear Embedding of PMU Data for Efficient Fault Detection in Power Systems. *IEEE Trans. Instrum. Meas.* **2022**, *71*, 1–4. [CrossRef]

170. Aloghareh, F.H.; Jannati, M.; Shams, M. A deep long Short-Term memory based scheme for Auto-Reclosing of power transmission lines. *Int. J. Electr. Power Energy Syst.* **2022**, *141*, 108105. [CrossRef]

171. Zhang, Y.; Wang, J.; Khodayar, M.E. Graph-based faulted line identification using micro-PMU data in distribution systems. *IEEE Trans. Smart Grid* **2020**, *11*, 3982–3992. [CrossRef]

172. Gilanifar, M.; Cordova, J.; Wang, H.; Stifter, M.; Ozguven, E.E.; Strasser, T.I.; Arghandeh, R. Multi-task logistic low-ranked dirty model for fault detection in power distribution system. *IEEE Trans. Smart Grid* **2019**, *11*, 786–796. [CrossRef]

173. Shadi, M.R.; Ameli, M.T.; Azad, S. A real-time hierarchical framework for fault detection, classification, and location in power systems using PMUs data and deep learning. *Int. J. Electr. Power Energy Syst.* **2022**, *134*, 107399. [CrossRef]

174. Sarangi, S.; Biswal, C.; Sahu, B.K.; Rout, P.K. Active Islanding Detection and Analysis of Total Harmonic Distortion for Inverter-Interfaced Microgrid Based on High-Frequency Signal Installation. In Proceedings of the 2022 1st International Conference on Sustainable Technology for Power and Energy Systems (STPES), Srinagar, India, 4–6 July 2022; IEEE: Piscataway, NJ, USA, 2022; pp. 1–6.

175. Kim, J.S.; Kim, C.H.; Oh, Y.S.; Cho, G.J.; Song, J.S. An islanding detection method for multi-RES systems using the graph search method. *IEEE Trans. Sustain. Energy* **2020**, *11*, 2722–2731. [CrossRef]

176. Kabalci, Y. A survey on smart metering and smart grid communication. *Renew. Sustain. Energy Rev.* **2016**, *57*, 302–318. [CrossRef]

177. Franco, R.; Sena, C.; Taranto, G.N.; Giusto, A. Using synchrophasors for controlled islanding—A prospective application for the Uruguayan power system. *IEEE Trans. Power Syst.* **2012**, *28*, 2016–2024. [CrossRef]

178. Sankar, A.; Sunitha, R. Synchrophasor data driven islanding detection, localization and prediction for microgrid using energy operator. *IEEE Trans. Power Syst.* **2021**, *36*, 4052–4065.

179. Samantaray, S.R.; Kamwa, I.; Joos, G. Phasor measurement unit based wide-area monitoring and information sharing between micro-grids. *IET Gener. Transm. Distrib.* **2017**, *11*, 1293–1302. [CrossRef]

180. Werho, T.; Vittal, V.; Kolluri, S.; Wong, S.M. A potential island formation identification scheme supported by PMU measurements. *IEEE Trans. Power Syst.* **2015**, *31*, 423–431. [CrossRef]

181. Brahma, S.; Kavasseri, R.; Cao, H.; Chaudhuri, N.R.; Alexopoulos, T.; Cui, Y. Real-time identification of dynamic events in power systems using PMU data, and potential applications—Models, promises, and challenges. *IEEE Trans. Power Deliv.* **2016**, *32*, 294–301. [CrossRef]

182. Barocio, E.; Pal, B.C.; Fabozzi, D.; Thornhill, N.F. Detection and visualization of power system disturbances using principal component analysis. In Proceedings of the 2013 IREP Symposium Bulk Power System Dynamics and Control-IX Optimization, Security and Control of the Emerging Power Grid, Madison, WI, USA, 29–31 May 2013; IEEE: Piscataway, NJ, USA, 2013; pp. 1–10.

183. Rafferty, M.; Liu, X.; Laverty, D.M.; McLoone, S. Real-time multiple event detection and classification using moving window PCA. *IEEE Trans. Smart Grid* **2016**, *7*, 2537–2548. [CrossRef]

184. Liu, X.; Laverty, D.M.; Best, R.J.; Li, K.; Morrow, D.J.; McLoone, S. Principal component analysis of wide-area phasor measurements for islanding detection—A geometric view. *IEEE Trans. Power Deliv.* **2015**, *30*, 976–985. [CrossRef]

185. Guo, Y.; Li, K.; Laverty, D.M.; Xue, Y. Synchrophasor-based islanding detection for distributed generation systems using systematic principal component analysis approaches. *IEEE Trans. Power Deliv.* **2015**, *30*, 2544–2552. [CrossRef]

186. Muda, H.; Jena, P. Phase angle-based PC technique for islanding detection of distributed generations. *IET Renew. Power Gener.* **2018**, *12*, 735–746. [CrossRef]

187. Gayathry, V.; Sujith, M. Machine learning based synchrophasor data analysis for islanding detection. In Proceedings of the 2020 International Conference for Emerging Technology (INCET), Belgaum, India, 5–7 June 2020; IEEE: Piscataway, NJ, USA, 2020; pp. 1–6.

188. Ali, W.; Ulasyar, A.; Mehmood, M.U.; Khattak, A.; Imran, K.; Zad, H.S.; Nisar, S. Hierarchical control of microgrid using IoT and machine learning based islanding detection. *IEEE Access* **2021**, *9*, 103019–103031. [CrossRef]

189. Kumar, G.P.; Jena, P. Pearson's correlation coefficient for islanding detection using micro-PMU measurements. *IEEE Syst. J.* **2020**, *15*, 5078–5089. [CrossRef]

190. Shukla, A.; Dutta, S.; Sadhu, P.K. An island detection approach by μ-PMU with reduced chances of cyber attack. *Int. J. Electr. Power Energy Syst.* **2021**, *126*, 106599. [CrossRef]

191. Chauhan, K.; Sodhi, R. A distribution-level PMU enabled Teager-Kaiser energy based islanding detector. *Electr. Power Syst. Res.* **2021**, *192*, 106964. [CrossRef]

192. Laverty, D.M.; Duggan, C.; Hastings, J.C.; Morrow, D.J. Islanding detection by phase difference method using a low cost quasi-PMU. *IET Gener. Transm. Distrib.* **2021**, *15*, 3302–3314. [CrossRef]

193. Dua, G.S.; Tyagi, B.; Kumar, V. Microgrid Differential Protection Based On Superimposed Current Angle Employing Synchrophasors. *IEEE Trans. Ind. Inform.* **2022**, 1–9. [CrossRef]

194. Som, S.; Dutta, R.; Gholami, A.; Srivastava, A.K.; Chakrabarti, S.; Sahoo, S.R. Dpmu-based multiple event detection in a microgrid considering measurement anomalies. *Appl. Energy* **2022**, *308*, 118269. [CrossRef]

195. Li, S.; Sun, Y.; Qin, S.; Shi, F.; Zhang, H.; Xu, Q.; Xie, W.; Zhang, Y. PMU-based harmonic phasor calculation and harmonic source identification. In Proceedings of the 2018 2nd IEEE Conference on Energy Internet and Energy System Integration (EI2), Beijing, China, 20–22 October 2018; IEEE: Piscataway, NJ, USA, 2018; pp. 1–6.

196. Frigo, G.; Derviškadić, A.; Pegoraro, P.A.; Muscas, C.; Paolone, M. Harmonic phasor measurements in real-world PMU-based acquisitions. In Proceedings of the 2019 IEEE International Instrumentation and Measurement Technology Conference (I2MTC), Auckland, New Zealand, 20–23 May 2019; IEEE: Piscataway, NJ, USA, 2019; pp. 1–6.

197. Chen, L.; Zhao, W.; Wang, F.; Huang, S. Harmonic phasor estimator for P-class phasor measurement units. *IEEE Trans. Instrum. Meas.* **2019**, *69*, 1556–1565. [CrossRef]

198. Bernard, L.; Goondram, S.; Bahrani, B.; Pantelous, A.A.; Razzaghi, R. Harmonic and interharmonic phasor estimation using matrix pencil method for phasor measurement units. *IEEE Sens. J.* **2020**, *21*, 945–954. [CrossRef]

199. Carta, A.; Locci, N.; Muscas, C. A PMU for the measurement of synchronized harmonic phasors in three-phase distribution networks. *IEEE Trans. Instrum. Meas.* **2009**, *58*, 3723–3730. [CrossRef]

200. Sun, Y.; Li, S.; Xu, Q.; Xie, X.; Jin, Z.; Shi, F.; Zhang, H. Harmonic contribution evaluation based on the distribution-level PMUs. *IEEE Trans. Power Deliv.* **2020**, *36*, 909–919. [CrossRef]

201. Castello, P.; Laurano, C.; Muscas, C.; Pegoraro, P.A.; Toscani, S.; Zanoni, M. Harmonic synchrophasors measurement algorithms with embedded compensation of voltage transformer frequency response. *IEEE Trans. Instrum. Meas.* **2020**, *70*, 9001310. [CrossRef]

202. Chen, L.; Farajollahi, M.; Ghamkhari, M.; Zhao, W.; Huang, S.; Mohsenian-Rad, H. Switch status identification in distribution networks using harmonic synchrophasor measurements. *IEEE Trans. Smart Grid* **2020**, *12*, 2413–2424. [CrossRef]

203. Ahmadi-Gorjayi, F.; Mohsenian-Rad, H. A Physics-Aware MIQP Approach to Harmonic State Estimation in Low-Observable Power Distribution Systems Using Harmonic Phasor Measurement Units. *IEEE Trans. Smart Grid* **2022**, *14*, 2111–2124. [CrossRef]

204. Aligholian, A.; Mohsenian-Rad, H. GraphPMU: Event Clustering via Graph Representation Learning Using Locationally-Scarce Distribution-Level Fundamental and Harmonic PMU Measurements. *IEEE Trans. Smart Grid* **2022**. [CrossRef]

205. Liang, G.; Weller, S.R.; Zhao, J.; Luo, F.; Dong, Z.Y. The 2015 Ukraine blackout: Implications for false data injection attacks. *IEEE Trans. Power Syst.* **2016**, *32*, 3317–3318. [CrossRef]

206. Pal, S.; Sikdar, B.; Chow, J.H. Classification and detection of PMU data manipulation attacks using transmission line parameters. *IEEE Trans. Smart Grid* **2017**, *9*, 5057–5066. [CrossRef]

207. Wang, X.; Shi, D.; Wang, J.; Yu, Z.; Wang, Z. Online identification and data recovery for PMU data manipulation attack. *IEEE Trans. Smart Grid* **2019**, *10*, 5889–5898. [CrossRef]

208. Xue, A.; Xu, F.; Xu, J.; Chow, J.H.; Leng, S.; Bi, T. Online pattern recognition and data correction of PMU data under GPS spoofing attack. *J. Mod. Power Syst. Clean Energy* **2020**, *8*, 1240–1249. [CrossRef]
209. Liang, H.; Zhu, L.; Yu, F.R.; Wang, X. A Cross-Layer Defense Method for Blockchain Empowered CBTC Systems Against Data Tampering Attacks. *IEEE Trans. Intell. Transp. Syst.* **2022**, *24*, 501–515. [CrossRef]
210. Yasinzadeh, M.; Akhbari, M. Detection of PMU spoofing in power grid based on phasor measurement analysis. *IET Gener. Transm. Distrib.* **2018**, *12*, 1980–1987. [CrossRef]
211. Shereen, E.; Dán, G. Model-based and data-driven detectors for time synchronization attacks against PMUs. *IEEE J. Sel. Areas Commun.* **2019**, *38*, 169–179. [CrossRef]
212. Gao, P.; Wang, M.; Ghiocel, S.G.; Chow, J.H.; Fardanesh, B.; Stefopoulos, G. Missing data recovery by exploiting low-dimensionality in power system synchrophasor measurements. *IEEE Trans. Power Syst.* **2015**, *31*, 1006–1013. [CrossRef]
213. Kamal, M.; Farajollahi, M.; Mohsenian-Rad, H. Analysis of cyber attacks against micro-PMUs: The case of event source location identification. In Proceedings of the 2020 IEEE Power Energy Society Innovative Smart Grid Technologies Conference (ISGT), Washington, DC, USA, 17–20 February 2020; IEEE: Piscataway, NJ, USA, 2020; pp. 1–5.
214. Li, Q.; Li, F.; Zhang, J.; Ye, J.; Song, W.; Mantooth, A. Data-driven cyberattack detection for photovoltaic (PV) systems through analyzing micro-PMU data. In Proceedings of the 2020 IEEE Energy Conversion Congress and Exposition (ECCE), Detroit, MI, USA, 11–15 October 2020; IEEE: Piscataway, NJ, USA, 2020; pp. 431–436.
215. Ghafouri, M.; Au, M.; Kassouf, M.; Debbabi, M.; Assi, C.; Yan, J. Detection and mitigation of cyber attacks on voltage stability monitoring of smart grids. *IEEE Trans. Smart Grid* **2020**, *11*, 5227–5238. [CrossRef]
216. Chakhchoukh, Y.; Lei, H.; Johnson, B.K. Diagnosis of outliers and cyber attacks in dynamic PMU-based power state estimation. *IEEE Trans. Power Syst.* **2019**, *35*, 1188–1197. [CrossRef]
217. Khalafi, Z.S.; Dehghani, M.; Khalili, A.; Sami, A.; Vafamand, N.; Dragičević, T. Intrusion detection, measurement correction, and attack localization of PMU networks. *IEEE Trans. Ind. Electron.* **2021**, *69*, 4697–4706. [CrossRef]
218. Elimam, M.; Isbeih, Y.J.; Azman, S.K.; ElMoursi, M.S.; Al Hosani, K. Deep Learning-Based PMU Cyber Security Scheme Against Data Manipulation Attacks With WADC Application. *IEEE Trans. Power Syst.* **2022**, *38*, 2148–2161. [CrossRef]
219. Martin, K.; Benmouyal, G.; Adamiak, M.; Begovic, M.; Burnett, R.; Carr, K.; Cobb, A.; Kusters, J.; Horowitz, S.; Jensen, G.; et al. IEEE standard for synchrophasors for power systems. *IEEE Trans. Power Deliv.* **1998**, *13*, 73–77. [CrossRef]
220. Martin, K.E.; Hamai, D.; Adamiak, M.G.; Anderson, S.; Begovic, M.; Benmouyal, G.; Brunello, G.; Burger, J.; Cai, J.Y.; Dickerson, B.; et al. Exploring the IEEE standard C37. 118–2005 synchrophasors for power systems. *IEEE Trans. Power Deliv.* **2008**, *23*, 1805–1811. [CrossRef]
221. Martin, K.E.; Brunello, G.; Adamiak, M.G.; Antonova, G.; Begovic, M.; Benmouyal, G.; Bui, P.D.; Falk, H.; Gharpure, V.; Goldstein, A.; et al. An overview of the IEEE standard C37. 118.2—Synchrophasor data transfer for power systems. *IEEE Trans. Smart Grid* **2014**, *5*, 1980–1984. [CrossRef]
222. Löper, M.; Trummal, T.; Kilter, J. Analysis of the applicability of PMU measurements for power quality assessment. In Proceedings of the 2018 IEEE PES Innovative Smart Grid Technologies Conference Europe (ISGT-Europe), Sarajevo, Bosnia and Herzegovina, 21–25 October 2018; IEEE: Piscataway, NJ, USA, 2018; pp. 1–6.
223. Huang, C.; Li, F.; Zhou, D.; Guo, J.; Pan, Z.; Liu, Y.; Liu, Y. Data quality issues for synchrophasor applications Part I: A review. *J. Mod. Power Syst. Clean Energy* **2016**, *4*, 342–352. [CrossRef]

*Article*

# Virtual Armature Resistance-Based Control for Fault Current Limiting in a High-Order VSG and the Impact on Its Transient Stability

Daniel Carletti, Thiago Amorim and Lucas Encarnação *

Department of Electrical Engineering, Federal University of Espírito Santo (UFES), Av. Fernando Ferrari, 514, Vitória 29075-910, Brazil; danielc.ufes@gmail.com (D.C.); t.s.amorim@hotmail.com (T.A.)
* Correspondence: lucas.encarnacao@ufes.br

**Abstract:** This article proposes a fault ride through (FRT) technique for a high-order virtual synchronous generator (VSG) that adjusts its virtual armature resistance. When a fault is detected by a dedicated algorithm, the proposed control adjusts the resistance parameter accordingly. The main contribution of this article is to adjust the virtual resistance directly in the machine model to limit the current during faults, unlike other techniques proposed in the literature that add another control loop to produce the virtual impedance effects. To validate the effectiveness of the proposed control, a hardware-in-the-loop real-time simulation platform was adopted using a Typhoon HIL 402 device and a Texas Instruments F28379D digital controller. The results demonstrate that the control effectively limits the converter's current while still contributing to raising the system's critical clearing time (CCT) and improving transient stability. The proposed FRT strategy is validated in a three-phase fault scenario in which a 500 kVA–480 V converter's peak fault current is reduced from 5 kA to 1.4 kA, depending on the resistance value adjusted. The transient stability is also analyzed in 30 different scenarios and the VSG support on the CCT is reduced by 23 ms on average. However, when compared to the baseline scenario without the VSG, the system still sees an increase in CCT with the current limiting control applied. Additionally, the control allows the VSG to smoothly transition to island mode in a scenario where the fault is cleared and the grid is disconnected by a protection system.

**Keywords:** fault ride through; grid-forming converter; power system stability; transient stability; virtual synchronous generator

**Citation:** Carletti, D.; Amorim, T.; Encarnação, L. Virtual Armature Resistance-Based Control for Fault Current Limiting in a High-Order VSG and the Impact on Its Transient Stability. *Energies* **2023**, *16*, 4680. https://doi.org/10.3390/en16124680

Academic Editors: Alon Kuperman and Mario Marchesoni

Received: 18 April 2023
Revised: 26 May 2023
Accepted: 8 June 2023
Published: 13 June 2023

## 1. Introduction

The increasing usage of renewable energy sources (RES) in the power system has raised concerns regarding its safe and stable operation. Most of this type of generation is connected to the system via small, distributed generation units, which are linked to the grid through voltage source converters (VSC). However, large-scale generation units, particularly photovoltaic and wind turbines, are becoming more prevalent in the energy market [1]. Due to their non-inertial nature, these energy sources do not contribute naturally to the system's angular and frequency stability, which can result in instability scenarios and blackouts. To address this, grid-forming (GFM) controls have been proposed for the VSC so that these energy sources, which are connected to the system by electronic converters, can also contribute to the system's stability by providing virtual inertia.

The GFM converter is characterized as a converter that is controlled in a way where its behavior is similar to that of a voltage source with low output impedance. Power control is accomplished by emulating the synchronization principle of a real synchronous machine [2]. A GFM control based on the direct emulation of the synchronous machine model is called a virtual synchronous generator (VSG). This can be realized through the application of high-order models, including a model of the electromagnetic behavior of the machine windings, as well as through low-order models, which apply only the swing equation for synchronization control and an external loop for voltage control [3–5].

Several authors have recently studied the transient stability of the system in the presence of these converters. As RES participation increases, and the system's equivalent inertia decreases, the system's stability margin for large disturbances (such as three-phase short circuits) eventually reduces. This leads to increased vulnerability to collapse due to an imbalance between instantaneous generation and demand. Additionally, the VSG, which has a GFM characteristic based on a voltage source with low output impedance, can produce high-magnitude fault currents that may damage the converter during voltage sag or short circuit situations [2]. Recent studies have demonstrated the contribution of VSGs to the power system's transient angular stability. For example, in [6], the Lyapunov direct method is used to investigate the transient stability of VSGs, while the effect of the model parameters of a low-order VSG on stability is analyzed. In [7], the transient stability of multiple GFM schemes, including the VSG, is examined, and design rules are proposed to enhance transient stability. The stability of paralleled VSGs in islanded microgrids is analyzed in [8]. In [9], adaptive control is proposed to improve the transient stability of a high-order VSG. Furthermore, several proposals have been discussed to improve the VSG's fault ride through (FRT). For instance, current saturation of the proportional integral (PI) controllers of the cascaded control loops [10], virtual impedances to limit the converter reference voltage [11], directly limiting the reference voltage in the GFM control structure [12], and FRT enhancement via a model predictive control (MPC) [13] are some of the main solutions proposed.

The current limitation of VSCs through a virtual impedance loop is discussed in [14–16]. The proposed methods for fault current limiting through virtual impedance involve the use of additional loops in the control scheme of distributed generation (DG) inverters. This technique effectively suppresses fault current and subsequent oscillations during faults as well as in the post-fault restoration process. The application of this fault current limitation strategy is specifically considered for the VSG in [17–19]. In [18], the authors proposed a dynamic virtual impedance (DVI) to provide damping torque for a low-order VSG. The DVI aims to emulate the transient and subtransient reactance's ability to provide damping and, consequently, to reduce the high damping coefficient normally used in low-order VSGs. A virtual resistance is considered for synchronous resonance damping in reference [17], but the negative effect of the virtual resistance on the transient stability is also mentioned. Increasing the resistance value can improve system stability by reducing power oscillations, but it may also result in increased losses and reduced efficiency [11]. On the other hand, a higher reactance value can enhance voltage regulation and reduce harmonic distortion, as mentioned in [11]. In [19], the authors presented design rules and different control strategies to optimize the output impedance of a VSG to improve fault current and large-disturbance stability. In [20], the authors propose a strategy based on virtual impedance to improve the low-voltage ride through of a low-order model VSG, aiming to regulate the current of the VSG under transient and steady-state conditions.

The strategies proposed and discussed earlier are based on additional control loops to adaptively adjust the value of virtual resistance and reactance at the output of the converter. This article proposes the use of a VSG implemented using a high-order model, with algebraic equations that also model the electrical part of a synchronous generator (SG). The high-order model provides equations for calculating the armature voltage of the virtual generator with its output impedance. The novelty of this work lies in the adjustment of the virtual armature resistance directly in the high-order model to limit the current during faults. This reduces the complexity of the control by not having an extra control loop to reshape the converter output impedance. The resistance variation is performed only during system faults, while in steady-state operation, the VSG operates with its original value for the armature resistance parameter. The resistance variation was adopted instead of reactance due to its direct relationship with the damping of power oscillations in the system, as described in [11]. By increasing the armature resistance, the VSG can provide more damping to the grid, which can help maintain the stability of the system during faults. An algorithm is designed to detect fault conditions, such as voltage sag and overcurrent, and the virtual resistance variation is implemented accordingly. As a result, the added resistance effectively limits the current during and after the fault, aiding in returning to steady-state operation.

The article is organized as follows: the structure of the implemented VSG is presented in Section 2. Section 3 presents in detail the current limiting strategy applied to the VSG presented in Section 2. Section 4 presents the system used for validation and discusses the results obtained in a hardware-in-the-loop (HIL) simulation platform. Finally, Section 5 is dedicated to the conclusions.

## 2. The High-Order Virtual Synchronous Generator

The VSG, which is a control mechanism that imitates the behavior of an SG, is controlled by a mathematical model of the SG. This model consists of a set of differential and algebraic equations, and its order and complexity depend on the number of differential equations used. It is common in VSG topologies to use a low-order model, as described in the literature. In this case, only the machine swing equation is employed to control the inverter synchronization, while an external loop regulates the output voltage [3,4]. This article proposes the development of a VSG using a third-order machine model, which includes the swing equation as a second-order differential equation, as well as a differential equation corresponding to the quadrature-axis transient voltage [21]. The swing equation is decomposed and presented in Equations (1) and (2) [21]:

$$P_m - P_e - D\Delta\omega = 2H\frac{d\omega}{dt} \tag{1}$$

$$\frac{d\delta}{dt} = \omega \tag{2}$$

where $P_m$ and $P_e$ are mechanical and electrical powers, respectively, $\omega$ is the mechanical speed, $\delta$ is the rotor angle, $D$ is the mechanical damping coefficient, $H$ is the rotor inertia constant, and $t$ is the time in seconds.

The transient voltage equation is shown in Equation (3) [21]:

$$T'_{do}\dot{E}'_q = E_{fd} - E'_q + i_d\left(X_d - X'_d\right) \tag{3}$$

where $T'_{do}$ is the quadrature-axis open circuit transient time constant, $X_d$ and $X_q$ are the synchronous reactances in the direct and quadrature axis, respectively, $X'_d$ and $X'_q$ are the transient reactances in the direct and quadrature axis, respectively, $I_d$ and $I_q$ are the current components in the direct and quadrature axis, respectively, $R_a$ is the machine armature resistance, $E'_q$ is the quadrature axis transient voltage, and $E_{fd}$ is the machine excitation voltage.

The model also has algebraic equations for calculating the electrical power $P_e$, presented in Equation (4), and equations for calculating the terminal voltages in the dq frame axis, shown in Equations (5) and (6) as $V_{td}$ and $V_{tq}$ [21]:

$$P_e = E'_q i_q + \left(X'_d - X_q\right)i_d i_q \tag{4}$$

$$V_{td} = -X'_q i_q - i_d R_a \tag{5}$$

$$V_{tq} = E'_q + X'_d i_d - i_q R_a \tag{6}$$

Figure 1 depicts the VSG control structure proposed in this paper. The measured three-phase voltages $v_{abc}$ and currents $i_{abc}$ at the point of common coupling (PCC) are transformed to the dq reference frame and fed into the machine model. The active power set-point $P_{ref}$, the reference frequency $f_n$, and the reference voltage $V_{ref}$ are all control inputs.

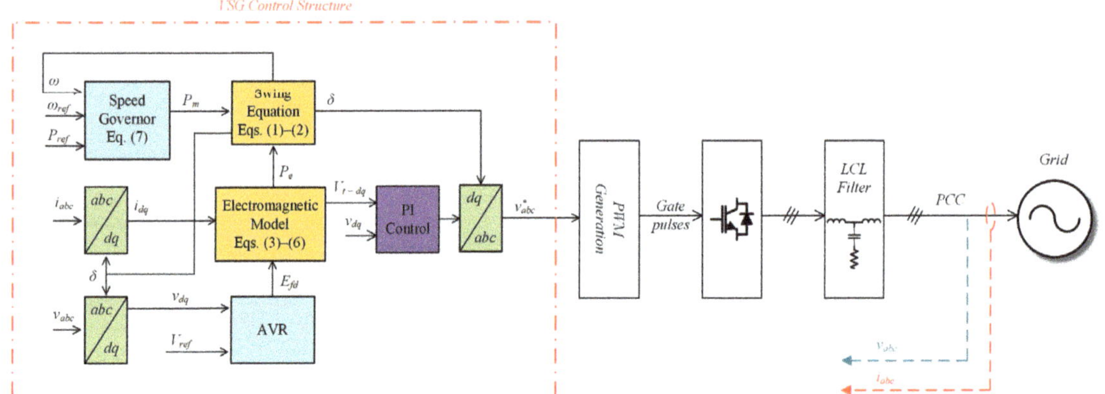


**Figure 1.** The control structure of the high-order VSG converter implemented.

The VSG speed governor is based on a P/f droop control and its characteristic equation is shown in Equation (7) [21]:

$$P_m = P_{ref} + D_p f_n (1 - \omega) \tag{7}$$

where $D_p$ is the frequency regulation droop coefficient.

The IEEE DC1A reference excitation system is used as an automatic voltage regulator (AVR) [22] but with a simplified structure by removing the saturation and voltage transducer. The output terminal voltages calculated in Equations (5) and (6) are the inputs to a double PI control that adjusts the output voltage of the converter. The PI control loop outputs are the pulse-width modulation (PWM) voltage references in the dq frame that are transformed back to three-phase signals ($v^*_{abc}$) before being used in the PWM switching signal generation.

In summary, the yellow blocks in Figure 1 implement the machine model, while the blue blocks serve as auxiliary controls for adjusting the VSG's voltage and frequency.

*Transient Stability Analysis*

Transient stability is a crucial aspect of power systems, as it determines the system's ability to maintain synchronism in the event of a severe disturbance. The response of the system during such an event is characterized by significant variations in the rotor angles of the generators, which are influenced by the non-linear power–angle relationship [23]. Although there have been extensive discussions on this type of stability in the literature, a detailed approach is beyond the scope of this work.

In general, several factors determine the transient stability of an SG connected to a system. These factors can be listed didactically when considering a simple grid, such as a single machine infinite bus (SMIB) system. In a three-phase SMIB system, stability analysis using the equal area criterion has shown that the fault CCT is directly related to factors such as generator loading, fault type and location, system reactances, voltage magnitudes in the system, and generator inertia [24]. The CCT is the maximum duration of a fault before the system reaches an unstable equilibrium point, and its calculation for a SMIB system is shown in Equation (8) [21]:

$$\text{CCT} = \sqrt{\frac{4H}{\omega P_m}(\delta_{cr} - \delta_0)} \tag{8}$$

where $\delta_{cr}$ represents the machine's critical angle and $\delta_0$ the machine's rotor angle in a pre-fault steady state.

Determining stability in a multi-machine system is more complex, although many of the factors that affect it remain the same. The difference lies in the interaction between the machines and their inertial contributions. To increase the total inertia of a system that experiences a percentage increase in RES compared to the total generation connected to it, a VSG is an option. A virtual inertia system is comprised of RES, short-term energy storage, and converters, which are controlled by a VSG algorithm. The DC-link capacitor stores the kinetic energy, which compensates for it. In this context, the energy stored in the capacitor is equivalent to the virtual rotational energy of the VSG.

However, the VSG proposed for this work and several other structures presented in the literature lack the inherent capacity to protect against fault overcurrent. As a result, the converter is exposed to destructive current levels, and it must be disconnected in these situations to avoid damage. This effectively prohibits the converter's potential contribution in fault situations. Therefore, it is crucial to suggest alternatives to limit the fault current injection of VSGs to maintain their contribution to system stability.

## 3. Fault Ride through (FRT) Strategy Based on Virtual Armature Resistance

The main objective of this work is to use an FRT technique to limit the current of high-order VSGs during three-phase faults. The goal is to validate not only the effectiveness of limiting current injection during the fault but also the VSG contribution to the system's transient stability. This will be investigated by comparing the CCT of the system with and without FRT control in a hybrid grid with a VSG operating in parallel to a conventional SG.

The adopted technique for limiting the fault current will consist of increasing the value of the virtual armature resistance in the high-order VSG model. The modification affects the model's Equations (5) and (6), which are used to calculate the reference terminal voltage for switching the VSG. To better understand the true effect of adding virtual resistance, one can visualize the model as a voltage source behind impedance, as shown in Figure 2 in an equivalent circuit representation. In Figure 2, $X_g$ and $R_g$ are the grid impedance parameters, $V_g$ is the grid voltage, $X_s$ and $R_a$ are the VSG armature impedance parameters, $E_a$ is the VSG transient voltage, and $I_t$ and $V_t$ are the VSG output current and voltage, respectively.

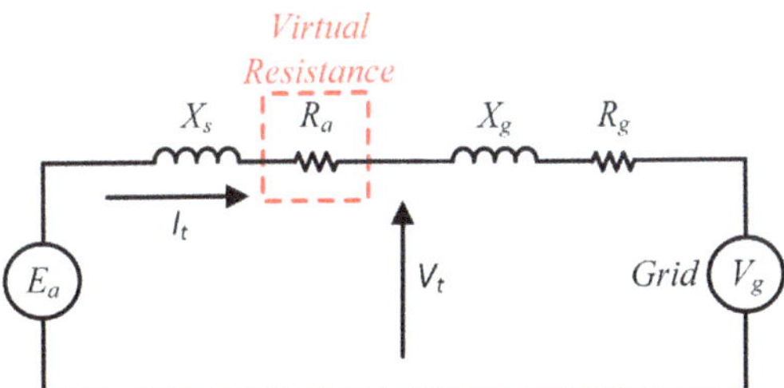


**Figure 2.** Voltage behind impedance equivalent model of the VSG.

The output current of the machine is directly proportional to the impedance of its armature, as shown in Figure 2 and Equations (5) and (6). The current flowing through the machine windings is limited by increasing the value of $R_a$ during transient operation. Increasing the virtual resistance parameter has the same effect for limiting the VSG's current, in this case by lowering the reference voltages generated for switching by Equations (5) and (6). As the output current increases, the voltage sag in the virtual armature circuit increases, lowering the values of $V_{td}$ and $V_{tq}$.

A fault detection algorithm is used to create the proposed virtual resistance variation. The algorithm adds resistance to the model if the voltage $V_t$ and terminal current $I_t$ levels of the VSG exceed certain limits. When a fault causes a voltage drop at the VSG terminals, the FRT control increases the resistance to limit the current during the fault. When the fault is cleared, the voltage rises, but the control continues to act in the post-fault period to limit the VSG's return current. The FRT control resets the resistance value to its nominal value when both sensing parameters return to steady-state levels. Typically, the synchronous machine's normal armature resistance value is so low that it is ignored. The resistance value used in the FRT strategy is comparable to the reactance in p.u The proposed FRT control algorithm is depicted in the flowchart in Figure 3.

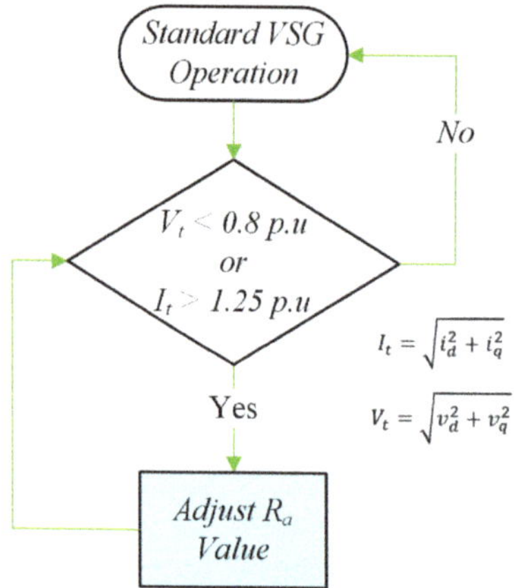


**Figure 3.** Flowchart of the proposed FRT control.

## 4. Case Studies

### 4.1. Case Description

The system shown in Figure 4 was used to validate the proposed FRT control. The studied VSG is connected in parallel to an SG and a PCC, which is connected to an infinite bus by a line impedance in the system depicted in Figure 4. The SG's speed governor and AVR are the same as those used in the VSG.

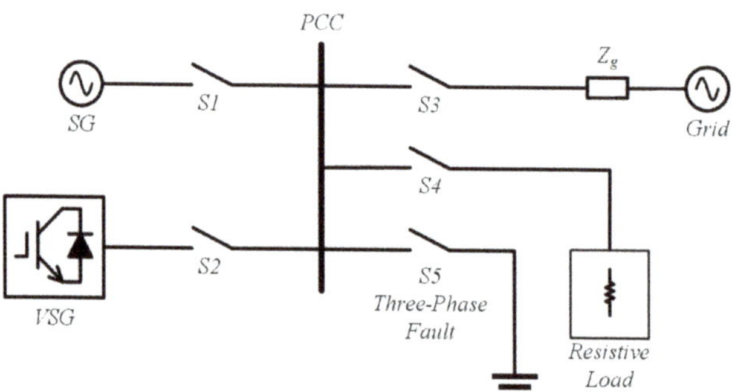


**Figure 4.** Structure of the simulated system.

A three-phase fault is applied to the PCC to evaluate the proposed control. For a VSG islanding test, a constant impedance load is also connected to the PCC. Switches S1 to S5 allow the system to be configured for different scenarios. Using the Typhoon HIL 402 device, the converter and grid/load system will be simulated in real-time in the HIL environment. Texas Instruments' digital signal processor TMSF28379D was used to implement the VSG control. A Yokogawa ScopeCorder DL350 was used to capture the VSG's output voltage and current. Figure 5 depicts the schematics for the test bench used to obtain the results. Figure 6 shows the HIL test bench with its components in operation.

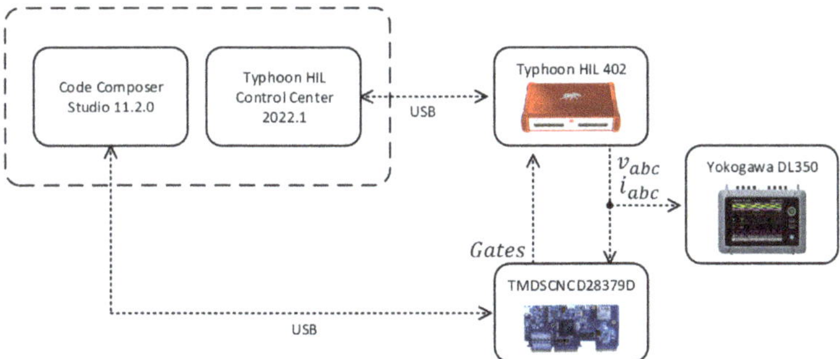


**Figure 5.** Schematics for the HIL test bench used for real-time simulation.

**Figure 6.** HIL test bench assembled.

The TMSF28379D DSP is connected to the Typhoon HIL 402 device via an interface card in charge of routing the analog signals from the simulation to the DSP and the PWM signals from the DSP to the real-time simulation. In this simulation, six analog outputs from the HIL 402 were used for the three-phase voltage and current, while six digital inputs were used to receive the PWM signals generated by the control embedded in the TMSF28379D.

The proposed FRT control's performance is evaluated using two metrics. The first metric is the control's ability to limit the VSG current during the fault. The SG and VSG are connected to the PCC along with the infinite bus for this analysis via switches S1, S2, and S3, respectively. The active power set-points $P_{ref}$ on both the VSG and the SG are set to 1.0 p.u Switch S5 is turned on for a set time to apply the fault to the PCC. The VSG output voltage and current curves are then analyzed for different values of virtual resistance added by the FRT control. Then, for a given fixed resistance adjustment, the system CCTs will be checked with and without the action of the FRT on the VSG.

Finally, a VSG islanding test is performed by disconnecting the synchronous generator (S1) and the infinite bus (S3) at the beginning of the fault. In this islanding scenario, the resistive load starts the test connected to the PCC through switch S4. The parameters used to simulate the model in the HIL test bench and the control system are given in Table A1 of Appendix A.

### 4.2. Current Limiting Results

In this case, the system was simulated with switches S1, S2, and S3 turned on. By activating switch S5, a three-phase fault is applied for 200 ms. This case was first simulated without the resistance adjustment to serve as a baseline and then for three different virtual resistance adjustment values: 0.5 p.u, 1.0 p.u, and 1.5 p.u.

The voltage and current graphs obtained are shown in Figures 7–10. In all figures, subfigure (A) shows the voltage curve of the VSG for 1 s, subfigure (B) shows the voltage curve of the VSG with a zoom for the short-circuit instant from 0.5 s to 0.7 s, the subfigure (C) shows the current curve of the VSG for 1 s, and subfigure (D) shows the voltage curve of the VSG zoomed in for the short-circuit instant from 0.5 s to 0.7 s. The vertical scale for current in all scenarios was kept the same to highlight the control's current-limiting feature. The fault was applied at 500 ms and cleared at 700 ms.

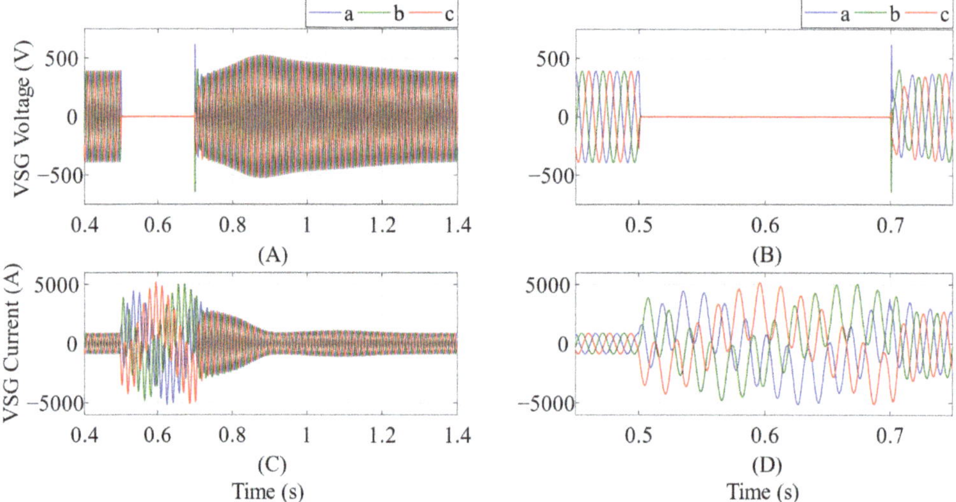

**Figure 7.** Voltage and current curves for the VSG without FRT control: (**A**) VSG voltage, (**B**) zoomed-in VSG voltage during the short-circuit instant, (**C**) VSG current, (**D**) zoomed-in VSG current during the short-circuit instant.

According to Figure 7, the VSG peak current exceeds 5.0 kA during the fault and remains high for approximately 200 ms after the disturbance has ended. This maximum current level is more than five times the nominal peak current of the VSG, which was approximately 850 A pre-fault and would undoubtedly damage the inverter if it continued to operate with the fault.

From Figures 8–10 the effect of the proposed FRT technique on the VSG current limitation is shown. Figure 8 shows that boosting the virtual resistance to 0.5 p.u reduces the peak current reached during the fault to 3.3 kA and also lowers the current levels during post-fault recovery. However, at this level, inverter operation is still unsafe. Therefore, Figures 9 and 10 present results for higher values of virtual resistance.

Figure 9 shows that the added resistance of 1.0 p.u was able to limit the peak current during the fault to 1.9 kA, resulting in better post-fault recovery behavior. Finally, Figure 10 depicts the result for a 1.5 p.u added resistance, with a maximum peak current of approximately 1.4 kA, a 64% increase over the VSG's nominal current. The results show that the proposed FRT control is effective in limiting the increase in VSG current during a three-phase short circuit situation. It is expected that the VSG will be able to remain connected to the grid even in the event of a severe fault, thereby contributing to the stability of the electrical system.

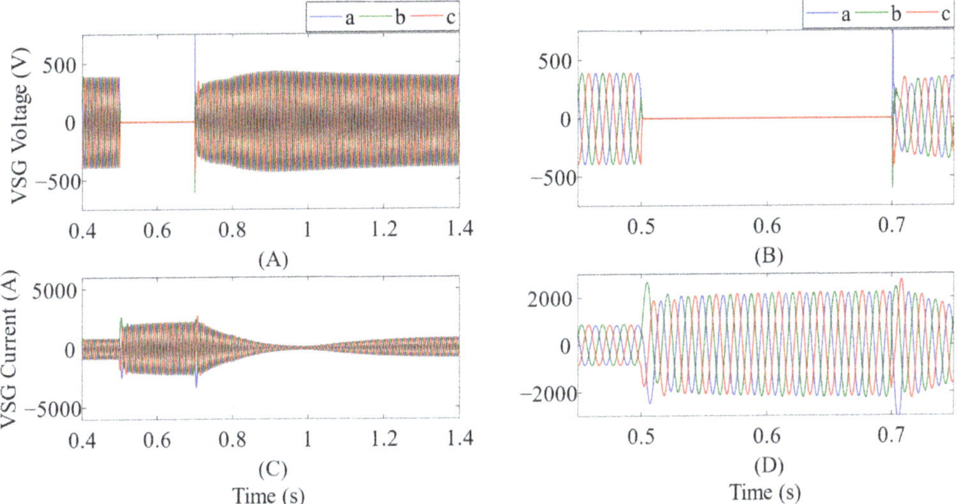

**Figure 8.** Voltage and current curves for the VSG with FRT control and 0.5 p.u virtual resistance: (**A**) VSG voltage, (**B**) zoomed-in VSG voltage during the short-circuit instant, (**C**) VSG current, (**D**) zoomed-in VSG current during the short-circuit instant.

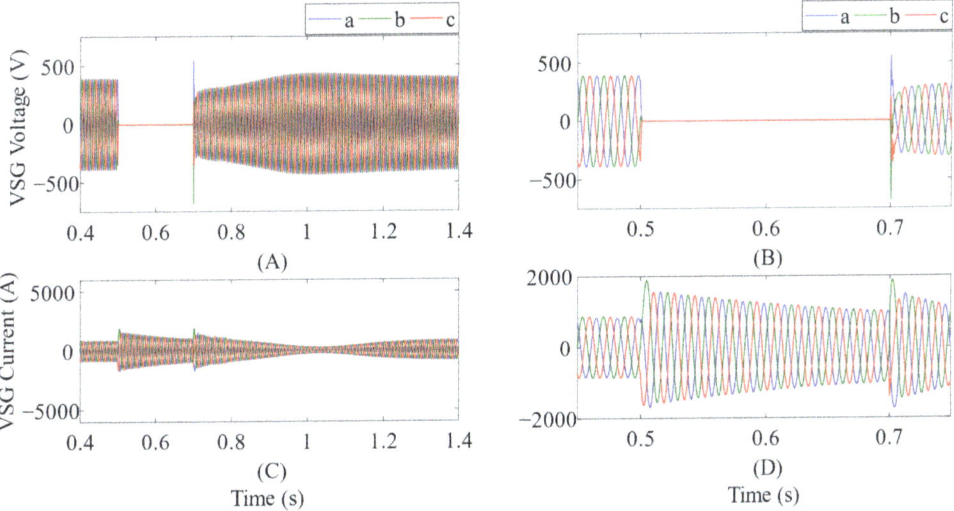

**Figure 9.** Voltage and current curves for the VSG with FRT control and 1.0 p.u virtual resistance: (**A**) VSG voltage, (**B**) zoomed-in VSG voltage during the short-circuit instant, (**C**) VSG current, (**D**) zoomed-in VSG current during the short-circuit instant.

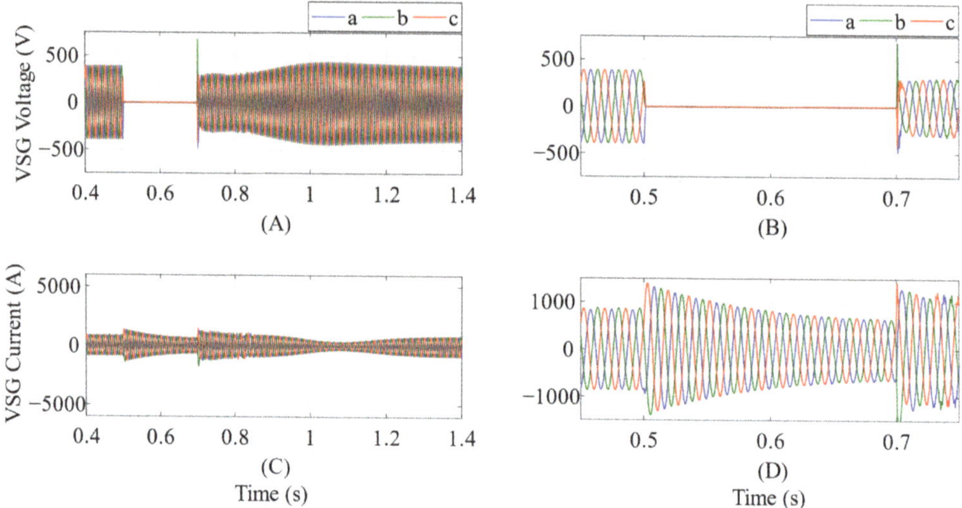

**Figure 10.** Voltage and current curves for the VSG with FRT control and 1.5 p.u virtual resistance: (**A**) VSG voltage, (**B**) zoomed-in VSG voltage during the short-circuit instant, (**C**) VSG current, (**D**) zoomed-in VSG current during the short-circuit instant.

### 4.3. Critical Clearing Time Results

In this case, it was determined whether the VSG continues to contribute to the system's transient stability despite the presence of the FRT control limiting its fault current. The system was first simulated without the FRT control for reference and then with the proposed control included. For this, the resistance added to the model by the FRT control was fixed at 1.5 p.u., and the same scenario as in the previous case was simulated by varying the fault time until finding the CCT or finding the maximum fault time before the system reaches instability. The loading of the VSG and the SG was also changed through their $P_{ref}$ set-points for a more comprehensive analysis. The VSG power was varied between 0.0 p.u and 1.0 p.u., and the power of the SG was varied between 0.2 p.u and 1.0 p.u., both in 0.2 p.u increments, to create a set of 60 simulation scenarios with 30 for each condition (with and without FRT control).

Tables 1 and 2 show the CCT results obtained. The reference CCTs of the SG operating alone can be obtained from scenarios with the VSG's power set to zero.

For a graphical representation of the results, surface planes were plotted with the generator's powers on the $x$ and $y$-axis and the CCT on the $z$-axis. Figure 11 shows the planes.

**Table 1.** CCT results in milliseconds without FRT control.

| $P_{VSG}$ (p.u) | $P_{SG}$ (p.u) | | | | |
|---|---|---|---|---|---|
| - | **1.0** | **0.8** | **0.6** | **0.4** | **0.2** |
| 0.0 | 253 | 301 | 346 | 389 | 429 |
| 0.2 | 359 | 405 | 441 | 484 | 517 |
| 0.4 | 347 | 391 | 431 | 469 | 501 |
| 0.6 | 332 | 376 | 415 | 452 | 485 |
| 0.8 | 317 | 359 | 392 | 424 | 465 |
| 1.0 | 298 | 341 | 374 | 408 | 438 |

**Table 2.** CCT results in milliseconds with FRT control.

| $P_{VSG}$ (p.u) | $P_{SG}$ (p.u) | | | | |
|---|---|---|---|---|---|
| - | 1.0 | 0.8 | 0.6 | 0.4 | 0.2 |
| 0.0 | 253 | 301 | 346 | 389 | 429 |
| 0.2 | 297 | 350 | 397 | 439 | 478 |
| 0.4 | 301 | 350 | 396 | 437 | 474 |
| 0.6 | 298 | 347 | 391 | 431 | 468 |
| 0.8 | 293 | 341 | 385 | 425 | 461 |
| 1.0 | 285 | 334 | 377 | 416 | 452 |

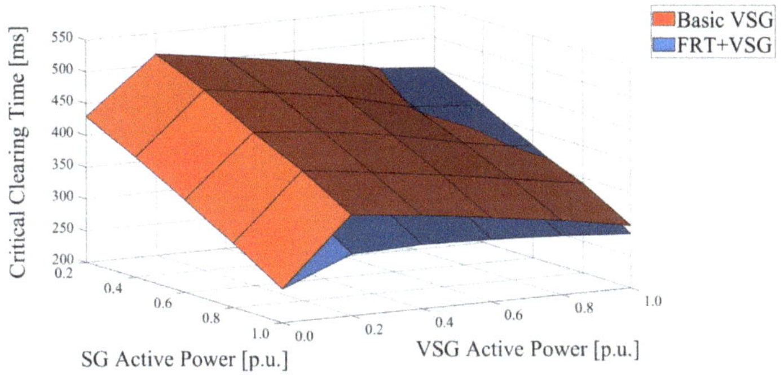


**Figure 11.** Visual representation of the CCT results.

It is possible to see that the CCT increases in all scenarios, with or without the FRT control in the VSG turned on. In other words, the VSG always contributes to increasing the CCT and, thus, the system's transient stability margin. This demonstrates that the proposed FRT control not only limits the VSG fault currents to safe levels, but it also maintains its inertial contribution to the system even when the fault is active. However, the results show that the performance of the proposed FRT control affects this contribution. This behavior is to be expected, especially given the change in VSG behavior during the post-fault recovery period. The presence of a virtual resistance in the armature circuit alters the characteristic of the synchronous machine's power-angle curve after fault clearing, resulting in this change in the stability margin. The only exception to this rule is when the VSG loading exceeds the SG. This effect is visible on the surfaces of Figure 11 when the blue plane representing the scenario with the FRT control has higher CCT values in some scenarios where the power delivered by the VSG is greater than the power delivered by the SG.

The CCT was determined by repeating the simulations with different fault times until the characteristic of power oscillation indicative of transient instability was confirmed, as shown in Figures 12 and 13. In all figures, subfigure (A) shows the voltage curve of the VSG for 1 s, subfigure (B) shows the voltage curve of the VSG with a zoom for the short-circuit instant from 0.5 s to 0.85 s, the subfigure (C) shows the current curve of the VSG for 1 s, and subfigure (D) shows the voltage curve of the VSG zoomed in for the short-circuit instant from 0.5 s to 0.85 s.

Figures 12 and 13 show simulations in which the generator and the VSG were at maximum load and for a fault extinction time of 350 ms. In both scenarios, it is possible to observe that the transient instability presents itself through post-fault power swings with or without FRT control.

### 4.4. Islanding Scenario

In all previous cases, the synchronous generator and the grid remained connected to the PCC during a fault so that an individual analysis of the VSG's behavior could be performed. However, this is not a realistic operating scenario. In a real-world power system, carefully programmed protection elements would isolate the fault and protect the generator and feeder. A VSG without overcurrent protection, such as an FRT control strategy, would invariably be disconnected from the PCC, causing loads connected to the same PCC to lose power.

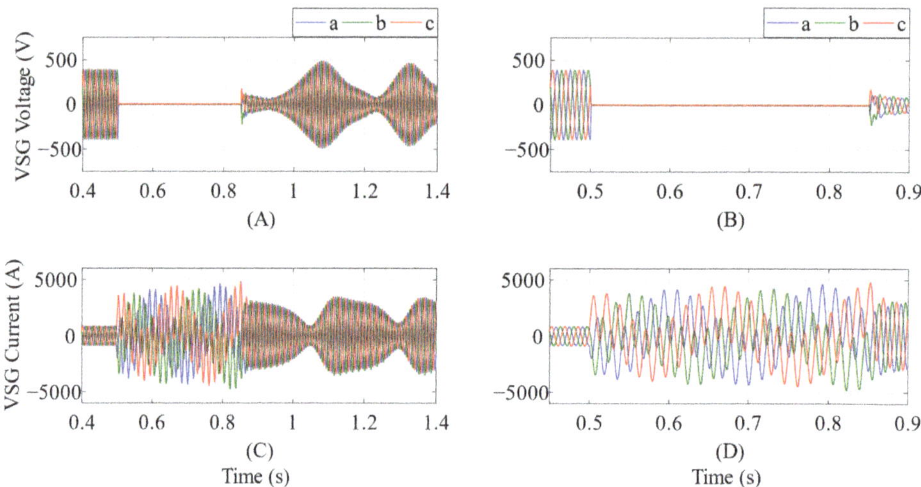

**Figure 12.** Transient instability characteristic of the VSG without FRT control and 350 ms of fault clearing time: (**A**) VSG voltage, (**B**) zoomed-in VSG voltage during the short-circuit instant, (**C**) VSG current, (**D**) zoomed-in VSG current during the short-circuit instant.

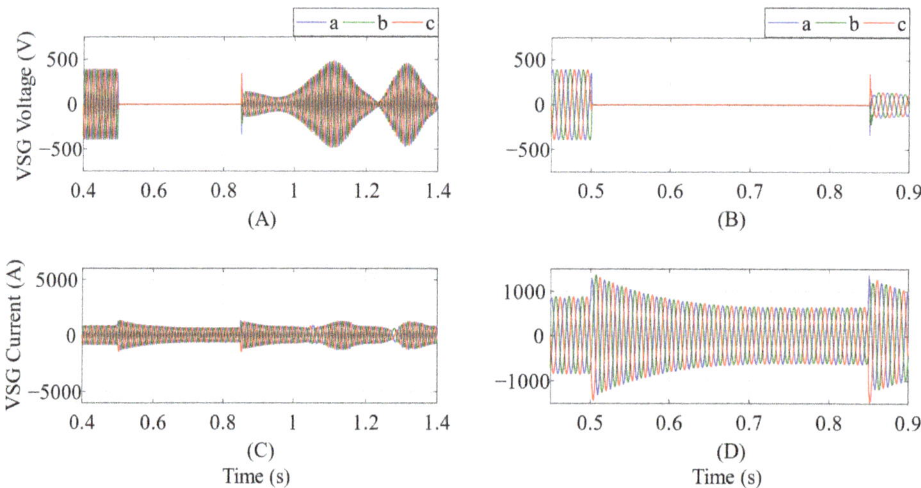

**Figure 13.** Transient instability characteristic of the VSG with FRT control and 350 ms of fault clearing time: (**A**) VSG voltage, (**B**) zoomed-in VSG voltage during the short-circuit instant, (**C**) VSG current, (**D**) zoomed-in VSG current during the short-circuit instant.

This final simulation case aims to demonstrate that the high-order VSG, in conjunction with the proposed FRT control, can survive the fault applied in the PCC and smoothly transition to island mode operation. The system is simulated with all switches turned on except the fault switch. A fault is applied by turning switch S5 on for 200 ms, and after 30 ms switches S1 and S3 are opened. The scenario is simulated for the VSG with and without the proposed FRT control. The results are shown in Figures 14 and 15. In Figure 14, subfigure (A) shows the voltage curve of the VSG for 1.6 s, subfigure (B) shows the voltage curve of the VSG with a zoom for the short-circuit instant from 0.5 s to 0.7 s, the subfigure (C) shows the current curve of the VSG for 1.6 s, and subfigure (D) shows the voltage curve of the VSG zoomed in for the short-circuit instant from 0.5 s to 0.7 s. In Figure 15, subfigures (A) and (C) shows the voltage and current for 2.6 s while subfigures (B) and (D) shows the same time period as the subfigures (B) and (D) from Figure 14.

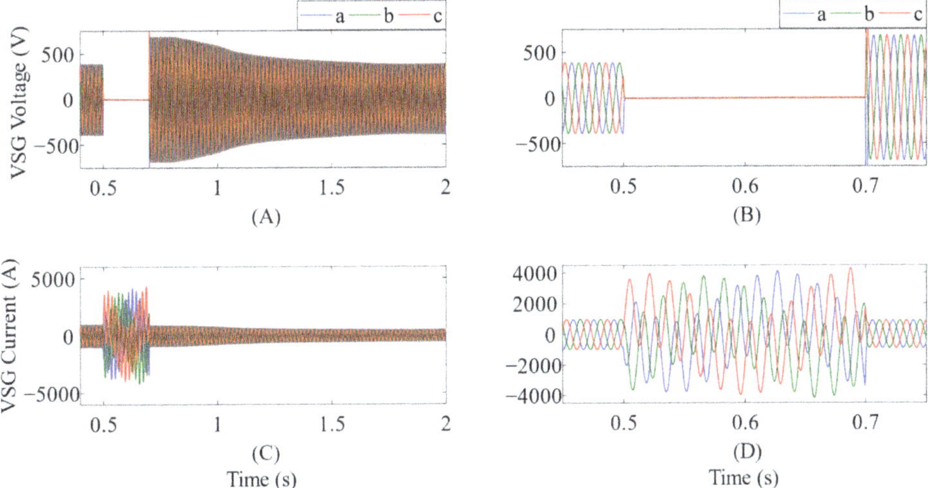

**Figure 14.** Islanding scenario for the VSG without FRT control: (**A**) VSG voltage, (**B**) zoomed-in VSG voltage during the short-circuit instant, (**C**) VSG current, (**D**) zoomed-in VSG current during the short-circuit instant.

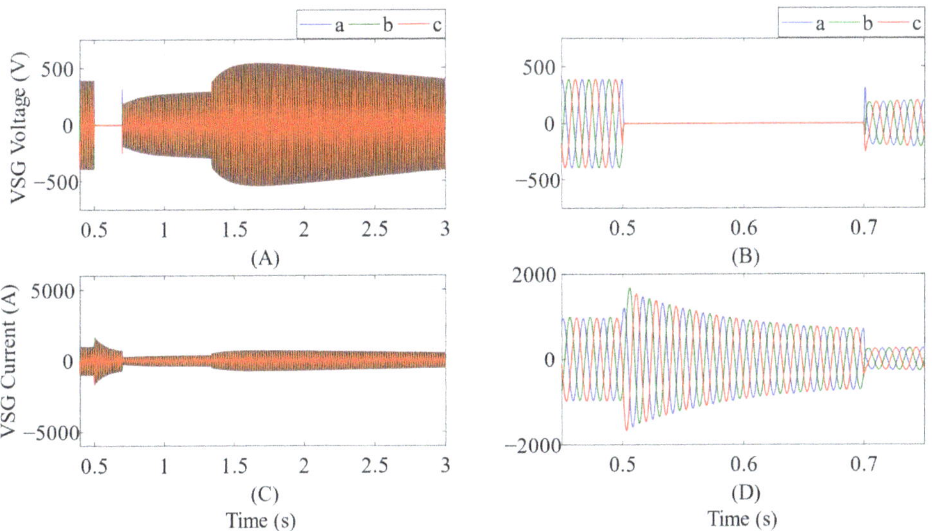

**Figure 15.** Islanding scenario for the VSG with FRT control: (**A**) VSG voltage, (**B**) zoomed-in VSG voltage during the short-circuit instant, (**C**) VSG current, (**D**) zoomed-in VSG current during the short-circuit instant.

After the fault is extinguished, the VSG could operate in island mode and feed the 300-kW load in both scenarios. However, without FRT control, the VSG is subject to high fault currents as shown in Figure 14, which would prevent it from operating.

As illustrated in Figure 15, the FRT control limits inverter current to safe levels and enables a smooth transition to an island operation mode.

## 5. Conclusions

This paper proposes a fault current limiting technique that adjusts the virtual armature resistance of high-order VSGs to improve their FRT ability. The aim of limiting the VSG's fault current is to preserve its inertial contribution to the system's transient angular stability. The primary contribution of this article lies in the way the adjustment control of virtual resistance was performed. The change is directly applied to the armature resistance parameter, which is an integral part of the synchronous machine model used to implement the VSG. Consequently, this adjustment does not increase the complexity of the VSG, as it eliminates the need for a separate control loop dedicated to the virtual impedance.

To evaluate the effectiveness of the proposed system, real-time simulations in hardware-in-the-loop were carried out. During a three-phase fault, the results demonstrated that increasing the resistance value led to a decrease in current until hitting diminishing returns. Without the proposed resistance adjustment, the peak value of the transient current reaches levels above 5 kA. By adjusting the resistance to 1.5 p.u, it was possible to limit the current to 1.4 kA, showing that the proposed control works. The CCT results show a reduction compared to the scenario without fault current limitation. For example, for the SG and VSG injecting 1.0 p.u of active power to the grid, the CCT without virtual resistance adjustment is 298 ms, while for the scenario with the proposed resistance adjustment the CCT is reduced to 285 ms. However, this CCT is still higher when compared to the case where only the SG is connected and operating with 1.0 p.u of power, which in this case is 253 ms. The results demonstrate the method's effectiveness in reducing the VSG's current during and after the fault's extinction in the PCC while maintaining its contribution to the system's transient stability with the presence of FRT control, albeit naturally diminished.

In an islanding scenario, the SG and grid are disconnected, leaving only the VSG to supply power to an active load after the fault. The proposed technique proves its ability to limit the islanding current transient by reducing it from 4 kA to 1.6 kA and smoothly transitioning the VSG to the island operating mode.

**Author Contributions:** Conceptualization, D.C. and L.E.; methodology, D.C.; software, D.C. and T.A.; validation, D.C.; formal analysis, D.C., L.E. and T.A.; investigation, D.C.; resources, L.E.; writing—original draft preparation, D.C.; writing—review and editing, D.C., L.E. and T.A.; visualization, D.C. and T.A.; supervision, L.E.; project administration, L.E.; funding acquisition, L.E. All authors have read and agreed to the published version of the manuscript.

**Funding:** This work was supported by the National Council for Scientific and Technological Development—CNPq (grant numbers 409024/2021-0 and 311848/2021-4) and Espírito Santo Research and Innovation Support Foundation—FAPES (grant numbers 514/2021, 668/2022 and 1024/2022).

**Data Availability Statement:** No new data were created or analyzed in this study. Data sharing is not applicable to this article.

**Conflicts of Interest:** The authors declare no conflict of interest.

## Appendix A. Hardware-in-the-Loop Simulation Parameters

Table A1 shows the parameters used to simulate the system in the HIL test bench.

**Table A1.** Simulation parameters.

| Subsystem | Symbol | Description | Value |
|---|---|---|---|
| Grid | $S_g$ | Base power | 1 MVA |
|  | $V_g$ | Base voltage (L-L) | 480 V |
|  | $f_n$ | Nominal frequency | 60 Hz |
|  | $Z_g$ | Line impedance | $(0.5 + j0.25)$ p.u |
|  | $P_L$ | Resistive load | 300 kW |
| VSG | $S_{VSG}$ | Nominal power | 500 kVA |
|  | $f_{sw}$ | Switching frequency | 10 kHz |
|  | $V_{dc}$ | DC-link voltage | 1200 V |
|  | $H$ | Inertia constant | 4 s |
|  | $D$ | Damping coefficient | 0.0 |
|  | $T'_{do}$ | Transient voltage time constant | 5.2 s |
|  | $X_d$ | d-axis synchronous reactance | 0.92 p.u |
|  | $X_q$ | q-axis synchronous reactance | 0.504 p.u |
|  | $X'_d$ | d-axis transient reactance | 0.3 p.u |
|  | $X'_q$ | q-axis transient reactance | 0.504 p.u |
|  | $R_a$ | Nominal armature resistance | 0.002 p.u |
| LCL filter | $L_i$ | Inverter side inductance | 0.665 mH |
|  | $L_g$ | Grid side inductance | 0.101 mH |
|  | $C_f$ | Filter capacitor | 0.287 mF |
|  | $R_d$ | Damping resistor | 0.1844 |
| Synchronous generator | $S_{VSG}$ | Nominal power | 500 kVA |
|  | $H$ | Inertia constant | 4 s |
| Speed governor and AVR | $D_P$ | P/f Droop coefficient | 2.0 |
|  | $V_{Ref}$ | Voltage reference | 1.0 p.u |
|  | $K_A$ | Voltage regulator gain | 100.0 |
|  | $T_A$ | Voltage regulator time constant | 0.001 s |
|  | $K_F$ | Feedback loop gain | 0.001 |
|  | $T_F$ | Feedback loop time constant | 0.1 s |
|  | $K_E$ | Exciter gain | 1.0 |
|  | $T_E$ | Exciter time constant | 0.1 s |

## References


1. Cabrera-Tobar, A.; Bullich-Massagué, E.; Aragüés-Peñalba, M.; Gomis-Bellmunt, O. Review of advanced grid requirements for the integration of large-scale photovoltaic power plants in the transmission system. *Renew. Sustain. Energy Rev.* **2016**, *62*, 971–987. [CrossRef]
2. Rosso, R.; Wang, X.; Liserre, M.; Lu, X.; Engelken, S. Grid-forming converters: Control approaches, grid-synchronization, and future trends—A review. *IEEE Open J. Ind. Appl.* **2021**, *2*, 93–109. [CrossRef]
3. Shadoul, M.; Ahshan, R.; Al-Abri, R.S.; Al-Badi, A.; Albadi, M.; Jamil, M. A Comprehensive Review on a Virtual-Synchronous Generator: Topologies, Control Orders and Techniques, Energy Storages, and Applications. *Energies* **2022**, *15*, 8406. [CrossRef]
4. Encarnação, L.F.; Carletti, D.; de Angeli Souza, S.; de Barros, O., Jr.; Broedel, D.C.; Rodrigues, P.T. Virtual Inertia for Power Converter Control. In *Advances in Renewable Energies and Power Technologies*, 1st ed.; Yahyaoui, I., Ed.; Elsevier Science: Amsterdam, The Netherlands, 2018; Volume 2, pp. 377–411.
5. Abubakr, H.; Guerrero, J.M.; Vasquez, J.C.; Mohamed, T.H.; Mahmoud, K.; Darwish, M.M.; Dahab, Y.A. Adaptive LFC incorporating modified virtual rotor to regulate frequency and tie-line power flow in multi-area microgrids. *IEEE Access* **2022**, *10*, 33248–33268. [CrossRef]
6. Shuai, Z.; Shen, C.; Liu, X.; Li, Z.; Shen, Z.J. Transient angle stability of virtual synchronous generators using Lyapunov's direct method. *IEEE Trans. Smart Grid* **2018**, *10*, 4648–4661. [CrossRef]
7. Pan, D.; Wang, X.; Liu, F.; Shi, R. Transient stability of voltage-source converters with grid-forming control: A design-oriented study. *IEEE J. Emerg. Sel. Top. Power Electron.* **2019**, *8*, 1019–1033. [CrossRef]
8. Cheng, H.; Shuai, Z.; Shen, C.; Liu, X.; Li, Z.; Shen, Z.J. Transient angle stability of paralleled synchronous and virtual synchronous generators in islanded microgrids. *IEEE Trans. Power Electron.* **2020**, *35*, 8751–8765. [CrossRef]
9. Carletti, D.; Amorim, A.E.A.; Amorim, T.S.; Simonetti, D.S.L.; Fardin, J.F.; Encarnacao, L.F. Adaptive armature resistance control of virtual synchronous generators to improve power system transient stability. *Energies* **2020**, *13*, 2365. [CrossRef]
10. Zamani, M.A.; Yazdani, A.; Sidhu, T.S. A control strategy for enhanced operation of inverter-based microgrids under transient disturbances and network faults. *IEEE Trans. Power Deliv.* **2012**, *27*, 1737–1747. [CrossRef]

11. He, J.; Li, Y.W. Analysis, design, implementation of virtual impedance for power electronics interfaced distributed generation. *IEEE Trans. Ind. Appl.* **2011**, *47*, 2525–2538. [CrossRef]
12. Natarajan, V.; Weiss, G. Synchronverters with better stability due to virtual inductors, virtual capacitors, and anti wind-up. *IEEE Trans. Ind. Electron.* **2017**, *64*, 5994–6004. [CrossRef]
13. Jongudomkarn, J.; Liu, J.; Ise, T. Virtual synchronous generator control with reliable fault ride-through ability: A solution based on finite-set model predictive control. *IEEE J. Emerg. Sel. Top. Power Electron.* **2019**, *8*, 3811–3824. [CrossRef]
14. Lu, X.; Wang, J.; Guerrero, J.M.; Zhao, D. Virtual-impedance-based fault current limiters for inverter dominated AC microgrids. *IEEE Trans. Smart Grid* **2018**, *9*, 1599–1612. [CrossRef]
15. Paquette, A.D.; Divan, D.M. Virtual impedance current limiting for inverters in microgrids with synchronous generators. *IEEE Trans. Ind. Appl.* **2014**, *51*, 1630–1638. [CrossRef]
16. Wang, X.; Li, Y.W.; Blaabjerg, F.; Loh, P.C. Virtual-impedance-based control for voltage-source and current-source converters. *IEEE Trans. Power Electron.* **2014**, *30*, 7019–7037. [CrossRef]
17. Xiong, X.; Wu, C.; Blaabjerg, F. Effects of virtual resistance on transient stability of virtual synchronous generators under grid voltage sag. *IEEE Trans. Ind. Electron.* **2021**, *69*, 4754–4764. [CrossRef]
18. Huang, L.; Xin, H.; Yuan, H.; Wang, G.; Ju, P. Damping effect of virtual synchronous machines provided by a dynamical virtual impedance. *IEEE Trans. Energy Convers.* **2020**, *36*, 570–573. [CrossRef]
19. Li, M.; Shu, S.; Wang, Y.; Yu, P.; Liu, Y.; Zhang, Z.; Hu, W.; Blaabjerg, F. Analysis and Improvement of Large-Disturbance Stability for Grid-Connected VSG Based on Output Impedance Optimization. *IEEE Trans. Power Electron.* **2022**, *37*, 9807–9826. [CrossRef]
20. Liu, Y.; Wang, Y.; Liu, H.; Xiong, L.; Li, M.; Peng, Y.; Xu, Z.; Wang, M. An LVRT strategy with quantitative design of virtual impedance for VSG. *Int. J. Electr. Power Energy Syst.* **2022**, *140*, 107661. [CrossRef]
21. Machowski, J.; Lubosny, Z.; Bialek, J.W.; Bumby, J.R. *Power System Dynamics: Stability and Control*, 2nd ed.; John Wiley & Sons: Hoboken, NJ, USA, 2008; p. 456.
22. *IEEE Std 421.5-2016 (Revision of IEEE Std 421.5-2005)*; IEEE Recommended Practice for Excitation System Models for Power System Stability Studies. IEEE: New York, NY, USA, 2016; pp. 1–207.
23. Kundur, P.; Paserba, J.; Ajjarapu, V.; Andersson, G.; Bose, A.; Canizares, C.; Vittal, V. Definition and classification of power system stability IEEE/CIGRE joint task force on stability terms and definitions. *IEEE Trans. Power Syst.* **2004**, *19*, 1387–1401.
24. Kundur, P.S.; Malik, O.P. *Power System Stability and Control*; McGraw-Hill Education: New York, NY, USA, 2022; pp. 827–954.

*Article*

# A Hybrid Physical Co-Simulation Smart Grid Testbed for Testing and Impact Analysis of Cyber-Attacks on Power Systems: Framework and Attack Scenarios

Mahmoud S. Abdelrahman, Ibtissam Kharchouf , Tung Lam Nguyen and Osama A. Mohammed *

Department of Electrical and Computer Engineering, Florida International University, Miami, FL 33174, USA; mabde046@fiu.edu (M.S.A.); ikhar002@fiu.edu (I.K.); tunguyen@fiu.edu (T.L.N.)
* Correspondence: mohammed@fiu.edu; Tel.: +1-305-348-3040

**Abstract:** With the deployment of numerous innovative smart grid technologies in modern power systems, more real-time communication and control are required due to the complexity and proliferation of grid-connected systems, making a power system a typical cyber-physical system (CPS). However, these systems are also exposed to new cyber vulnerabilities. Therefore, understanding the intricate interplay between the cyber and physical domains and the potential effects on the power system of successful attacks is essential. For cybersecurity experimentation and impact analysis, developing a comprehensive testbed is needed. This paper presents a state-of-the-art Hybrid Physical Co-simulation SG testbed at FIU developed for in-depth studies on the impact of communication system latency and failures, physical events, and cyber-attacks on the grid. The Hybrid SGTB is designed to take full advantage of the benefits of both co-simulation-based and physical-based testbeds. Based on this testbed, various attack strategies are tested, including man-in-the-middle (MitM), denial-of-service (DoS), data manipulation (DM), and setting tampering (change) on various power system topologies to analyze their impacts on grid stability, power flow, and protection reliability. Our research, which is based on extensive testing on several testbeds, shows that using hybrid testbeds is justified as both practical and effective.

**Keywords:** smart grid; cyber-physical power system; hybrid testbed; cyber-attacks; ns-3; OPAL-RT

**Citation:** Abdelrahman, M.S.; Kharchouf, I.; Nguyen, T.L.; Mohammed, O.A. A Hybrid Physical Co-Simulation Smart Grid Testbed for Testing and Impact Analysis of Cyber-Attacks on Power Systems: Framework and Attack Scenarios. *Energies* **2023**, *16*, 7771. https://doi.org/10.3390/en16237771

Academic Editor: José Matas

Received: 25 October 2023
Revised: 17 November 2023
Accepted: 19 November 2023
Published: 25 November 2023

## 1. Introduction

The smart grid is one of the fundamental technologies supporting sustainable economic and societal growth. Today's energy infrastructure is shifting significantly across all generation, transmission, and distribution systems to provide dependability and sustainability to the power system network. Therefore, electric power systems have evolved into densely interconnected cyber-physical systems that rely heavily on advanced communications due to the networked physical and electronic sensing, monitoring, and control devices connected to a control center in the energy control and protection system [1]. As a result, this expansion has increased the vulnerability of power systems to various cyber-attacks, which might have various negative consequences and cascading failures, from the destruction of interconnected critical infrastructure to the loss of life [2]. The power grid's primary goal is to deliver electricity to load centers with reliability. The physical layer of large electric power networks is coupled with a cyber system of information and communication technologies, including complex devices and systems like supervisory control and data acquisition (SCADA) systems and intelligent electronic devices (IEDs), to operate effectively and dependably. By using these devices and systems to control circuit breakers, transformers, capacitor banks, and other equipment, power systems are particularly vulnerable to cyber-attacks. The protection and control systems, which are known as the backbones of power networks because the former detects abnormal conditions and quickly corrects them, and the latter maintains the integrity of the system and stabilizes it in the wake of physical disturbances, are the main beneficiaries of these communication-based schemes and components. Power system controllers in the SCADA hub process the data gathered and communicate the necessary commands to the appropriate actuators.

Cybercriminals recently launched a wave of expertly coordinated and sophisticated attacks. Utility grid operators deal with cyber incursions daily, including denial of service, physical systems, malicious intent, and authentication threats. The risk is substantially higher when it comes to SCADA systems, security, control, and protection equipment recognized vulnerabilities. Multiple cyber-attacks against smart grid systems have occurred in recent years. During the 2015 cyber-attack on the Ukrainian electricity grid, circuit breakers were opened and closed without the control centers' knowledge [3]. By simulating cyber-attacks in an Aurora generator test, researchers have examined the severity of malicious switching and DoS attacks on protective digital relays for the Ukrainian power grid [4]. A new discussion about limiting remote access to circuit breakers and managing them locally was triggered by this cyber-attack. However, restricting breakers' direct remote access does not guarantee that an online attack will not send them dangerous commands. Protective relays can control breakers locally, frequently relying on communication networks susceptible to cyber-attacks [5]. As a result, the corresponding breaker has been indirectly targeted if an attack can change the data sent to a relay so that the relay issues a false trip instruction. The automatic generation controller (AGC) controls the power exchange between adjacent areas at predetermined levels and maintains the system frequency within acceptable ranges by altering the load reference set-point. Through a communication system, all inputs and outputs are sent and received. AGC systems are, however, susceptible to cyber-attacks because they rely on the communication infrastructure. The grid's frequency, stability, and profitable functioning can all be directly impacted by cyber-attacks introducing false control or measurements into the AGC data stream [6]. A successful attack on a protection system may involve interfering with the measurements taken and judgments made by a remote relay in communication with a local relay [7]. According to [8], there are three main categories of research on the cybersecurity of protection systems in AC and DC systems. The first group concentrates on attack modeling and risk assessment in protection systems, a description of the framework for modeling coordinated switching attacks over circuit breakers and relays, and an assessment of the effect of bus and transmission line protection techniques on the cybersecurity of the power system [9–11]. The cybersecurity of substations has received a lot of attention in the second group of studies. In [12], the authors introduced a technique for detecting and thwarting cyber-attacks on substation automation systems. The third group of studies has created a mechanism that uses power networks' physical and digital characteristics to identify attacks and distinguish them from faults [9]. Therefore, a power network's cybersecurity of its protection and control mechanisms should be upgraded in addition to its physical security. This upgrade must be previously tested and validated before implementation.

The ability of energy assets to identify and comprehend such innovative and intelligent threats and system vulnerabilities to build smart, effective countermeasures while maintaining the real-time functioning of the energy system in a secure and reliable condition is a critical challenge. This requires efforts and endeavors from grid operators and researchers to study the system operation under these abnormal conditions with extensive testing and impact analysis without harming the system's security and reliability. However, the high cost of implementing and utilizing actual system hardware and software for testing purposes and the possible harm caused by simulating cyber-attacks on live power systems are major obstacles [13]. In addition, real measurements (such as voltages, currents, and frequency) and information and communications technology (ICT) data (such as communication protocols and security logs) are unavailable due to power companies' security concerns. Therefore, for research and demonstration purposes, one of the most important tools for investigating the cyber-physical security of power systems is a cyber-physical testbed, which is a useful choice for gathering accurate data from physical and cyber systems [14]. Research initiatives targeted at enhancing the resilience of the electrical system can be conducted and evaluated in testbeds that can successfully incorporate both the cyber and physical components of the smart grid [15]. It is seen as a viable approach to investigating and addressing cybersecurity challenges. From the architecture viewpoint, the development of a cyber-physical smart grid and the accompanying testbeds, which include a variety of testing paradigms with related challenges and their solutions, were thoroughly reviewed by Smadi et al. [16]. According to this study, the authors classified and explained three categories of testbeds: hardware-based, software-based, and hybrid-based.

The cyber-physical testbed offers a realistic setting for studying how a sophisticated power and ICT systems interact through simulation and modeling. It is crucial to research the cause-and-effect correlations of cyber intrusions, the susceptibility and resilience of power systems, and the performance and dependability of applications in a testbed's realistic environment. Power, communication, and control/protection systems are the fundamental elements of the SG testbeds in the electric power sector. In the power system model, the measurements and status information needed for substation situational awareness are obtained. The communication system uses a variety of industry-standard protocols and communication media to communicate data between the substation and the control center, including measurements and commands necessary for the efficient operation of the power system. The control and protection systems encompass all the apparatus used and deployed in the control center. To the best of our knowledge, research in this area has been sparse, and the field has not been fully investigated despite the growing concern regarding the benefits and technical challenges of smart grid vulnerability analyses. Investigations into attack simulations and assessments of their effects on the infrastructure have not received much attention. Therefore, creating a new framework to facilitate smart grid attack simulation, emulation, and even real attacks is crucial.

Several smart grid testbeds have been created. Each testbed has particular characteristics and purposes of its own from the viewpoints of both architecture and functions. In [16], the authors provided a comprehensive list of existing cyber-physical smart grid testbeds from various research institutes. The majority of these testbeds are either simulation-based or use a co-simulation platform. Accordingly, the cyber-attack model may be simulated in the power system layer by some assumptions or emulated through the communication layer if the communication network is modeled. There are not many testbeds that are entirely hardware-based. Even though fully hardware-based testbeds are hard to implement, they can guarantee fidelity. Some of these testbeds [17–24] are listed and compared with the developed Hybrid SGTB in Table 1 of this paper. This study describes the cutting-edge Hybrid Physical Co-simulation SG Testbed at Florida International University, which was created for in-depth research on the effects of communication system latency and failures, physical events, and cyber-attacks on the grid. In the physical testbed part, the physical and network layers are deployed using real hardware and industrial-grade software, providing a high-fidelity model that closely mimics the real CPS. Since it is pretty challenging to reconfigure such testbeds for different research endeavors, the co-simulation testbed, which is mainly constructed from OPAL-RT as a modern real-time simulator that can accurately mimic the response of an actual physical system in real time and ns-3 for communication network emulation, is developed to overcome this challenge. This testbed can provide a more flexible configuration, easier expansion to large-scale CPSs, easier testing of evolving cyber-attacks, and full instrumentation through software probes to determine exactly what happened at every component of the CPS. It is designed to take full advantage of the benefits of co-simulation- and physical-based testbeds. Hence, the power system can either be an actual power system component of the physical testbed or simulated in real time, as in the OPAL-RT using the co-simulated testbed. Also, the measurements and actuation commands are either sensed directly from a physical device in the physical testbed or simulated and transmitted over the network emulation in the co-simulated testbed. To assess the effects of these cyber-attacks on grid stability, power flow, and protection dependability, a variety of attack tactics, including data manipulation (DM), setting change, man-in-the-middle (MitM), and denial-of-service (DoS), are tested on the basis of this testbed on various power system topologies by real attackers and virtual attackers. Our study demonstrates that the usage of hybrid testbeds is justified as both feasible and efficient.

**Table 1.** Taxonomy of cyber-physical smart grid testbeds.

| Testbed | Power System | | Communication Network Model | Commercial Devices | Attack Model | |
|---|---|---|---|---|---|---|
| | Practical | Simulation | | | Real Agent | Emulation |
| [17] | N/A | RTDS | ns-3 | SEL IEDs | N/A | DaterLab |
| [18] | N/A | RTDS | ISEAGE | IEDs/PLC | N/A | ISEAGE |
| [19] | N/A | OPAL-RT | OMNeT++ | SEL IEDs | N/A | N/A |
| [20] | N/A | RTDS | WANE | N/A | N/A | WANE |
| [21] | N/A | PSCAD | OMNeT++ | N/A | N/A | OMNeT++ |
| [22] | N/A | GridLAB-D | ns-3 | N/A | N/A | ns-3 |
| [23] | N/A | OPAL-RT | Ethernet-based | SEL IEDs | N/A | Simulated |
| [24] | N/A | OPAL-RT | Exata CPS | SEL IEDs | N/A | Exata CPS |
| Hybrid SG-TB | Reduced scale power system | OPAL-RT | Ethernet-based and ns-3 | ABB/SEL IEDs | Real PCs | ns-3/ Docker container |

To summarize, the key contributions of the authors are as follows:

- We developed a hybrid smart grid testbed (SGTB) as a comprehensive environment for testing and impact analysis studies of cyber-attacks on the power grid, comprising a fully physical testbed and co-simulation testbed.
- In the physical testbed, we built a reduced-scale power system including generation, transmission, and loads with the full connection of instrument transformers and commercial devices to provide protection and control applications through a real Ethernet network.
- We developed a real-time cyber-physical co-simulation testbed on three different machines to fully represent the different layers in the CPS using OPAL-RT and ns-3 integrated with docker containers.
- We demonstrated the impact of different attacks on the tested grid using real agents (PCs) connected through the communication network in the physical testbed.
- We developed virtual attack models using the docker containers with the ns-3 model to emulate or closely imitate the attacker behavior in the co-simulation testbed.

The rest of this paper is organized as follows. In Section 2, we describe the architecture of the hybrid SGTB. Section 3 explains cyber-attacks in cyber-physical energy power systems and different techniques for testing these attacks. Cyber threats in digital substation protection are introduced with a brief description of relay configuration and its data model in Section 4. Implementations of two attack scenarios in the physical testbed are described in detail in Section 5. Section 6 discusses our proposed co-simulation testbed set-up in detail, including OPAL-RT, ns-3, and docker container. We discuss the attack model of DoS and MitM in the co-simulation platform in Section 7. In Section 8, we conclude this work and propose future evolution and enhancement of the platform.

## 2. Hybrid SGTB Architecture

A testbed is an experimental environment outfitted with state-of-the-art equipment and technology assembled to produce a test system or equipment. Through testbeds, it is possible to validate cutting-edge concepts for digital transformation and digitalization as well as products, systems, and technology. They are frequently used for instructive and demonstrative applications and in research, development, and invention projects. Most testbeds are created primarily for the assessment and validation of certain tasks. Few testbeds offer complete hardware and software assessment policies for research purposes; however, certain testbeds offer insights for particular study disciplines. This section examines the design and execution of a complete cyber-power testbed, including simulation, emulation, and real devices in a modular approach. Commercial hardware, software, and simulated devices make up the data measuring and collection system. This testbed also has hardware-in-the-loop (HIL) and controller-in-the-loop (CIL) features enabling one to connect to and communicate with real hardware controllers and relays. However, with no power amplifier in the laboratory, it does not provide power hardware-in-the-loop (PHIL). Protection relays and every other control component in a microgrid may be evaluated early since they are real. Most current cyber-physical power system testbeds utilize this paradigm since commercial products provide both efficient ICT network integration and a broad degree of system-in-the-loop tests. The Hybrid SGTB at FIU is a composite of a physical testbed with everything real as a reduced-scale power system, and the co-simulation testbed is built using a combination of real-time simulators. The general architecture of the testbed is shown in Figure 1.

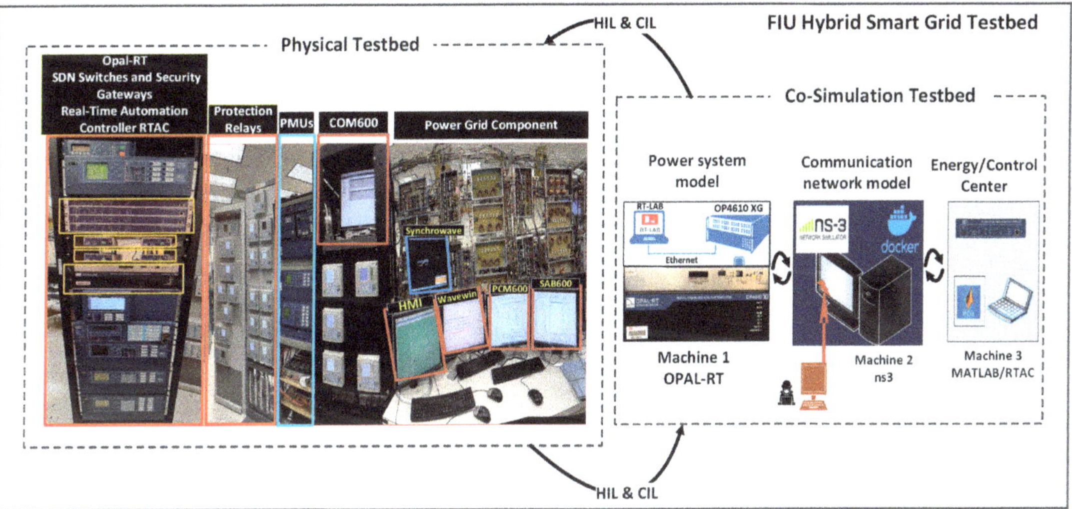


**Figure 1.** Hybrid SGTB architecture at FIU.

### 2.1. Physical Testbed

Considering the smart grid broadly, it is defined by its capabilities and operational traits rather than by the use of any specific technology. The smart grid testbed at FIU is a cutting-edge research facility that was created as an integrated hardware-based AC/DC system. This state-of-the-art system's hardware, software, and communication components can all conceptualize a comprehensive cyber-physical smart grid framework. The physical testbed is a unique, reduced-scale low-voltage testbed that provides researchers with a way to evaluate the operation and cybersecurity of power systems. While all system components run at voltages below 230 volts, they imitate functions at higher voltages and larger sizes. As shown in Figure 2a, the components of the physical testbed system are interconnected. It features four AC generators set up in a ring configuration, and supply loads are linked to the load buses via line models.

Additionally, this system has connections to the available DAQs monitoring system's communication infrastructure and the wide-area communications network to implement the equipment for power system control and the SCADA system. Figure 2b depicts the control virtual instrument (VI) used in the LabVIEW environment to manage and display input/output data as data, curves, or indicators. The detailed parameters of power system components (generators, transmission lines, and loads) were discussed and illustrated in [25]. The platform is being used to create microgrid technologies, such as converters based on power electronics, energy storage, communications protocols, and integration of DERs and vehicles [26]. To simulate various power grid schemes, the configuration of the testbed may also be dynamically changed. This physical part of the testbed has been modified several times with a new infrastructure to mimic the requirements and challenges of a smart grid regarding integrating renewables and energy storage devices and connecting various commercial IED types. Various protection relays and phasor measurement units (PMUs) from different vendors, including Schweitzer and ABB have been installed at different points through voltage and current sensors based on their functions, such as generator protection, line protection, and motor protection.

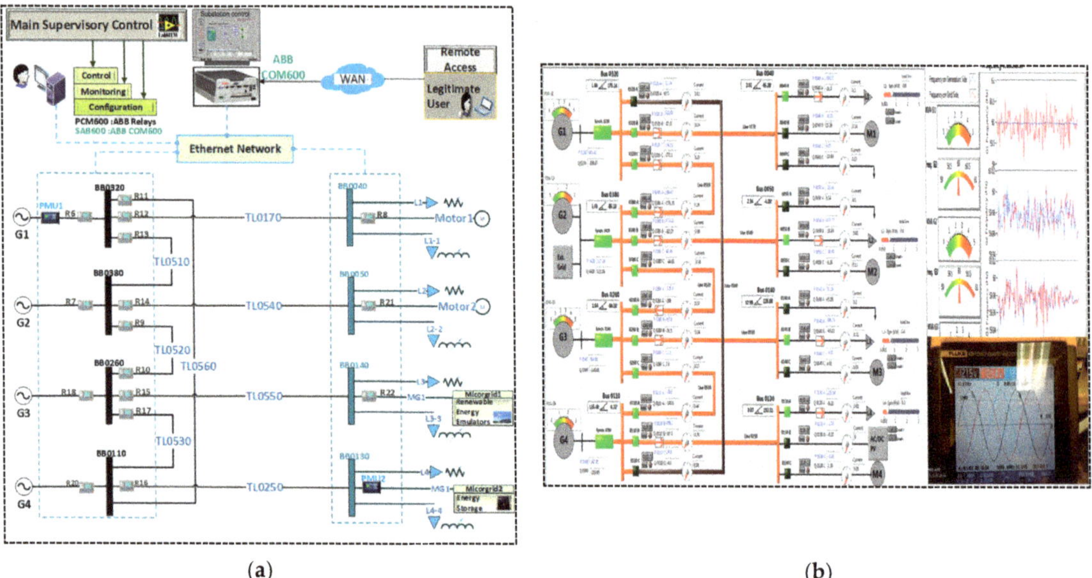

(a)       (b)

**Figure 2.** Physical testbed infrastructure: (**a**) SLD and protection relays deployment; (**b**) main supervisory control panel using LabVIEW.

*2.2. Co-Simulation Testbed*

The desire for co-simulation environments that embody several domains' characteristics is growing. Therefore, studies of the effects of data transfers on control systems can be carried out with better precision with network emulation capabilities. Regarding facilitating data connectivity on the testbed, Ethernet is a popular and easily accessible data communications technology that supports a wide range of protocols, including TCP/IP and UDP. As shown in Figure 3, the FIU testbed implements and builds a comprehensive cyber-physical power system of three-domain modeling and simulation of three different machines: (1) Machine 1 is OPAL-RT for power system domain modeling and simulation in real time, (2) Machine 2 is a Linux computer for communication network domain modeling and emulation using ns-3 and docker containers, and (3) Machine 3 is for power system application domain implementation which can be industrial-grade devices such as protection relays, real-time automation controllers, or simulation tools such as MATLAB/Simulink.

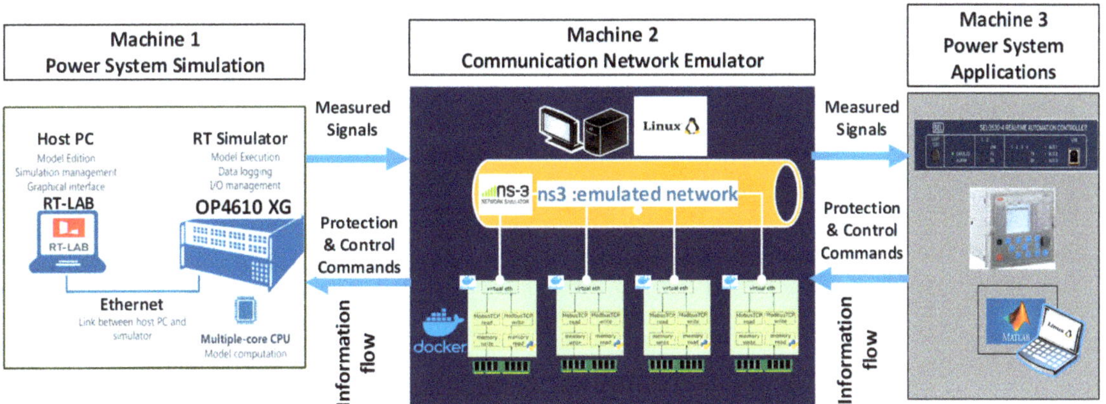


**Figure 3.** Co-simulation testbed architecture and main components.

## 3. Modeling and Testing of Cyber-Attacks in Energy Power Systems

### 3.1. CPS Layers and Attack Modeling

Typical cyber-physical systems (CPSs) combine computational and communication components to control, protect, and manage physical objects. Understanding how cyber and physical components interact is necessary to research cyber-physical challenges. The three components of a power system are generation, transmission, and distribution, which are sensed and actuated through IoT devices. Sensors communicating with field devices (generators, transmission lines, etc.) send measurements to control centers using dedicated communication protocols. In the physical layer, the measurements $y(t)$ may refer to quantities such as voltage, current, and frequency. These measurements are processed by computational protection and control algorithms running in the control center to make operational decisions. Actuators are then given the decision $u(t)$ to alter the field devices. Figure 4 fully represents the interaction between physical and cyber layers. To better study the impacts of cyber-physical events such as cyber-attacks and/or communication failure, the communication layer in the middle must be modeled and appropriately simulated with a high level of flexibility.

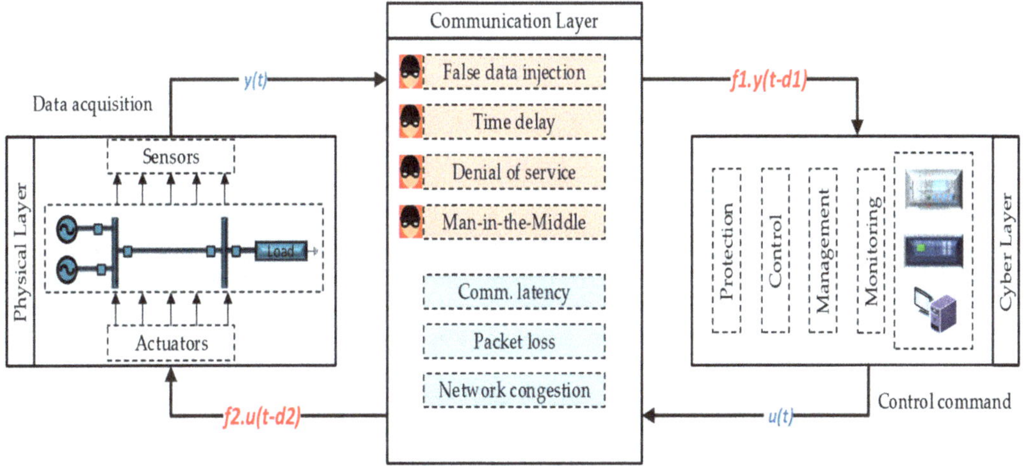


**Figure 4.** Cyber-physical system layer data and interaction.

An adversary could build attack templates intended to alter the content of or impose a time delay or denial in the communication of these control/measurement signals by exploiting weaknesses along the communication channels [27]. In this case, the sensor measurements utilized as an input to the control or protection algorithm will differ from the actual condition by $f1.y(t - d1)$. Similarly, the control or protection decision output from the control or protection system will be deviated from the correct one by $f2.u(t - d2)$. As these attacks can potentially seriously compromise the security and dependability of the power system, it is crucial to research and understand their effects. These effects can be quantified in terms of load loss, frequency and voltage violations, and their subsequent effects. Attack studies will also aid the development of defenses against attacks or ways to lessen their effects. Algorithms for attack-resistant control and bad data detection are examples of countermeasures [28].

### 3.2. Testing Methods of Cyber-Attacks

The smart grid needs to be cyber-physically analyzed; hence, it is important to establish adequate modeling and simulation approaches for the different domains in the smart grid. These domains, which have traditionally been the focus of power system modeling, can be modeled and simulated using various techniques. Three basic types of experimental settings exist simulations, data from real-world systems (such as data obtained from a power company), and real-world testbeds, which are real-world systems used just for testing and not for actual use. Each of them has benefits as well as disadvantages. For instance, using simulations makes it feasible to examine the effects of attacks without posing any safety risks, but doing so ignores many of the issues that real-world practitioners face. Deploying attacks in a testbed, however, offers a more realistic investigation of the system's weaknesses but also carries the danger of causing costly equipment damage or even personal injury. However, deploying attacks in a testbed allows for a more realistic investigation of the system's vulnerabilities, but it also entails the risk of expensive equipment damage or even personal injury.

Based on the modeling technique used, the attack can be simulated, emulated, or even a real agent as shown in Figure 5a. Simulation is a powerful means of studying networking problems because it gives the flexibility to model a wide range of scenarios, has a low usage and deployment cost, and provides reproducible experimentation. A single simulator can be used to represent and simulate the power networks in these scenarios such as Matlab, DIgSILENT, PSSE, etc., representing the communication infrastructure as directly connected links. In this scenario, the attacker will be modeled and simulated as a manipulation block with a generalized function to modify the targeted signal for the attack as shown in the left block in Figure 5a. This type of simulation has been improved recently by using the same software to simulate both the power and communication networks. Power System Computer Aided Design (PSCAD), for instance, is a tool for power system simulation that can be utilized for cyber-physical simulation with the introduction of communication network components, as detailed in [29]. Applications might be testable but not actual devices because they are often built utilizing software packages. In addition, these models must undergo extensive testing and validation. It is a more practical and feasible solution to maintain the simulation of power and communication systems in separate simulators and integrate them through a common framework to function together as a co-simulation system [30]. The necessary data flow interface and time synchronization of the two simulators are realized using the shared framework. The key benefit is that a cyber-physical simulation environment may be created using industrial-grade commercial devices and tools. By emulating the attacker model through the communication network, this attack emulation may be made to look realistic as shown in the right block in Figure 5a. On the other hand, as shown in the middle of Figure 5a, real attackers can be settled at the testbed to launch attacks targeting the protection or control agents in the physical testbed through a PC connected to the Ethernet network, which means testing with a high-fidelity model and demonstration with accurate results.

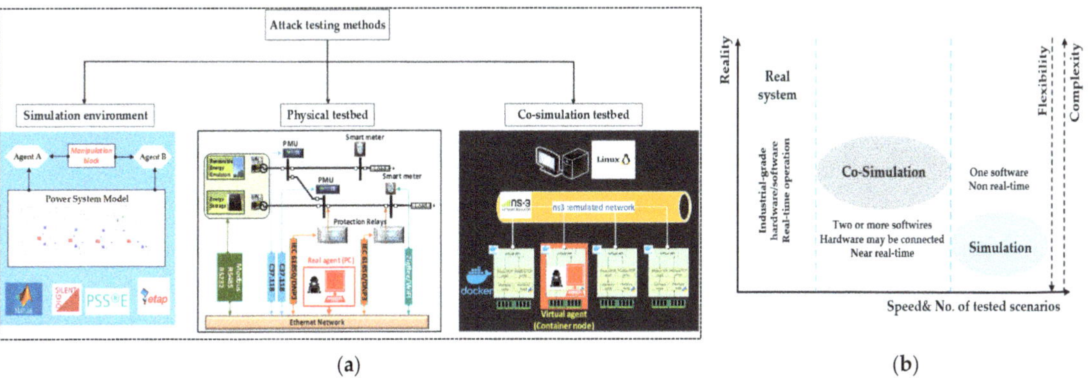

(a)                    (b)

**Figure 5.** Testing of cyber-attacks on power systems: (**a**) methods of testing; (**b**) degrees of testing methods in terms of reality, flexibility, and complexity.

Figure 5b depicts the key differences in testing methods in terms of the number of scenarios that can be conducted, considering flexibility, complexity, and how close they are to reality. In this work, we focus on the cyber-attacks targeting the protection and automation system of the power grid, especially cyber-attacks at digital substations using the physical testbed. In addition, the co-simulation testbed will be used to virtualize the cyber-attacks associated with the distributed control scheme. Therefore, real attackers/agents will be used with the physical testbed with detailed descriptions and studies in Sections 4 and 5, while Sections 6 and 7 include the implementations of attack emulators in the co-simulation testbed with different scenarios.

## 4. Cyber-Physical Substation

The substation is a crucial part of the power grid, which plays a critical role in connecting power generation sources into the grid and transmitting energy to the end-user. Substation measurements are utilized as input to various physical and cyber devices that can be physically or electrically coupled for monitoring, protection, and control applications. The transformation of the power system to a smart grid is being driven by automation of the power system and communication standards. One such standard that is an extensively used standard for substation automation and protection is IEC 61850. It permits crucial substation automation and protection components in digital substations to communicate and exchange data in real time. However, the Sampled Values (SV) and Generic Object-Oriented Substation Event (GOOSE) protocols of IEC 61850 may be vulnerable to cyber-attacks. The standard does not implement any encryption due to complex real-time requirements of trip signals for protection systems, typically in the range of 3–4 ms. The exploit of GOOSE protocol vulnerabilities within IEC 61850 is demonstrated in [31,32].

Figure 6 depicts the testbed's fundamental four-tier architecture. Four levels make up this system [1]: the process layer, the bay layer, the substation layer, and the network layer. The process layer comprises components such as circuit breakers, circuit transformers, voltage transformers, current transformers, actuators, sensors, and main and secondary switchgear. The use of input/output terminal equipment decreases the amount of hardwiring in the process. Station-level Ethernet switches connect IEDs in the bay layer for control and protection. The IEDs carry out operations such as bay control, protection, monitoring, and fault recording upon receiving communication commands from the station level. The station layer contains modems, Ethernet switches, HMI computers, and global positioning system (GPS) receivers. It is used for data storage, automation, archiving, and bay-level management. The network layer mainly allows remote access exchange of information and control. The substation and network layers are the main sources of cyber-attacks through the substation's insider or remote access. Blackouts may result from several cascading events by simultaneous cyber-attacks on crucial substations. Therefore, to increase the resilience of power grids, it is essential to improve the cybersecurity of substations and assess both physical and cybersecurity as one integrated structure.

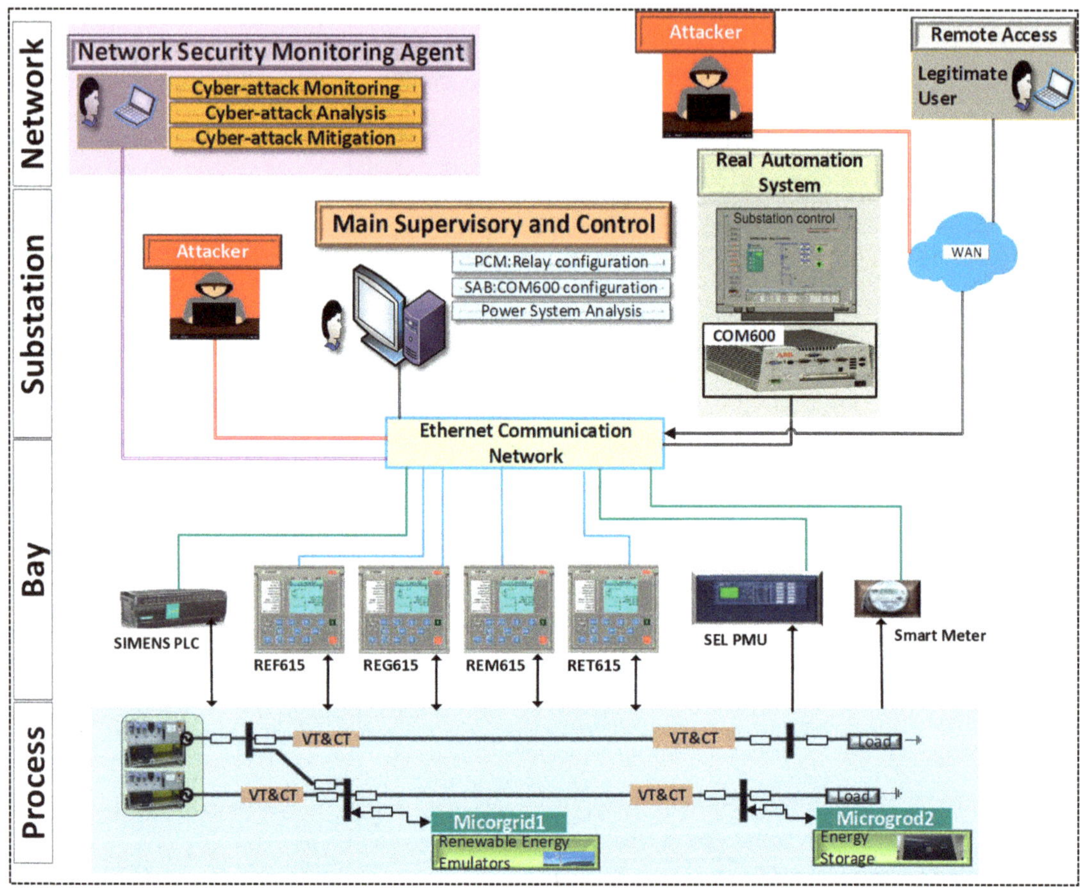


**Figure 6.** Cyber-physical substation layers.

*4.1. Cyber Intrusions in Protective Relays*

Digital substation protection systems are vulnerable to cyber-attacks that could disconnect lines and generation, leading to cascading failures in the power grid. Protective relays were initially electro-mechanical switches, but as they developed into sophisticated networked digital devices with enormous processing power, they became intelligent electronic devices (IEDs). As a result, both IT network and control system vulnerabilities are making IEDs cyber-vulnerable. Not every IED is critical, but some must be protected.

According to the attack tree in [12], opening CB as the attacker's primary goal may be achieved due to four malicious actions that can originate from different points, either by the station's insiders or remote access. However, some attackers may want to leave the CB closed to activate other relays to trip their associated CBs, negatively impacting the protection system operation and consequently resulting in service disruptions. Suppose the attackers have gained access to the station communication bus through the internal communication network or remote access from an external network. In that case, they might jeopardize the communication protocols and the station's hardware (protective relays or user interfaces). For instance, they might open circuit breakers using malicious commands or control. Another possible attack scenario is to modify the protective relay's settings so that the relay trips during normal operation or mal-operates due to miss-coordination, which could lead to a cascading issue. As shown in Figure 7, according to the attacker's goals, the protective relay under attack may send an unnecessary trip signal or not a necessary trip signal to the associated CB, causing inappropriate operation of the protection system. During normal operation (no-fault), the unnecessary trip signal from the primary protection, which may result from direct malicious command or changing the relay setting, will impact the system operation by service interruption and may lead to cascading failures. During the abnormal operation (fault condition), the prevented necessary trip signal from the primary protection, which may result from a DoS attack or relay setting altering, will impact the system operation by expanding the service interruption area, consequently leading to cascading failures. These severe consequences are mainly due to several trips through activating the backup protection system or the miss-coordination resulting from the attack.

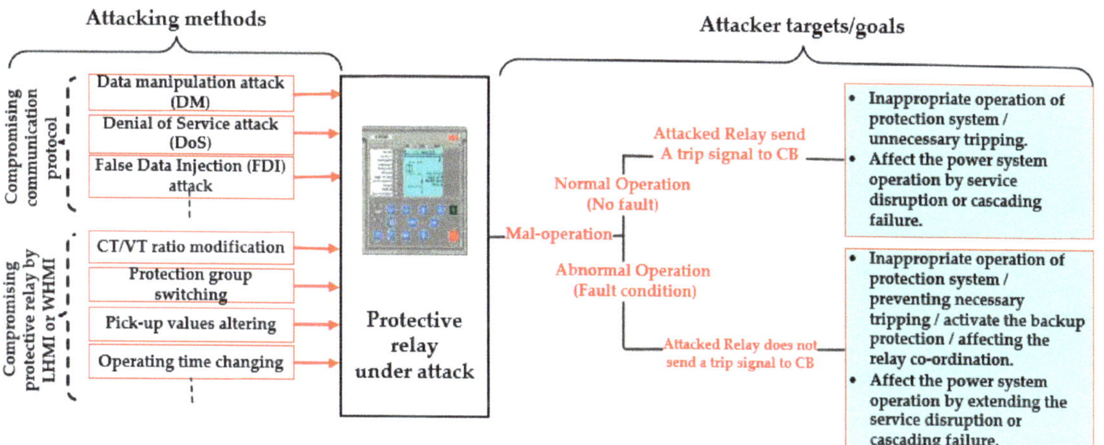


**Figure 7.** Attacker targets and attacking methods in protective relays.

### 4.2. Digital Relay Configuration and Data Model

The digital overcurrent relay, which is the major target of cyber-attacks in this work, is explained in this section along with a set of protective configurations. It is a potential target for several types of cyber-attacks, such as DoS, message suppression (MS), replay, man-in-the-middle, false data injection (FDI), data manipulation (DM), etc. Figure 8a depicts a simplified schematic diagram according to the IEC61850 definition of the digital overcurrent relay. A source of continuous measurement of current is each set of current signals at each node, which are derived from the current transformer (CT), which can be acquired as follows:

$$I^{meas} = [I_{R1}, I_{R2}, I_{R3}, I_{R4}, \dots I_{Rn}] \tag{1}$$

where $I^{meas}$ is the measured three-phase analog current signals from CTs, and $I_{R1} : I_{Rn}$ are the currents for Relay-1 through Relay-n. The relay protection block compares the measured values with the threshold or pickup values after receiving the values of the processed current signals from the CT measurements. Carefully choosing the pickup value requires considering the protection relay's operation's minimum fault current and maximum allowable load current. The typical relay pickup current ($I^{pu}$) can be represented as [33] and typically falls between the system's maximum load currents and minimum fault currents. The Protection and Control IED Manager PCM600 makes it easy to configure these relays, and the application configuration tool allows one to design, choose, and configure the necessary functionalities. Device configuration data are contained in the XML-based Substation Configuration Language (SCL) as a common language to achieve device interoperability. The configured SCL file is imported by IEC 61850-based devices over the ICT network, negating the requirement for human configuration. Multi-vendor interoperability is improved through standardized communication protocols and logical nodes. According to the standard, as shown in Figure 8a, the hierarchical model of the physical devices is composed of five layers, starting from the physical device (PD) layer to the data attribute (DA) layer [34]. According to IEC 61850-7-4, each PD in this data model has a group of logical devices (LDs), and each LD has several logical nodes (LNs) that define specific power system functions. All LNs have a class name consisting of four letters, the first of which denotes a group of LNs. Additionally, groups of data objects (DOs) are assembled into the LNs to gather data values, control outputs, and parameters. Figure 8b shows that the relays were configured and programmed using PCM600 2.10 32bit, with different IP addresses and technical keys.

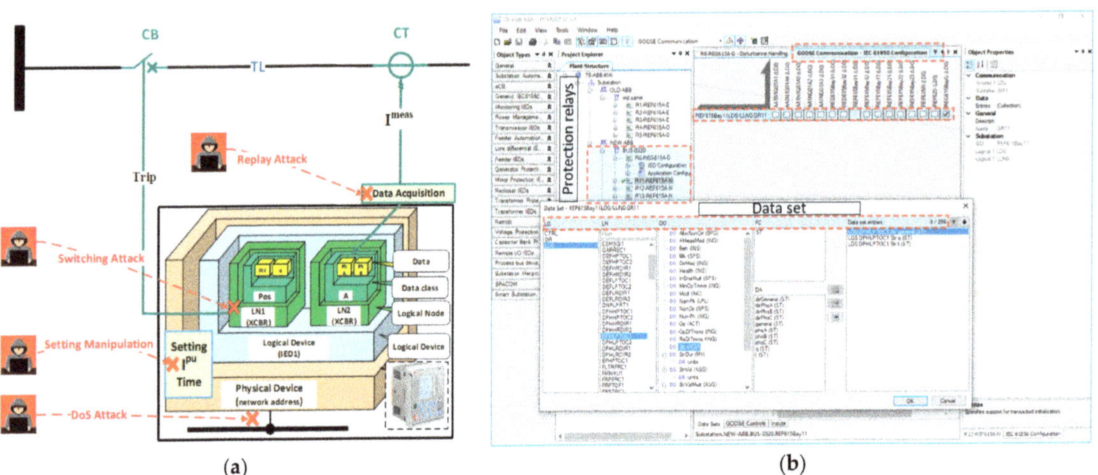


**Figure 8.** Protection relay data model and configuration: (**a**) digital relay schematic diagram with the data model; (**b**) relay configuration using PCM.

### *4.3. IEC 61850 Cybersecurity Threats*

Cybersecurity is a major concern with the growing integration of communication technologies aimed at enhancing the performance of protection systems. Within this framework, three fundamental security objectives focus on confidentiality, integrity, and availability. Communication systems must fulfill these three security objectives. Despite the protective measures introduced by IEC 62351 [35] to secure IEC 61850 against certain types of attacks, the IEC 61850 communication standard remains susceptible to a range of potential threats. Consequently, it is essential to identify potential attacks that may target protection systems and assess their potential impact. The origin of attacks on protection systems can vary from low- to high-skilled attackers. Skilled intruders may cause big damage to the utility because they may have good knowledge of the standard and the message content. Therefore, they may cause undesirable tripping even before the intrusion is detected. In this paper, we will test and assess the impact of a DM attack (switching attack) on the relay, causing the targeted relay to subscribe to the malicious GOOSE message and opening its associated circuit breaker.

## 5. Physical Testbed Attack Scenarios

Users can research plausible scenarios of cyber-attacks and defense techniques using a real-time cyber-physical testbed. This section will address situations that could occur, such as potential entry points to the substation systems and substation-compromising cyber-attacks by analyzing two different attacks. It is assumed that the attacker has gained access to the network data and IEDs by breaching the private network to emulate the attack in the testbed.

### 5.1. Switching Attack Scenario

The set-up consists of five IEDs as shown in Figure 9a: two for incomers (generators G1 and G3) and three for outgoing feeders. Communication between these IEDs and the substation control system is through the IEC 61850 GOOSE protocol. In this scenario, the attacker is represented by an agent (PC) connected to the local network, as shown in the figure. Suppose this intruder can successfully spoof a GOOSE message frame. In that case, it can increment the stNum, reset the sqNum, alter the Boolean data field, and then multicast the malicious message to subscribing IEDs. If the receiving IED accepts the manipulated GOOSE message, they will initiate the opening of the circuit breaker. The circuit breaker emulation was implemented with the help of the I/Os on the IEDs and enables a realistic emulation of the switching process.

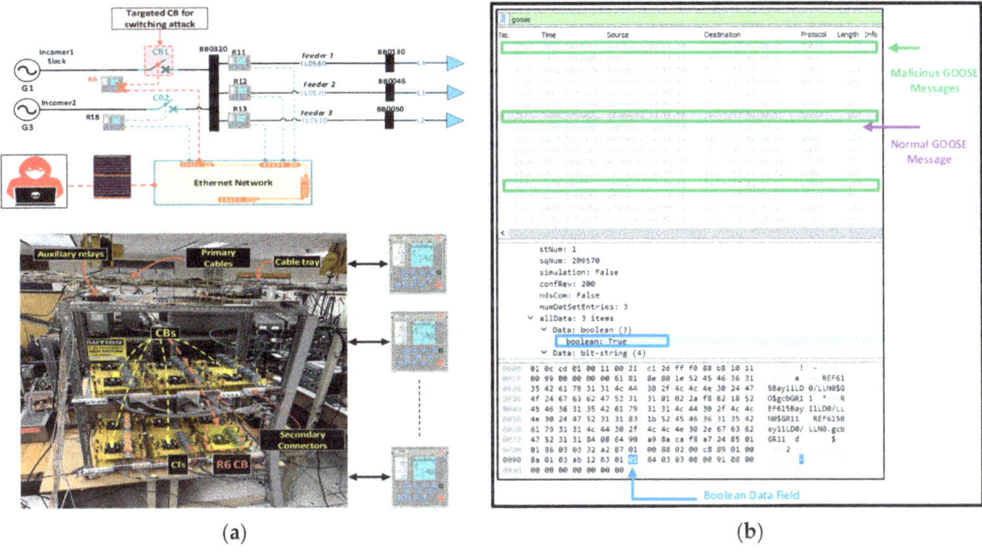


(a)         (b)

**Figure 9.** Switching attack scenario: (**a**) power system and relay connections; (**b**) switching attack technique.

### 5.1.1. Attack Model

Goose messages are designed for fast communication, adhering to a 4 ms transmission time specified by the IEC 61850 standard, where it does not employ encryption algorithms. Consequently, an eavesdropper with access to the network can intercept these unsecured GOOSE packets and efficiently collect the plaintext information. A familiar intruder with the message's content can manipulate the data field effectively. The attacker is implemented on a desktop computer that hosts Ubuntu 16.04.7 LTS on it and has 8 GB of RAM and 4 CPU cores. Based on the relay data model described in Section 5.2, each data name is specified by the standard and functionally correlates to a particular power system component. The LN with the name XCBR is used to mimic a circuit breaker with various data elements including, for example, *Pos*, which contains two attributes: *ctlVal*, which represents the opening and closing command, and *stVal*, which identifies the position. In the scenario we assessed, the intruder changed the Boolean data field of *ctlVal* from false to true, resulting in the tripping of the circuit breaker. In our adversarial model, we assume the attacker can intercept network traffic, extract essential fields from GOOSE messages, and execute a multicast attack by setting the Boolean data field to true, initiating the corresponding IED circuit breaker opening, as shown in Figure 9b.

### 5.1.2. Testing Scenario

During normal operation, the two generators supply the three loads, as shown in the first portion of Figure 10. The first incomer is slack, while the second generator is controlled with a specific input torque, which means G3 can supply a limited power at a certain frequency. Accordingly, if the load demand increases or the power output from G3 decreases, G1 will provide the increased power. When R6 attacked and opened the associated CB1, the system lost power coming from the main source; this led to an increase in the requested power from G3 and islanded G1 with zero load. In this case, the scenario was performed with no additional protection functions such as frequency protection to effectively demonstrate the impact of such attacks on power system operation. A significant overload may result from generators exceeding the allowed power limit due to a general rise in power. As shown in the figure, the power generated from G3 is increasing divergently with a severe drop in the frequency, which may result in catastrophic damage to the generator. In this scenario, G3 will provide the power demand with a very low frequency, as shown in the figure, which may also affect the connected loads. The frequency protection function must be activated with proper under-frequency settings to trip the CB if the frequency is below the threshold ($\cong$ 59.7 Hz). When the attacker opens R6 CB, the relay R18 connected to G2 will pick up as under frequency and, accordingly, send a trip signal to CB2. No power source will feed the three loads connected to the load buses. As a result of these attacks, a severe impact can be observed either through the operation under low frequency or with the blackout when the whole loads served by this substation lose supply. By increasing the grid connection, cascaded outages are expected.

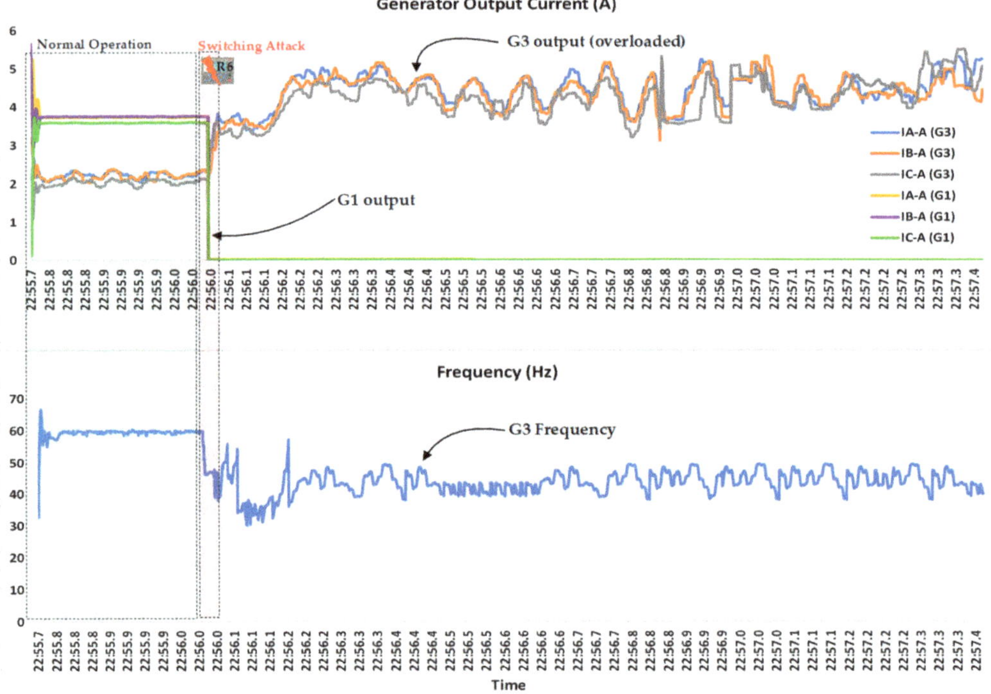


**Figure 10.** Switching attack results.

### 5.2. Relay Setting Tampering (Change) Attack Scenario

The protection system aims to compute the system state, measure signals such as voltages, currents, and frequencies using sensors, and take corrective action if any deviation from normal operation is identified based on appropriate configurations and settings. The faulted area can be isolated by switching actions when a fault occurs at any component inside the substation after picking up a certain time. The relay can be accessed and configured using PCM600, LHMI, or WHMI. The substation's communication protocol does not affect communication between the IED and PCM600 or WHMI. It might be seen as an additional communication channel using Ethernet and TCP/IP protocol. The attack on the protection scheme prevents us from achieving this goal. One of the most frequent attacks in protection relays is the setting tampering (setting change) attack; the attacker may maliciously change the relay configurations or setting functions, preventing the relay operation during a fault or forcing the relay to operate normally. In this scenario, these attackers could severely impact the security and stability of the power system and the communication between protective equipment. Therefore, studying the impacts of such types of attacks on the protection system is crucial.

#### 5.2.1. Relay Configuration and Attack Model

This section focuses mostly on cyber-attacks against digital overcurrent relays. In overcurrent protection, the protection relays first receive the current measured by the current transformers, subsequently comparing it with the preset threshold values. Relays trip under fault current, changing their output from an open contact to a closed state to clear the fault after a certain time based on the selected overcurrent characteristics. Time overcurrent (TOC) or inverse time over current denotes that the relay's trip time is inversely proportional to the applied fault current. Modern digital relays are programmable; thus, curve shapes can easily be changed without needing replacement. Nearly any mix of definite-time, inverse-time, and instantaneous elements can be applied. In general, the trip time for each standardized relay protection curve will be determined using the IEC 60255 or IEEE C37.112 [36] formulas as in (2),

$$t(I^{meas}) = TD * \left( \frac{K}{P^x - 1} + C \right) \tag{2}$$

where $t$ is the relay trip time, $TD$ is the time dial setting, $P$ is the ratio of measuring current or fault current to pick up current ($I^{meas}/I^{pu}$), and $K$, $C$, and $x$ are constants depending on the curve types. In this work, IEC standard inverse curves were selected such that $C = 0$, $k = 0.14$, and $x = 0.02$. Typically, a certain principle guides the selection of decision thresholds or pickup values, which are represented by (3) between the maximum load current of that feeder and the system's minimum short-circuit fault current.

$$I_{\text{max}-load} \leq I_{pu} \leq I_{min-fault} \tag{3}$$

However, in this work, some assumptions were considered due to the practical component limitations and avoiding harmful actions in testing fault scenarios. Therefore, the $I_{pu}$ was selected only depending on the maximum loading conditions. Hence, pick up current settings for R11, R12, and R13 should be above their associated feeder load currents. The standards state that the threshold value set should be twice (or equal to 200%) the nominal current flowing through lines TL0560, TL0170, and TL0510 for proper detection. IED settings are calculated for various operation conditions in advance and assigned to various configuration groups. The IED application or a manual menu selection can be used to alter the active configuration group. In addition to giving read/write access to executable files that perform monitoring, configuration, and essential operating tasks, reading configuration files reveals which services are currently active. The researchers claim that an attacker can take advantage of the flaw to access private data, including usernames and passwords, which they can then use to take complete control of the intended device.

To establish a WHMI connection to the protection relay, after opening the explorer, the protection relay's IP address must be typed in the address bar, and then the username and password must be typed. The IED settings can now be changed maliciously because the attacker has access to them. The attacker reconfigured the IED to operate under high-load conditions or fail to sense the fault currents, especially low fault currents. Therefore, under normal circumstances, feeder relays are anticipated to trip the respective breakers when a fault occurs in their lines. But if even one of these breakers or relays malfunctions due to a physical or digital abnormality, this leads to incorrect operation of the protection system.

### 5.2.2. Testing Scenario

The cyber-physical system comprises a reduced-scale power system, a communication network, and IEDs. The physical ABB relays were connected through an Ethernet communication network. The relays used in this section are the REF615 for feeder protection and REG615 for generator protection. In this work, both relays (IEDs) offer a protection unit PTOC with "time over-current" as its primary protection feature. Each relay is configured with two OC stages: stage I> is a standard inverse, and stage I >> is definite time as shown in Figure 11a. In normal operation, with the proper setting configuration of the five relays in either low- or high-loading conditions, physical faults will lead to normal protection system operation. When a fault occurs on a transmission line, the feeder relays R11, R12, and R13 trip, open the breakers, and send a GOOSE block to the generators' relays R6 and R18 to prevent a false operation in the unlikely event that one of them is detected as a second stage I >> [37]. However, we assumed low loading conditions such that G3 is out of service and only G1 supplies the total power to the loads L1, L2, and L3. By abruptly raising the load L1, an imitation of the three-phase fault on feeder TL0560 is applied. The stage PTOC.str picked up right away according to the setting of R11 I $\geq$ 3.8 A with an approximate time delay of 735 msec, according to the selected inverse characteristic. In addition, R6 was picked up as a first stage I $\geq$ 7.5 A; however, it was not trip due to the relay coordination scheme. In the case of high-fault conditions, R6 may have been picked up as a second stage I >> and trip before R11, and therefore, a GOOSE message was transmitted instructing the incomer relay R6 to stop stage I >> operation. R11 issued a trip order to the circuit breaker at a time delay of 735 msec, and the breaker was opened at 750 msec. As a result, the current through R11 decreased to zero, while the incoming current from G1 decreased to about 3.5 A to feed the other loads L2 and L3 through the healthy feeders. Individual disturbance recordings must be uploaded from the IED using PCM600 to monitor the recorded data as shown in Figure 11b. All disturbance recordings can be found in the C:\COMTRADE directory.

On the other hand, the protective mechanism was not functioning correctly due to the setting tampering (change) attack R11. A setting change attack was carried out on the suggested protective method using almost identical loading conditions and fault scenarios. Relay R11 was maliciously altered, as shown in Figure 12a, putting the protective relay's coordination and the protection system's dependability at risk. With the manipulation in the threshold values of the relays by the attacker through moving the threshold values upward, the relay will probably not detect the short-circuit fault, and the circuit breaker will not work under this abnormal condition or isolate the faulted line. Under the same fault conditions, the overcurrent relay R11 cannot detect the fault current caused by the fault, while stage I> of R6 was picked up almost simultaneously. In this case, the I> stage of R6 as a backup protection for R11 was activated, and the relay issued a trip signal to the linked CB after the calculated time on the relay characteristic, which is indicated by the red lines as shown in Figure 12b. Consequently, all the generator's power was lost on busbar BB0320, which rendered the remaining loads connected to the functional feeders unusable.

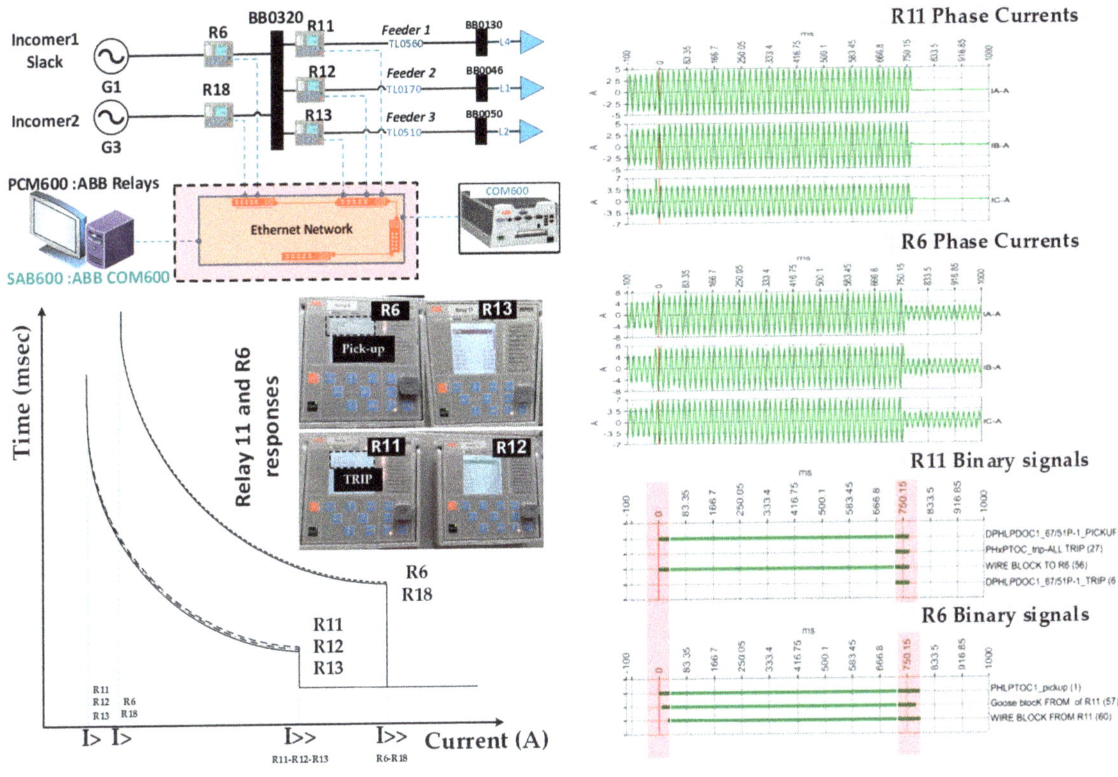

**Figure 11.** Normal protection system operation: (**a**) power system and relay setting configuration; (**b**) disturbance recorders of relays' analog and binary signals.

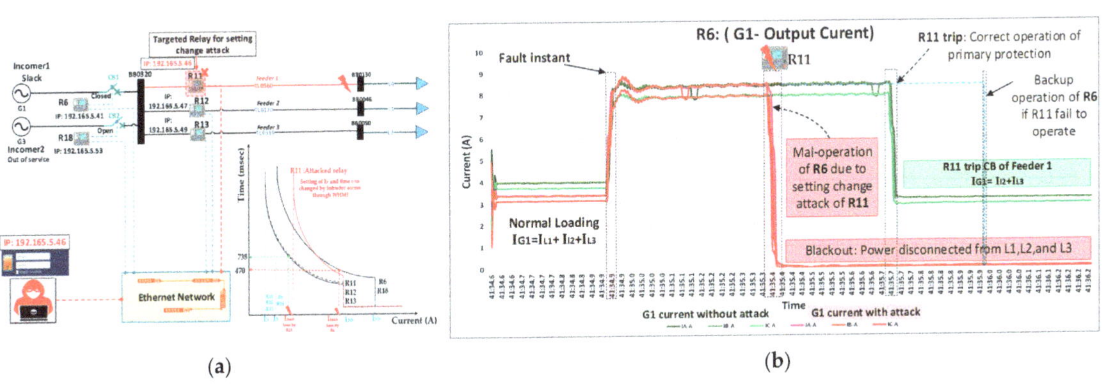

**Figure 12.** Protection system response under setting change attack: (**a**) relay coordination is altered by attacker; (**b**) generator current without/with attack.

A power protection system's reliability is a measure of the extent of certitude that it will work as intended and is made up of two components: dependability and security. Dependability is a measure of the degree of certainty that a protective system will function successfully when required and at the intended speed when the fault is in the protected zone. Security is a measure of the degree of assurance that the protection system will not act erroneously or quicker than intended when the fault is outside the protective zone. The correction operation of the protection system is operated correctly since R6 acted as a backup protection for R11. However, from the perspective of power system protection dependability, the system is unreliable because the primary protection (R11) did not operate correctly, which means the system is undependable, and R6 issued a trip signal to the associated CB, which means the system is insecure. From the reliability of the power system point of view, the loads connected to the healthy feeders lost power, and consequently, this was a complete blackout. Figure 12b shows the G6 current output in both cases, with green lines without attack and red lines if R11 is under attack. Conversely, if the threshold values are lowered because of the attack, any rise in the system load could potentially be regarded as a fault, and the relay will transmit the trip command under normal operation.

## 6. Cyber-Physical Co-Simulation Testbed Implementation and Set-up
### 6.1. System Overview

Despite the higher degree of reality introduced by the physical testbed in terms of testing and analysis of cyber-attack impacts on power system controls and protection, there is a lack of flexibility in conducting different attack scenarios and power system topologies. Therefore, building a real-time cyber-physical co-simulation testbed is crucial. The cyber-physical co-simulation system simulates the three domains of real-time power system operation, with information flowing continuously across a simulated communication network utilizing the ns3 tool between the power system domain and the control and energy management domain. The experimental platform was created to more broadly assess how the communications network and controls perform when used for grid control and protection applications. Any additional controllable device—a "smart grid device"—can be integrated into this platform in the future with only modest set-up adjustments. Figure 13 illustrates the implementation and construction of a comprehensive three-domain modeling and simulation cyber-physical power system on various machines. Machine 1 is the real-time power system simulator (OPAL-RT: OP4610XG), a compact mid-range simulator, while the third machine is used for implementing the power system control or protection scheme. Both machines are connected to machine 2 through the Ethernet network as shown on the right and left of the figure.

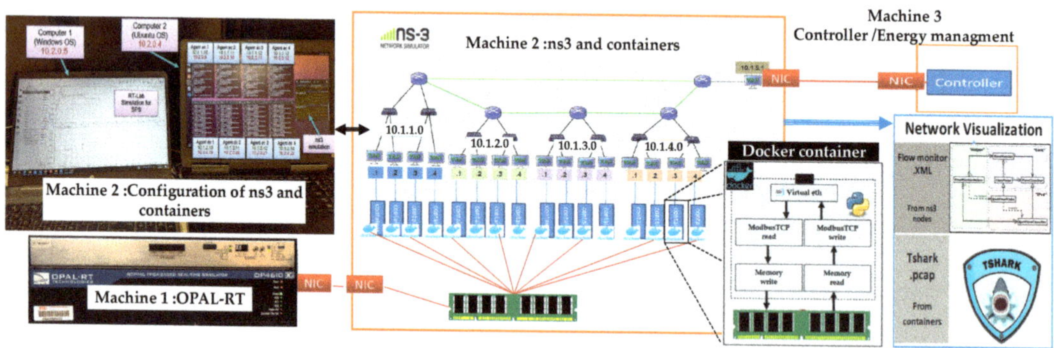


**Figure 13.** Cyber-physical co-simulation testbed.

### 6.2. Communication Network Emulation and Docker Containers

Regarding the set-up of machine 2, which is the core element of this research work as an excellent environment for testing and emulating not only cyber-attacks but also different communication issues such as latency, packet loss, and losing network links, it is running with Linux OS to host two major parts of the communication network infrastructure as shown in the middle of the figure. The first part is ns-3 for communication network emulation. ns-3 is a free, extendable, advanced network simulation framework [38]. An extensive library of network model protocols, including those for multicasting, IP-based applications (TCP, UDP), routing, and wireless and wired networks, is accessible on top of the ns-3 architecture. The ns-3 core, a time sync module, a simulated communication network, and a network application module are the four primary ns-3 components available in an ns-3 process to support all additional simulator aspects. The second part is docker containers that serve as intermediaries to convey data between the network nodes in ns-3 and the OPAL-RT simulation. These containers can be considered as local/primary controllers or agents that receive commands from a secondary/tertiary controller or as local sensors that transmit measurements. A docker is a software development tool and virtualization technology that makes it easier to develop, deploy, and manage programs utilizing containers [39]. A container is a small, independent executable software package that includes all the libraries, dependencies, configuration files, and other components required to run a program. Multiple containers can run concurrently on a single host since containers are secure due to their isolation. In this work, the docker containers are designed for interfacing between the ns-3 nodes and their corresponding device in the power system model, which runs in the OPAL-RT. Based on the structure of a container, as illustrated in Figure 14, agents within containers can connect with the power system modeled on OPAL-RT, and they can communicate with other containers using ns-3 by two virtual network interfaces: veth1 and veth2. veht1 is used to exchange data with other containers via ns-3 through Linux bridge and tapping device connection by the host operating system, while veth2 is used to send/receive measurements or control signals from/to OPAL-RT. In this work, the application inside the container is configured to act as an agent representing the sensor or controller. However, one of these containers or additional containers can be inserted into the network to behave like an attacker receiving information and sending it after some modifications.

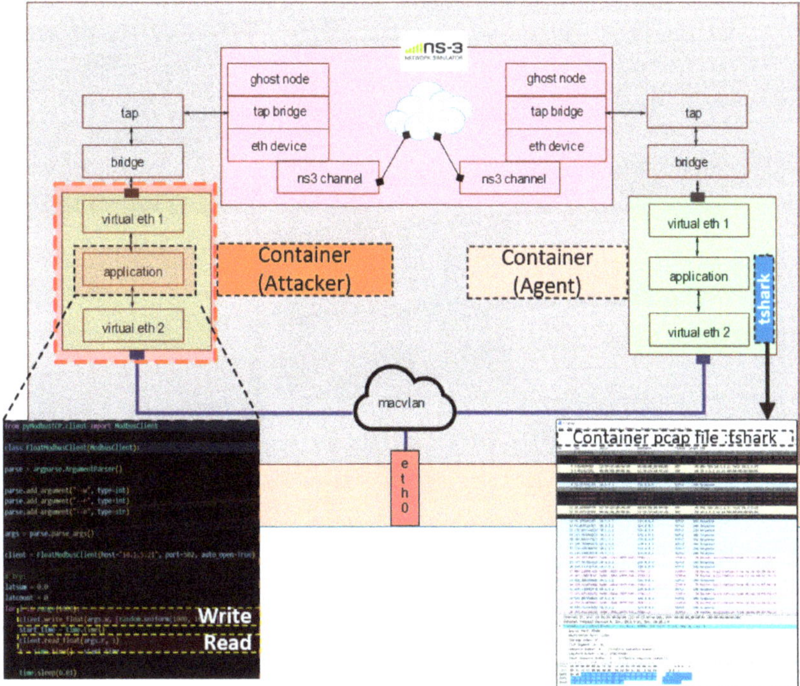


**Figure 14.** Container/ns-3 connectivity.

## 6.3. Network Performance Tests and Visualization

The communication network is a crucial system for ensuring the dependability of control and protection applications, and its condition depends on where the application is applied. The network's behavior and performance can considerably impact the system's availability, stability, and performance as a whole. Communication network tests are necessary to reproduce and evaluate these impacts before implementation, including network performance and communication protocols. In [40], the network latencies are stochastically characterized by natural probabilistics to evaluate the network performance for the shipboard power system application. Different communication protocols may be utilized in power systems since the controllers/agents may be supplied by multiple vendors and employed for a variety of control/protection functions. With the proposed platform, the communication between agents in docker containers through ns-3 using different protocols such as UDP/IP, TCP/IP, DDS RTPS, and IEC61850 GOOSE was conducted by running applications in two different containers to send and receive messages. Testing of the communication using TCP/IP and UDP/IP by assigning one container as a server and the other as a client, with checking the results using the Wireshark is shown in Figure 15. In addition, to visualize, monitor, and analyze the communication network model, some additional tools have been installed recently in the second machine such as FlowMonitor and tshark. The FlowMonitor module is a core feature of ns-3 that facilitates the collection of a common set of network performance measurements of packet-related data, such as throughput, loss ratio, packet delay, bit rate, and round-trip time. It saves them to permanent storage in XML files describing the flow of information between all the system's nodes in ns-3. Moreover, to evaluate the communication system's performance in real time, a network packet analyzer tool (tshark) was installed in all the containers (nodes) in the system to capture and analyze network packets through the generated PCAP files as shown in Figure 14.

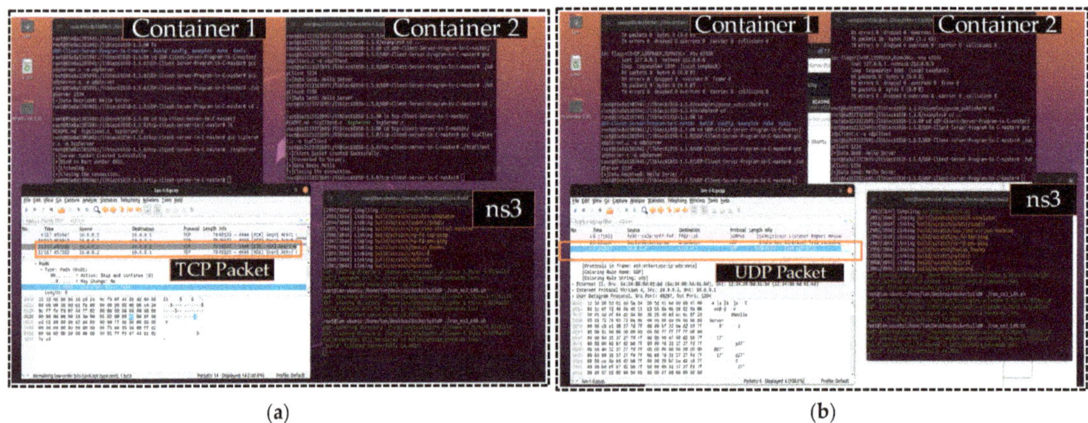


(a)  (b)

**Figure 15.** Container-based testing protocols: (**a**) TCP protocol testing; (**b**) UDP protocol testing.

## 7. Co-Simulation Testbed Attack Scenarios

The objective of the test case in this section is to assess the control system's performance under cyber-attacks. Using the well-developed co-simulation testbed, potential cyber risks with proper attack models through emulation of their behaviors will be studied. This section will develop and discuss two types of attacks: DoS attack and MitM attack. By carrying out actions specifically designed to target the system under investigation and to power system protocols, the adversary can carry out DoS and MiTM attacks that have tangible consequences. The threat model we present and implement in this work is based on emulating the attacker behavior through ns-3 and docker containers; however, the adversary is restricted by the available resources in the Linux containers in ns-3 and docker containers.

### 7.1. DoS Attack Scenario through ns-3

This section evaluates the resilience of the MVAC/DC ship power system under DoS attack. To conduct the evaluation, the network emulator ns-3 is used to simulate the ship communication network in real time, as illustrated in Figure 16a. The considered network consists of two local networks corresponding to the system's AC and DC sides. Docker containers coexist with the ns-3 network emulation system on a Linux host computer and serve as interfaces between the power system simulation in Opal-RT and ns3. Specifically, each container links a device in Opal-RT to a network node in ns-3. The goal of the control system is to maintain stability in the ship power system under uncertainties in the overall system. It ensures that the voltage and frequency remain at their designated reference values. A distributed control strategy [41] is employed, where individual local controllers or agents are implemented within containers. These agents communicate with each other through the provided communication network. The proper functioning of the communication network is crucial for the system to operate effectively.

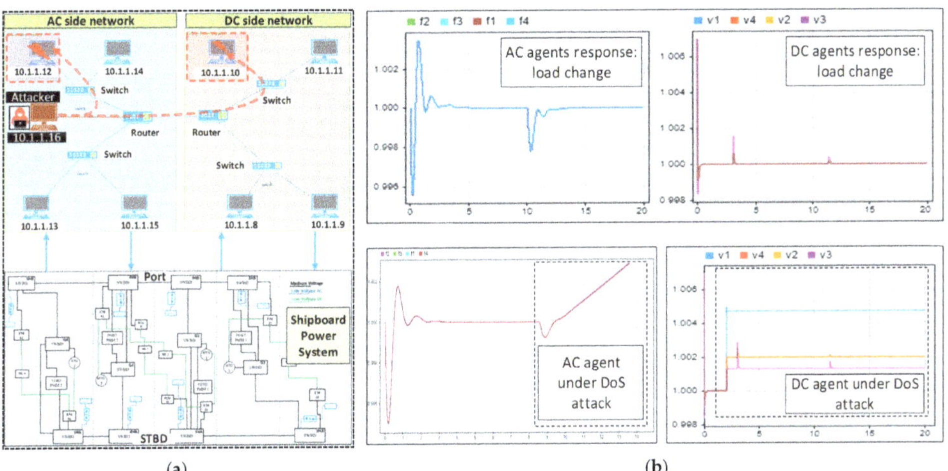


**Figure 16.** System study under DoS attack: (**a**) SPS under study and agent's implementation in ns-3; (**b**) agent's response under normal operation, DoS attack of an AC agent, and DoS attack of a DC agent.

The investigation assumes that an attacker can gain access to communication. There are numerous techniques for launching denial of service (DoS) attacks, such as ping of death (PoD), Internet control message protocol (ICMP) flood, and user datagram protocol (UDP) flood. All DoS attack techniques aim to interfere with the targeted node's communication channels, even though they employ various Open Systems Interconnection (OSI) layers, including application, presentation, session, transport, network, data link, and even physical layer protocols [42]. In our threat model, an additional container is used to act as the attacker and is connected to the ns-3 network using open-source tools. In the test case, the attacker employs the hping3 tool to launch a DoS attack by flooding the target with traffic. This command-line tool generates DoS attacks that overload the network or the application layer, causing delayed message delivery. The flooding attack can take various forms depending on the network protocol used. While simple to perform, this type of attack can cause significant disturbances. The test results depicted in Figure 16b show that the cyber-attack affects the AC system when the attacker is attached to the AC local network. As a result, the target agent or controller becomes unavailable, rendering the control system non-functional. The frequency of the AC system is no longer maintained at the reference value, and in the worst case, the system becomes unstable, resulting in a blackout event for the entire ship power system. Similarly, if the attacker is targeting a DC agent in the DC side network, this will impact the DC voltage, as shown in the figure, when compared to the normal operation without attack.

### 7.2. MitM Attack Scenario through ns-3

In a similar approach, a MiTM attack is carried out using the co-simulation part of the platform, and Wireshark is used to observe the network flow to discover the attack behavior and impacts. A MiTM attack is a type of attack in which an intruder positions themselves between two communicating agents to intercept and/or alter data traveling between them. By embedding themselves within a conversation, the intruder can eavesdrop or impersonate one of the devices, allowing them to perform false data injection (FDI) and false command injection (FCI) attacks that can compromise power system operations. As shown in Figure 17, the additional container is used to simulate the intruder in the network. This container uses the Address Resolution Protocol (ARP) spoofing technique to link its MAC address with the IP address of the victim container. When a packet is sent from the source agent in the AC side network, it is routed through the MitM attacker before reaching the destination agent. The evidence of the MiTM attack is determined by analyzing the average round trip time (RTT), retransmission rate, and average processing time of packets. When a packet is sent, the sender starts a variable-length retransmission timer and waits for the acknowledgment. If no acknowledgment is received before the timer expires, the sender assumes the packet is lost and retransmits it. To detect if there is a MitM attack, the ping command can calculate round-trip times and packet loss statistics and display a summary on completion. As depicted in the upper right figure, the test results indicate that when the packet is sent through ns-3 via appropriate nodes, the recorded time is t = 1.28 ms, marked by the yellow dotted lines. However, when the MitM attacker captures the message to modify it, the recorded time increases to t = 2.29 ms (almost doubled). This significant discrepancy in the recorded time strongly suggests the presence of a MiTM attack, which can be used as an indicator to detect the presence of attackers in the system.

**Figure 17.** The effect of MitM attack on an AC agent in the ship power system.

## 8. Conclusions

The effective way to comprehend security threat events and their effects on the power grid is through cyber-physical testbeds, which can help facilitate grid resiliency to cyber threats. The FIU Hybrid SGTB offers a realistic testing environment with real power system components, controls, and protective software. While this offers the best testing conditions, many research projects find it impractical to test modern communication systems or large-scale power systems due to their complexity and flexibility. Integrating a co-simulation-based testbed to the physical testbed can make testing and validation more convenient and flexible, enabling the incorporation of both real and virtual components.

In this study, the Hybrid SGTB introduced a comprehensive framework for running and simulating various power system topologies' physical and cyber components by using components like industrial-grade devices, real-time simulators, and various automation tools for experiment orchestration, data collection, and visualization. The attack model, impact on system dynamics, and cascading failures are experimentally proven through a suggested cyber-physical experimental framework that closely replicates real-world conditions within a digital substation, including IEDs and protection measures. Various experimental scenarios were used to implement cases of data manipulation and setting change attacks by real agents (attackers) using the physical testbed. In addition, two emulated attacks on the shipboard power system model using the co-simulation testbed, MitM and DoS, were performed through virtual agents using the integration of ns-3 and docker containers. In the future, a network security monitoring agent as a vulnerability scanner component will be implemented in the physical testbed for monitoring, analysis, and intrusion detection. In addition, a real-time automation controller will be integrated into the co-simulation testbed for different control applications. Moreover, the communication network will be fully modeled using Exata CPS running on OPAL-RT to evolve testing attack scenarios and help implement intrusion detection techniques.

**Author Contributions:** Conceptualization, M.S.A., and O.A.M.; Methodology, M.S.A., and T.L.N.; Software, I.K., and T.L.N.; Validation, M.S.A., and T.L.N.; Formal analysis, M.S.A., and I.K.; Investigation, O.A.M.; Resources, O.A.M.; Writing—original draft, M.S.A., and I.K.; Writing—review & editing, T.L.N., and O.A.M.; Visualization, M.S.A., and I.K.; Supervision, O.A.M. All authors have read and agreed to the published version of the manuscript.

**Funding:** The work in this article was partially supported by grants from the Office of Naval Research, ESRDC, and the National Science Foundation.

**Data Availability Statement:** Data are contained within the article.

**Acknowledgments:** We acknowledge the collaborative discussions, equipment support, and software provided by ABB and Schweitzer Engineering Laboratories.

**Conflicts of Interest:** The authors declare no conflict of interest.

## References

1. Buchholz, B.M.; Styczynski, Z. *Smart Grids-Fundamentals and Technologies in Electricity Networks*; Springer: Berlin/Heidelberg, Germany, 2014; pp. 225–275.
2. Krause, T.; Ernst, R.; Klaer, B.; Hacker, I.; Henze, M. Cybersecurity in Power Grids: Challenges and Opportunities. *Sensors* **2021**, *21*, 6225. [CrossRef] [PubMed]
3. Whitehead, D.E.; Owens, K.; Gammel, D.; Smith, J. Ukraine Cyber-Induced Power Outage: Analysis and Practical Mitigation Strategies. In Proceedings of the 2017 70th Annual Conference for Protective Relay Engineers (CPRE), College Station, TX, USA, 3–6 April 2017; pp. 1–8.
4. Liang, G.; Weller, S.R.; Zhao, J.; Luo, F.; Dong, Z.Y. The 2015 Ukraine Blackout: Implications for False Data Injection Attacks. *IEEE Trans. Power Syst.* **2017**, *32*, 3317–3318. [CrossRef]
5. Ward, S.; O'Brien, J.; Beresh, B.; Benmouyal, G.; Holstein, D.; Tengdin, J.T.; Fodero, K.; Simon, M.; Carden, M.; Yalla, M.V.V.S.; et al. Cyber Security Issues for Protective Relays; C1 Working Group Members of Power System Relaying Committee. In Proceedings of the 2007 IEEE Power Engineering Society General Meeting, Tampa, FL, USA, 24–28 June 2007; pp. 1–8.
6. Tan, R.; Nguyen, H.H.; Foo, E.Y.; Yau, D.K.; Kalbarczyk, Z.; Iyer, R.K.; Gooi, H.B. Modeling and mitigating impact of false data injection attacks on automatic generation control. *IEEE Trans. Inf. Forensics Secur.* **2017**, *12*, 1609–1624. [CrossRef]

7.   Rahman, M.S.; Mahmud, M.A.; Oo, A.M.T.; Pota, H.R. Multi-agent approach for enhancing security of protection schemes in cyber-physical energy systems. *IEEE Trans. Ind. Inform.* **2017**, *13*, 436–447. [CrossRef]
8.   Ameli, A. Application-Based Measures for Developing Cyber-Resilient Control and Protection Schemes in Power Networks. Ph.D. Thesis, UWSpace, Waterloo, ON, Canada, 2019.
9.   Liu, S.; Mashayekh, S.; Kundur, D.; Zourntos, T.; Butler-Purry, K. A framework for modeling cyber-physical switching attacks in smart grid. *IEEE Trans. Emerg. Top. Comput.* **2013**, *1*, 273–285. [CrossRef]
10.  Liu, X.; Shahidehpour, M.; Li, Z.; Liu, X.; Cao, Y.; Li, Z. Power system risk assessment in cyber-attacks considering the role of protection systems. *IEEE Trans. Smart Grid* **2017**, *8*, 572–580. [CrossRef]
11.  Touhiduzzaman, M.; Hahn, A.; Srivastava, A. A diversity-based substation cyber defense strategy utilizing coloring games. *IEEE Trans. Smart Grid* **2018**, *10*, 5405–5415. [CrossRef]
12.  Hong, J.; Nuqui, R.F.; Kondabathini, A.; Ishchenko, D.; Martin, A. Cyber Attack Resilient Distance Protection and Circuit Breaker Control for Digital Substations. *IEEE Trans. Ind. Inform.* **2018**, *15*, 4332–4341. [CrossRef]
13.  Ani, U.D.; Watson, J.M.; Green, B.; Craggs, B.; Nurse, J. Design Considerations for Building Credible Security Testbeds; A Systematic Study of Industrial Control System Use Cases. *J. Cyber Secur. Technol.* **2021**, *5*, 71–119. [CrossRef]
14.  Yang, Y.; Mclaughlin, K.; Littler, T.; Sezer, S.; Im, G.; Yao, Z.Q.; Pranggono, B.; Wang, H.F. Man-in-the-middle attack test-bed investigating cyber-security vulnerabilities in Smart Grid SCADA systems. In Proceedings of the International Conference on Sustainable Power Generation and Supply (SUPERGEN 2012), Hangzhou, China, 8–9 September 2012; pp. 1–8.
15.  Siaterlis, C.; Garcia, A.P.; Genge, B. On the use of Emulab testbeds for scientifically rigorous experiments. *IEEE Commun. Surv. Tutor.* **2012**, *15*, 929–942. [CrossRef]
16.  Smadi, A.A.; Ajao, B.T.; Johnson, B.K.; Lei, H.; Chakhchoukh, Y.; Abu Al-Haija, Q. A Comprehensive Survey on Cyber-Physical Smart Grid Testbed Architectures: Requirements and Challenges. *Electronics* **2021**, *10*, 1043. [CrossRef]
17.  Liu, R.; Vellaithurai, C.; Biswas, S.S.; Gamage, T.T.; Srivastava, A.K. Analyzing the Cyber-Physical Impact of Cyber Events on the Power Grid. *IEEE Trans. Smart Grid* **2015**, *6*, 2444–2453. [CrossRef]
18.  Hahn, A.; Ashok, A.; Sridhar, S.; Govindarasu, M. Cyber-Physical Security Testbeds: Architecture, Application, and Evaluation for Smart Grid. *IEEE Trans. Smart Grid* **2013**, *4*, 847–855. [CrossRef]
19.  Nelson, A.; Chakraborty, S.; Wang, D.; Singh, P.; Cui, Q.; Yang, L.; Suryanarayanan, S. Cyber-physical test platform for microgrids: Combining hardware, hardware-in-the-loop, and network-simulator-in-the-loop. In Proceedings of the 2016 IEEE Power and Energy Society General Meeting (PESGM), Boston, MA, USA, 17–21 July 2016; pp. 1–5.
20.  Zhang, H.; Ge, D.; Liu, J.; Zhang, Y. Multifunctional cyber-physical system testbed based on a source-grid combined scheduling control simulation system. *IET Gener. Transm. Distrib.* **2017**, *11*, 3144–3151. [CrossRef]
21.  Wei, M.; Wang, W. Greenbench: A benchmark for observing power grid vulnerability under data-centric threats. In Proceedings of the IEEE INFOCOM 2014-IEEE Conference on Computer Communications, Toronto, ON, Canada, 27 April–2 May 2014; pp. 2625–2633.
22.  Duan, N.; Yee, N.; Salazar, B.; Joo, J.Y.; Stewart, E.; Cortez, E. Cybersecurity Analysis of Distribution Grid Operation with Distributed Energy Resources via Co-Simulation. In Proceedings of the 2020 IEEE Power & Energy Society General Meeting (PESGM), Montreal, QC, Canada, 2–6 August 2020.
23.  Gupta, K.; Sahoo, S.; Panigrahi, B.K.; Blaabjerg, F.; Popovski, P. On the Assessment of Cyber Risks and Attack Surfaces in a Real-Time Co-Simulation Cybersecurity Testbed for Inverter-Based Microgrids. *Energies* **2021**, *14*, 4941. [CrossRef]
24.  Chamana, M.; Bhatta, R.; Schmitt, K.; Shrestha, R.; Bayne, S. An Integrated Testbed for Power System Cyber-Physical Operations Training. *Appl. Sci.* **2023**, *13*, 9451. [CrossRef]
25.  Salehi, V.; Mohamed, A.; Mazloomzadeh, A.; Mohammed, O.A. Laboratory-Based Smart Power System, Part I: Design and System Development. *IEEE Trans. Smart Grid* **2012**, *3*, 1394–1404. [CrossRef]
26.  Hussein, H.; Aghmadi, A.; Nguyen, T.L.; Mohammed, O. Hardware-in-the-loop implementation of a Battery System Charging/Discharging in Islanded DC Micro-grid. In Proceedings of the SoutheastCon 2022, Mobile, AL, USA, 26 March–3 April 2022; pp. 496–500.
27.  Huang, Y.L.; Cárdenas, A.A.; Amin, S.; Lin, Z.S.; Tsai, H.Y.; Sastry, S. Understanding the physical and economic consequences of attacks on control systems. *Int. J. Crit. Infrastruct. Prot.* **2019**, *2*, 73–83. [CrossRef]
28.  Deng, W.; Yang, Z.; Xun, P.; Zhu, P.; Wang, B. Advanced Bad Data Injection Attack and Its Migration in Cyber-Physical Systems. *Electronics* **2019**, *8*, 941. [CrossRef]
29.  Menike, S.; Yahampath, P.; Rajapakse, A. Implementation of Communication Network Components for Transient Simulations in PSCAD/EMTDC. In Proceedings of the International Conference on Power Systems Transients (IPST2013), Vancouver, BC, Canada, 18–20 July 2013.
30.  Le, T.D.; Anwar, A.; Loke, S.W.; Beuran, R.; Tan, Y. GridAttackSim: A Cyber Attack Simulation Framework for Smart Grids. *Electronics* **2020**, *9*, 1218. [CrossRef]
31.  Hoyos, J.; Dehus, M.; Brown, T.X. Exploiting the GOOSE protocol: A practical attack on cyber-infrastructure. In Proceedings of the 2012 IEEE Globecom Workshops, Anaheim, CA, USA, 3–7 December 2012; pp. 1508–1513.
32.  Youssef, T.A.; El Hariri, M.; Bugay, N.; Mohammed, O.A. IEC 61850: Technology standards and cyber-threats. In Proceedings of the 2016 IEEE 16th International Conference on Environment and Electrical Engineering (EEEIC), Florence, Italy, 7–10 June 2016; pp. 1–6.

33. Amin, B.; Taghizadeh, S.; Rahman, M.S.; Hossain, M.J.; Varadharajan, V.; Chen, Z. Cyber-attacks in smart grid–dynamic impacts, analyses and recommendations. *IET Cyber-Phys. Syst. Theory Appl.* **2020**, *5*, 321–329. [CrossRef]
34. Azeem, A.; Jamil, M.; Qamar, S.; Malik, H.; Thokar, R.A. Design of Hardware Setup Based on IEC 61850 Communication Protocol for Detection & Blocking of Harmonics in Power Transformer. *Energies* **2021**, *14*, 8284.
35. Hussain, S.S.; Ustun, T.S.; Kalam, A. A review of IEC 62351 security mechanisms for IEC 61850 message exchanges. *IEEE Trans. Ind. Inform.* **2019**, *16*, 5643–5654. [CrossRef]
36. Benmouyal, G.; Meisinger, M.; Burnworth, J.; Elmore, W.A.; Freirich, K.; Kotos, P.A.; Leblanc, P.R.; Lerley, P.J.; McConnell, J.E.; Mizener, J.; et al. IEEE standard inverse-time characteristic equations for overcurrent relays. *IEEE Trans. Power Deliv.* **1999**, *14*, 868–872. [CrossRef]
37. Abdelrahman, M.S.; Kharchouf, I.; Alrashide, A.; Mohammed, O.A. A Cyber-Physical Smart Grid Testbed for Validation of GOOSE-Based Protection Strategies. In Proceedings of the 2022 IEEE Industry Applications Society Annual Meeting (IAS), Detroit, MI, USA, 9–14 October 2022; pp. 1–12.
38. NS3, NS3 Homepage. Available online: https://www.nsnam.org/ (accessed on 25 August 2023).
39. Wang, W. Research on Using Docker Container Technology to Realize Rapid Deployment Environment on Virtual Machine. In Proceedings of the 2022 8th Annual International Conference on Network and Information Systems for Computers (ICNISC), Hangzhou, China, 16–19 September 2022; pp. 541–544.
40. Abdelrahman, M.S.; Nguyen, T.L.; Mohammed, O.A. Stochastic Characterization-Based Performance Analysis of an Emulated Communication Network for Cyber- Physical Shipboard Power Systems. In Proceedings of the 2023 IEEE Electric Ship Technologies Symposium (ESTS), Alexandria, VA, USA, 1–4 August 2023; pp. 528–533.
41. Yoo, H.J.; Nguyen, T.T.; Kim, H.M. Consensus-based distributed coordination control of hybrid AC/DC microgrids. *IEEE Trans. Sustain. Energy* **2019**, *11*, 629–639. [CrossRef]
42. Kalluri, R.; Mahendra, L.; Kumar, R.S.; Prasad, G.G. Simulation and Impact Analysis of Denial-of-Service Attacks on Power SCADA. In Proceedings of the 2016 National Power Systems Conference (NPSC), Bhubaneswar, India, 19–21 December 2016; pp. 1–5.

 *energies*

*Article*

# Enhanced Efficiency on ANPC-DAB through Adaptive Model Predictive Control

Adriano Nardoto [1], Lucas Encarnação [2,*], Walbermark Santos [2], Arthur Amorim [1], Rodrigo Fiorotti [1], David Molinero [3] and Emilio Bueno [3]

[1] Department of Electrical Engineering, Federal Institute of Espírito Santo (IFES), BR101 Km 58, São Mateus 29932-540, Brazil; adriano.nardoto@ifes.edu.br (A.N.); arthur.amorim@ifes.edu.br (A.A.); rodrigo.fiorotti@ifes.edu.br (R.F.)

[2] Department of Electrical Engineering, Federal University of Espírito Santo (UFES), Av. Fernando Ferrari, 514, Vitória 29075-910, Brazil; walbermark.santos@ufes.br

[3] Department of Electronics, Alcalá University (UAH), Plaza San Diego S/N, 28801 Madrid, Spain; david.molinero@uah.es (D.M.); emilio.bueno@uah.es (E.B.)

* Correspondence: lucas.encarnacao@ufes.br

**Abstract:** This work studies the DC-DC conversion stage in solid-state transformers (SST). The traditional two- or three-level dual active bridge (DAB) topology faces limitations in microgrid interconnection due to power and voltage limitations. For this reason, the use of multilevel topologies such as active neutral point clamped (ANPC) is a promising alternative. Additionally, the efficiency of the SSTs is a recurring concern, and reducing losses in the DC-DC stage is a subject to be studied. In this context, this work presents a new control technique based on an adaptive model-based predictive control (AMPC) to select the modulation technique of an ANPC-DAB DC-DC converter aimed at reducing losses and increasing efficiency. The single-phase shift (SPS), triangular, and trapezoidal modulation techniques are used according to the converter output power with the aim of maximizing the number of soft-switching points per cycle. The performance of the proposed control technique is demonstrated through real-time simulation and a reduced-scale experimental setup. The findings indicate the effectiveness of the AMPC control technique in mitigating voltage source perturbations. This technique has low output impedance and is robust to converter parameter variations. Prototyping tests revealed that, in steady-state, the AMPC significantly improves converter efficiency without compromising dynamic performance. Despite its advantages, the computational cost of AMPC is not significantly higher than that of traditional model predictive control (MPC), allowing for the allocation of time to other applications.

**Keywords:** Dual Active Bridge (DAB) converter; active neutral point clamped (ANPC); model predictive control (MPC); power electronics; switching losses; adaptive control

**Citation:** Nardoto, A.; Encarnação, L.; Santos, W.; Amorim, A.; Fiorotti, R.; Molinero, D.; Bueno, E. Enhanced Efficiency on ANPC-DAB through Adaptive Model Predictive Control. *Energies* **2024**, *17*, 12. https://doi.org/10.3390/en17010012

Academic Editor: Ahmed Abu-Siada

Received: 14 November 2023
Revised: 4 December 2023
Accepted: 8 December 2023
Published: 19 December 2023

## 1. Introduction

The solid-state transformer (SST) is becoming an important technology and is foundational to applications that include traction systems, offshore energy generation, DC grids and especially microgrids. Its advantages, such as accurate output-voltage regulation, short-circuit current limitation, power factor compensation, reduced weight and volume in comparison with traditional transformers, and voltage-dip immunity under certain limitations, enable the construction of a new paradigm in electric power systems [1].

The future of the electric power system seems to be a smart grid with power-flow management, incorporated energy storage systems, quick transient response and high integration of renewable energy [2]. All of these features can be incorporated through the utilization of SST. However, this device still has opportunities for improvement, mainly related to cost reduction, reliability, and efficiency [3].

The usual topology of the SST is presented in Figure 1. It shows the AC-DC conversion (Stage 1), DC-DC conversion (Stage 2), and DC-AC conversion (Stage 3), with the high-voltage utility grid on the left and the low-voltage microgrid on the right and all stages allowing bidirectional power flow. Stage 2 traditionally employs a DC-DC converter through a resonant magnetic link to connect high-voltage DC to low-voltage DC, and the efficiency of SST is closely related to losses in this stage.

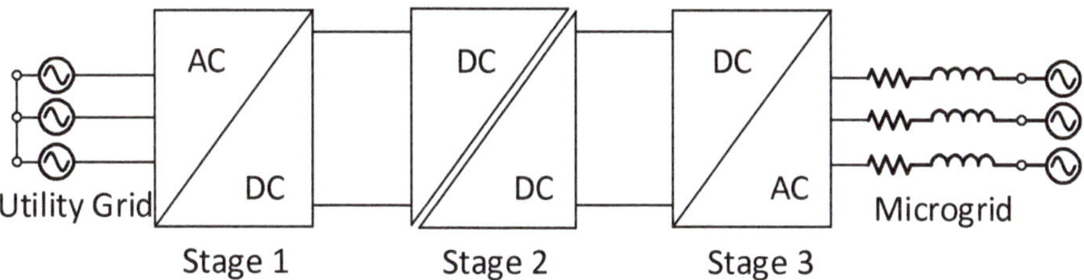


**Figure 1.** SST Topology.

The dual active bridge (DAB) is the general term for the isolated converter formed by the union of two active bridges through an inductor and a high-frequency transformer.

The requirements for building a DAB controller are a fast transient response, minimum steady-state error and reduced losses on the power converter. Hence, an adaptive model predictive control (AMPC) is proposed for a DAB converter in this work. The topology adopted in this work is shown in Figure 2 and operates with an active neutral point clamped (ANPC) converter on the high-voltage side and a H-bridge on the low-voltage side. In this work, it is referred to as ANPC-DAB. Its performance is evaluated through a 8 kW prototype.

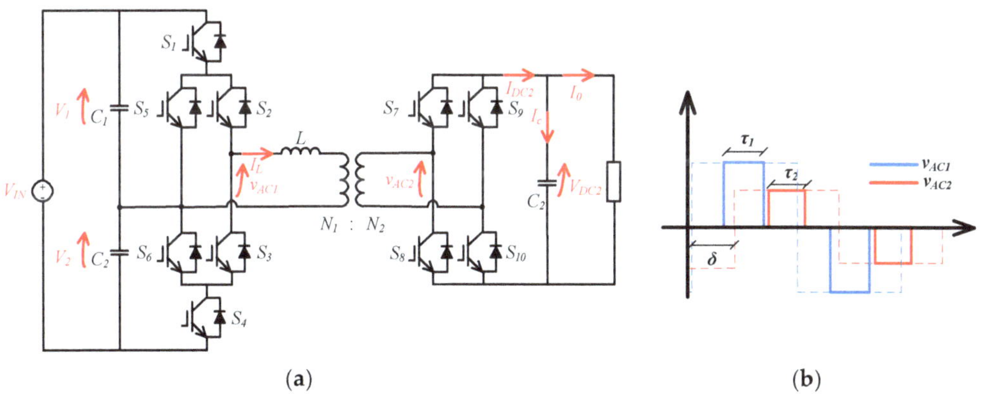


(**a**)                           (**b**)

**Figure 2.** DAB Converter (**a**) ANPC-DAB topology (**b**) Three-level voltages, $v_{AC1}$ and $v_{AC2}$.

### 1.1. Literature Review

#### 1.1.1. DAB Converter Topology

The H-bridge two-level DAB is the most commonly used topology for Stage 2 of an SST. Despite the many advantages of this structure, it is unable to deal with high power and high voltage, being limited by semiconductor technology.

The main solution to this difficulty is the series and parallel connection of the converters; however, additional challenges arise in equalizing voltages and sharing currents. Another alternative is the cascade connection of multiple DAB converters, which increases the number of power switches and the complexity of the control scheme. Hence, the utilization of multilevel structures seems to be the best solution. Many multilevel structures have been widely used with success. Various three-level (3L) topologies have been applied to the DAB converter: flying capacitor, neutral point clamped (NPC), and ANPC [4].

The NPC topology is still widely employed in the industrial sector, notably for operating medium-voltage electric motors [5]. However, the 3L-NPC has some limitations that affect its performance. Some of the switches must withstand higher voltages than others [6], and the power loss in each semiconductor is highly unbalanced [6].

The ANPC converter was proposed to overcome these problems by replacing the clamping diodes with active switches generating redundant switching stages [7]. These redundant stages are used to equalize the voltage and reduce unbalanced losses in each switch, increasing the power-processing capacity of the converter.

### 1.1.2. Modulation Techniques for Loss Reduction

Traditionally, the single-phase shift (SPS) modulation is used in DAB converters. The main advantage of this strategy is that it allows for zero voltage switching (ZVS) over a wide operating range [8]. However, when the relationship between the input and output voltage is far from unity, especially under small loads, these soft-switching techniques are not possible. Additionally, the converter experiences significant backflow-power in this scenario, leading to high circulating currents and increased stress on the switches and conduction losses. In conclusion, the SPS modulation may not ensure high efficiency of the DAB converter over its entire operating range.

A previous work has shown the improvements obtained through the utilization of trapezoidal and triangular modulation techniques [9,10]. These techniques can naturally extend the DAB's soft-switching operation and consist of the utilization of pulse widths different from 50% [11]. Their proper application can increase the number of soft commutations to four (for trapezoidal modulation) or six (for triangular modulation) out of a total of eight in a cycle, even during operation with reduced loads.

In the AC-DC conversion literature, various balancing control strategies have been proposed to optimize the loss distribution of the ANPC circuit [7,12,13]. The voltage synthesized by the rectifier takes on $V_{DC}$ ($P$) or zero ($Z$) during the positive cycle of the electrical grid, resulting in $P \leftrightarrow Z$ type commutations for half of the grid period. During the negative cycle of the electrical grid, the voltage synthesized by the rectifier takes on $-V_{DC}$ ($N$) or zero ($Z$), resulting in $N \leftrightarrow Z$ type commutations for the other half of the grid period. However, in the case of the DAB converter, the ANPC bridge losses distribution is carried out at each sampling time, with $P \leftrightarrow Z \leftrightarrow N$ switches occurring within each sampling period. As a result, all six switches are used within each switching period of the DAB converter.

The literature includes studies on ANPC-DAB switching for loss distribution among the converter switches. Some states involving the transition between the current path through the clamping diode and through the active switch of the converter are presented in [4]. In [14], a modulation is proposed to minimize the number of switching actions, realizing the ZVS turn-ON for all switches and avoiding hard switching turn-OFF in specific switches. In [15], the electromagnetic interference (EMI) caused by zero voltage switching in the ANPC-DAB converter was evaluated.

### 1.2. Contributions

The literature review revealed that work on DAB converters can be divided into two groups. The first group involves the development of modulation strategies to enhance the efficiency of the power converter, while the second group deals with novel control techniques to improve the transient and steady-state response. In this paper, both goals are pursued simultaneously through a novel scheme pf control and modulation.

The control proposal was developed to manage the DAB converter and is based on model predictive control (MPC), which incorporates a variable phase-shift adjustment depending on the error between the desired value and the measured variable. However, it has been enhanced for AMPC control. In this new configuration, the control is capable of selecting operation modes (trapezoidal and triangular) in order to minimize losses and thus enhance converter efficiency. The proposed AMPC is applied to an ANPC-DAB, and the results are validated through an experimental setup.

Going beyond previously published work found in the literature, this analysis includes the following: the utilization of a realistic switching frequency (20 kHz); a compensation scheme for the AMPC delay; tests with load disturbance rejection; an improved cost function that considers capacitor discharge; and a quantitative study of the influence of parameter variation.

### 1.3. Paper Organization

The paper is organized as follows: Section 2 presents the mathematical modeling of the SPS and the triangular and trapezoidal modulations. The proposed control scheme is detailed in Section 3. Section 4 shows the results of the computational simulation. The experimental results are shown and discussed in Section 5. Finally, Section 6 shows the conclusions of this work.

## 2. ANPC-DAB Topology and Modulations

The DAB converter is formed by two active bridges connected through a high-frequency transformer. In this study, an ANPC converter is used on the high-voltage side, operating as an inverter. There is an H-bridge on the low-voltage side, operating as a rectifier. Together, they form the ANPC-DAB topology, as shown in Figure 2a.

Figure 2b shows a generic representation of the three-level voltages $v_{AC1}$ and $v_{AC2}$ synthesized by the active bridges using a centralized PWM. The dashed line represents the voltage generated by SPS modulation. The phase shift $\delta$ and inner phase shift $\tau_1$ and $\tau_2$ are defined in Equation (1):

$$\begin{cases} -90° < \delta < 90° \\ 0 < \tau_1 < 180° \\ 0 < \tau_2 < 180° \end{cases} \tag{1}$$

One of the fundamental characteristics of the DAB converter is its ability to operate with ZVS in a large part of its operating range when operating in SPS modulation [8,16], and this ability can be extended to the entire operating range when using modulations with greater degrees of freedom [17,18].

The ZVS conditions of the ANPC-DAB are consistent with the ZVS conditions of a dual H-bridge DAB [4,14]. Thus, when they use the same phase-angle shift modulation method, the ANPC-DAB, DNPC-DAB, and H-bridge DAB circuits have the same phase-shift ranges for the ZVS of all switches [19].

The next section analyzes the switching used in the ANPC-DAB converter. Section 2.2 contains an analysis of the operating points at which the ZVS occurs in the SPS modulation. Section 2.3 contains an analysis of the ZVS involving triangular and trapezoidal modulations, the basis of AMPC.

### 2.1. ANPC-DAB State Switching

The 3L-ANPC topology has one active switch connected in parallel with each clamp diode. These two switches allow the AC terminal to be connected to the midpoint DC link (neutral point) in more than one switching state. The four resultant zero states are presented in Table 1 and described as $Z_{OH1}$, $Z_{OH2}$, $Z_{OL1}$, and $Z_{OL2}$. This configuration allows for a more even distribution of switch losses in the converter, increasing its energy-processing capacity.

**Table 1.** Switching sequences of the 3L-ANPC converter.

| AC Side | Switching State | Switch Sequence | | | | | |
|---|---|---|---|---|---|---|---|
| | | $S_1$ | $S_2$ | $S_3$ | $S_4$ | $S_5$ | $S_6$ |
| $V_{DC1}$ | P | 1 | 1 | 0 | 0 | 0 | 1 |
| 0 | $Z_{OH1}$ | 0 | 1 | 0 | 1 | 1 | 0 |
| 0 | $Z_{OH2}$ | 0 | 1 | 0 | 0 | 1 | 0 |
| 0 | $Z_{OL1}$ | 1 | 0 | 1 | 0 | 0 | 1 |
| 0 | $Z_{OL2}$ | 0 | 0 | 1 | 0 | 0 | 1 |
| $-V_{DC2}$ | N | 0 | 0 | 1 | 1 | 1 | 0 |

The switching scheme used in this work is shown in Figure 3. The $v_{AC1}$ and $v_{AC2}$ voltage synthesized and the trigger pulses on the switches for both sides of converter are own, highlighting the positive cycle in red and the negative cycle in blue.

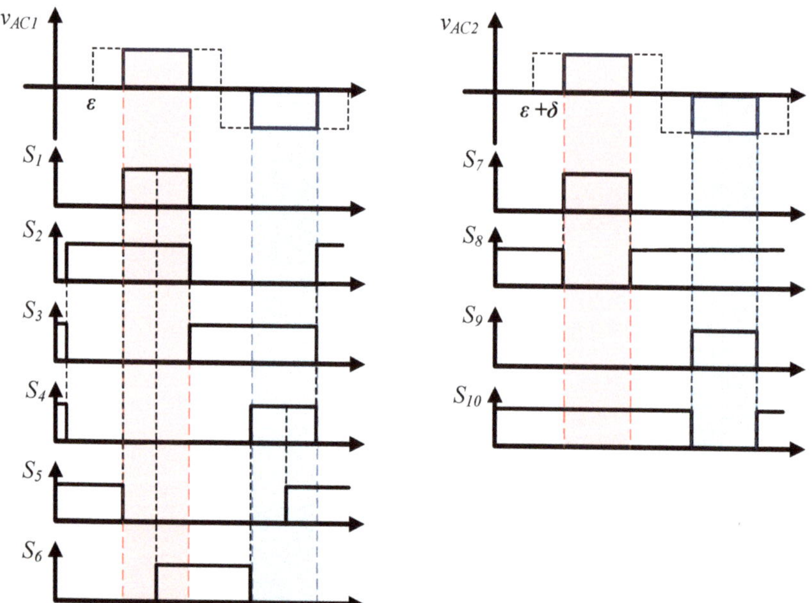


**Figure 3.** Voltage and the trigger pulses on the switches for both sides of the converter. The red color indicates the moments in time when $v_{AC1}$ and $v_{AC2}$ are positive, while the blue color indicates the moments when $v_{AC1}$ and $v_{AC2}$ are negative.

### 2.2. SPS Modulation

SPS modulation is the main modulation used by the ANPC-DAB converter. The only control variable is the phase shift, that is, $\tau_1 = \tau_2 = 180°$. In this work, the voltages on the ANPC-DAB input capacitors will be the same, that is, $V_1 = V_2 = V_{DC1}$. The power transferred using SPS modulation, for a given operating point, is shown in (2).

$$P_{SPS} = \frac{4 \cdot n \cdot V_{DC1} \cdot V_{DC2} \cdot sin\delta}{\pi^3 \cdot f_{sw} \cdot L} \tag{2}$$

where $L$ is power transfer inductance; $n$ is the transformer ratio ($n = N_1/N_2$); $f_{sw}$ is the switching frequency of the converter; and $V_{DC1}$ and $V_{DC2}$ are the input and output voltage on capacitors, respectively.

The power boundaries for the ZVS actuation of the primary and secondary active bridges are presented in (3). If the DAB's power is above $P_{ZVS2}$ or below $P_{ZVS1}$, the DAB converter cannot operate with ZVS [20].

$$\begin{cases} P_{ZVS1} = \left( \dfrac{V_{DC1}^{2} \cdot \delta}{2 \cdot \pi \cdot f_{sw} \cdot L \cdot \Phi} \right) \cdot \left( \dfrac{\pi - |\delta|}{\pi - 2 \cdot |\delta|} \right) \\[2ex] P_{ZVS2} = \left( \dfrac{V_{DC1}^{2} \cdot \delta}{2 \cdot \pi \cdot f_{sw} \cdot L \cdot \Phi} \right) \cdot \left( \dfrac{\pi - |\delta|}{\pi} \right) \cdot \left( \dfrac{\pi - 2 \cdot |\delta|}{\pi} \right) \end{cases} \tag{3}$$

where $\Phi$ is the rated phase shift.

The DAB transformation ratio $d$ is defined in (4). When $d = 1$, the DAB operates in ZVS over the whole power range. When $d > 1$, the primary bridge operates with hard switching. When $d < 1$, the secondary bridge operates with hard switching.

$$d = \frac{N_1 \cdot V_{DC2}}{N_2 \cdot V_{DC1}} \tag{4}$$

Figure 4 shows the DAB power curve as a function of various values of $d$ to highlight the ZVS region. The solid blue and red lines indicate the power limits within which the primary and secondary bridges, respectively, can operate in soft switching. The solid black line represents $d = 1$, where the converter operates in soft switching. The dashed lines show the operating points for $d \neq 1$. For this analysis, we assumed that $\Phi = 45°$.

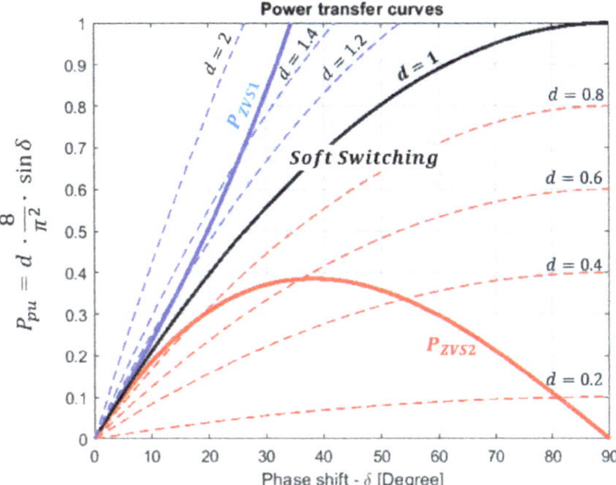


**Figure 4.** Power transfer curves and limit powers for ZVS operation.

These curves show that small phase shifts and significant deviations from unity in the DAB transformation ratio make it difficult to operate the converter in ZVS and decrease its efficiency at those points. Thus, SPS modulation is insufficient to ensure high-efficiency operation of the DAB converter over its entire operational range.

### 2.3. Triangular and Trapezoidal Modulation

An alternative method by which to extend soft switching to regions where SPS modulation does not allow ZVS operation is to apply triangular and trapezoidal modulation.

Triangular modulation has the advantage of allowing soft switching on six transitions, for a total of eight switchings. However, it cannot be used over the entire operating range of the converter because its power transmission capacity is limited, and it is possible to transfer power only when $d \neq 1$.

When $d < 1$, the phase shift $\delta$ and inner phase shift $\tau_1$ and $\tau_2$ are calculated so that the current in the inductor $I_L$ is zero when there are simultaneous changes of state: $Z \rightarrow P$ in the primary and secondary bridge and $P \rightarrow Z$ in the secondary bridge. When $d > 1$, the phase shift $\delta$ and inner phase shift $\tau_1$ and $\tau_2$ are calculated so that the current in the inductor $I_L$ is zero when there are simultaneous changes of state: $Z \to P$ in the primary bridge and $P \to Z$ in the primary and secondary bridges. In the negative half-cycle, the procedure is repeated, replacing $P$ with $N$. Figure 5 shows the waveforms of the voltages $v_{AC1}$ and $n \cdot v_{AC2}$ on the active bridges and the inductor current $I_L$.

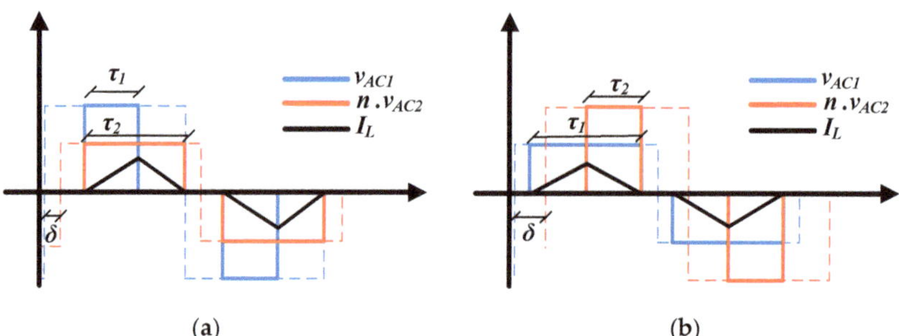

(a)  (b)

**Figure 5.** Voltage and current waveforms for triangular modulation (**a**) $d < 1$ (**b**) $d > 1$.

The values of the inner phase shift $\tau_1$ and $\tau_2$ for the triangular modulation are calculated from the voltages and phase shift. When $d < 1$, the calculation is done as shown in (5). When $d > 1$, the calculation is done as shown in (6).

$$\begin{cases} \tau_1 = 2 \cdot \delta \cdot \left( \dfrac{n \cdot V_{DC2}}{V_{DC1} - n \cdot V_{DC2}} \right) \\ \tau_2 = 2 \cdot \delta \cdot \left( \dfrac{V_{DC1}}{V_{DC1} - n \cdot V_{DC2}} \right) \end{cases} \tag{5}$$

$$\begin{cases} \tau_1 = 2 \cdot \delta \cdot \left( \dfrac{n \cdot V_{DC2}}{n \cdot V_{DC2} - V_{DC1}} \right) \\ \tau_2 = 2 \cdot \delta \cdot \left( \dfrac{V_{DC1}}{n \cdot V_{DC2} - V_{DC1}} \right) \end{cases} \tag{6}$$

When triangular modulation cannot supply the necessary power for a given operating point, trapezoidal modulation is used. Trapezoidal modulation is an extension of triangle modulation, which ensures smooth transition between modulations [11]. In addition, soft switching occurs in four out of the eight switchings per cycle. Moreover, this modulation is possible for any value of $d$. Figure 6 shows the waveforms of the voltages $v_{AC1}$ and $n \cdot v_{AC2}$ on the active bridges and the inductor current $I_L$ for the trapezoidal modulation.

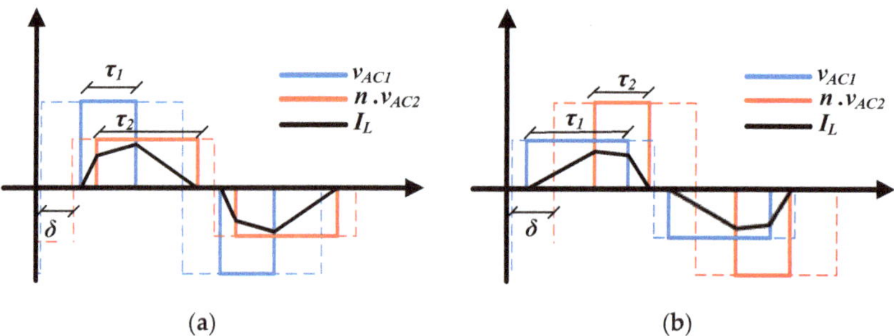

(a)  (b)

**Figure 6.** Voltage and current waveforms for trapezoidal modulation (**a**) $d < 1$ (**b**) $d > 1$.

The values of the inner phase shift $\tau_1$ and $\tau_2$ for the trapezoidal modulation are calculated from the voltages and phase shift, as shown in (7).

$$\begin{cases} \tau_1 = 2 \cdot (\pi - \delta) \cdot \left( \dfrac{n \cdot V_{DC2}}{n \cdot V_{DC2} + V_{DC1}} \right) \\ \tau_2 = 2 \cdot (\pi - \delta) \cdot \left( \dfrac{V_{DC1}}{n \cdot V_{DC2} + V_{DC1}} \right) \end{cases} \tag{7}$$

The natural transition between triangular and trapezoidal modulations is shown in Figure 7a, where the power transferred by the converter is plotted as a function of the phase shift, $\tau_1$ and $\tau_2$. Figure 7b shows the values of $\tau_1$ and $\tau_2$ as a function of $\delta$.

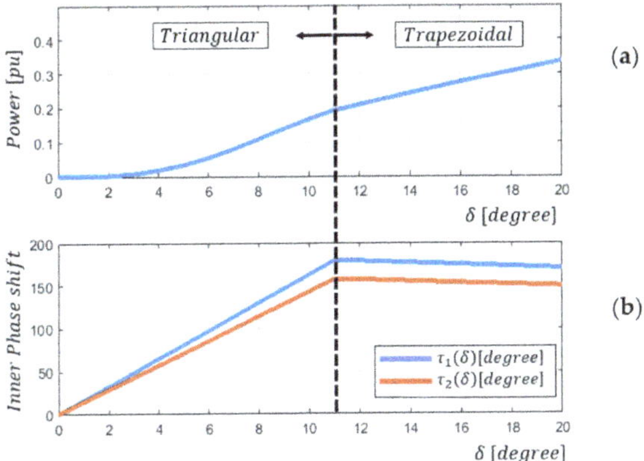


**Figure 7.** Natural transition between triangular and trapezoidal modulations (**a**) Power transferred by the converter as a function of the phase shift $\delta$ (**b**) Values of $\tau_1$ and $\tau_2$ as a function of delta $\delta$.

## 3. Proposed Adaptive Model Predictive Control (AMPC)

In power electronics, contemporary control methods are frequently employed. MPC is one notable example.

Control methodologies that aim to reduce switching losses in the converter usually require complex algorithms to define the three control variables ($\delta$, $\tau_1$ and $\tau_2$) to achieve the desired output voltage. In a deviation from these methods, the AMPC control uses triangular and trapezoidal modulations to reduce switching losses in the converter. The control defines the modulation to be used according to the converter operating point. The great advantage of this strategy is that there is no need for complex algorithms to define the optimal operating point. In addition, the power transfer in the converter happens smoothly and continuously during switching between triangular and trapezoidal modulations.

### 3.1. Mathematical Modeling

The power transferred by the converter depends on the voltages synthesized by the active bridges. In general, phase shift $\delta$ and inner phase shift, represented by $\tau_1$ and $\tau_2$, may be present in the pulses generated by bridges. In this case, either the phase shift or the variation in the inner phase shift may be used to regulate the power flow between the bridges. In this work, the voltages synthesized by the active bridges $v_{AC1}$ and $v_{AC2}$ were generated through a centralized PWM, as depicted in Figure 2b.

The average value of ANPC-DAB output current $I_{DC2}$ is obtained from phasor theory and is calculated as in (8).

$$\langle I_{DC2} \rangle_{T_s} = \frac{4.n.V_{DC1} \cdot sin\frac{\tau_1}{2} \cdot sin\frac{\tau_2}{2} \cdot sin\delta}{\pi^3 \cdot f_{sw} \cdot L} \tag{8}$$

where $T_s$ is sample time and $f_{sw} = 1/T_s$ is the switching frequency.

The average output capacitor quasi-instantaneous voltage is calculated as in (9).

$$C_2 \frac{d\langle V_{DC2} \rangle_{T_s}}{dt} = \langle I_{DC2} \rangle_{T_s} - \langle I_0 \rangle_{T_s} \tag{9}$$

where $I_0$ and $C_2$ are the measure output load current and the output capacitance of ANPC-DAB converter, respectively.

The discretization of (9), using the Euler method, defines the predicted voltage output (10).

$$V_{DC2}(k+1) = V_{DC2}(k) + \frac{I_{DC2} - I_0(k)}{C_2 \cdot f_{sw}} \tag{10}$$

where $I_0(k)$ and $V_{DC2}(k)$ are measured at time $k$ and $I_{DC2}$ is the defined current output derived from (8).

At instant $(k+1)$, there will be a synthesized voltage due to the state of the switches defined by the control at instant $k$. Between instants $k$ and $(k+1)$, the calculation will estimate the voltages for the next cycle, i.e., at time $(k+2)$. Therefore, this calculation must be done using the state voltage at time $(k+1)$, when the control will decide the new state of the switches. To this end, a delay compensation is determined by calculating the voltage that will be synthesized at time $(k+1)$ so that all possible predictions for instant $(k+2)$ can be analyzed. Thus, it is necessary to calculate references for two time steps because predicting variables for only one time step is insufficient to account for the processing delay of signals.

Assuming that a sampling period's load current does not change significantly, i.e., $I_0(k) = I_0(k+1)$, the $N$ possible output voltages at time $(k+2)$ are obtained from $N$ output current possibilities $I_{DC2}^N$, as shown in (11).

$$V_{DC2}^N(k+2) = V_{DC2}(k+1) + \frac{I_{DC2}^N - I_0(k)}{C_2 \cdot f_{sw}} \tag{11}$$

*3.2. Definition of Phase Shift and Inner Phase Shift*

In this work, the main objective of the AMPC strategy is to control the output voltage $V_{DC2}$ ensuring the lowest losses in the active bridges.

The first step is to define the value of the phase shift. The procedure consists of discretizing the entire operating range of the DAB converter into three phase-shift options, where one must be selected. The phase shift adopted is a compromise between the controller precision and the system transient response. These three phase-shift options are set based on the last angle taken by the control, as shown in (12).

$$\delta^N = \left[ \left( \delta_{old} - \delta_{step} \right); \delta_{old}; \left( \delta_{old} + \delta_{step} \right) \right] \tag{12}$$

where $\delta_{old}$ is the phase shift applied at the previous instant and $\delta_{step}$ is dynamically evaluated according to (13), the result of which determines whether to increase or decrease the transferred power [21].

$$\delta_{step} = \delta_{min} \cdot \left( 1 + \alpha \cdot V_{adp} \right) \tag{13}$$

where

$$V_{adp} = \begin{cases} |V^* - V_{DC2}|; & if \; |V^* - V_{DC2}| \leq Vm \\ Vm; & if \; |V^* - V_{DC2}| > Vm \end{cases} \tag{14}$$

where $V^*$ is the value the control will track, $\alpha$ is a gain, the lowest phase-shift angle is $\delta_{min}$ and the greatest value for $V_{adp}$ is $Vm$.

The dynamic choice of the $\delta_{step}$ value ensures that when the controlled variable's error is high, the response is accelerated by using the larger phase shift. On the other hand, small angles are used to increase precision when the error is small. The advantages of utilizing this adaptive technique include a low computational burden.

The next step is to define the modulation, i.e., the values of $\tau_1$ and $\tau_2$. The control system will verify whether it is possible to operate with triangular modulation at the current operating point of the DAB converter, considering that this modulation promotes six soft switchings of eight commutations in one cycle. The maximum power transferred by triangular modulation occurs when the value of $\tau_1$ reaches $180°$. If it is not possible to transfer the required power to the DAB converter through triangular modulation, trapezoidal modulation is used instead. Trapezoidal modulation can transfer more power than triangular modulation, but in each cycle (eight commutations), there are four soft-switching events. The values of the inner phase shift will be calculated for each value of $\delta^N$, generating the sets $\tau_1^N$ and $\tau_2^N$.

The modulation and inner phase-shift selection algorithm are shown in Figure 8.

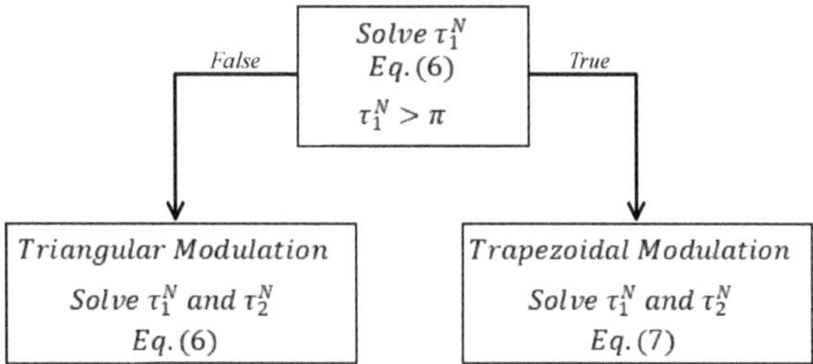

**Figure 8.** Modulation-selection algorithm.

*3.3. Loop Compensation and Cost Function*

Normally, the application of predictive control requires additional techniques to compensate for prediction errors in the controlled variables. One way to reduce these errors is to improve the mathematical model used in the control [22,23]. However, this solution increases the computational burden because it uses high-order models. Another method involves feedback compensation. This method has been used in several works [24–27]. In this work, the error compensation $V^*$ is shown in (15). The difference between the desired and measured values of the DAB converter output voltage is added to the desired value, resulting in a compensated reference value that reduces the steady-state inaccuracy [28].

$$V^* = V_{ref} + \left( V_{ref} - V_{DC2} \right) \tag{15}$$

where $V_{ref}$ is the desired voltage.

Finally, the optimal phase shift and its inner phase shift are defined by the smallest value of the cost function through the comparison between the desired value and the $N = 3$ values predicted by the AMPC. The cost function $G$ is defined in (16).

$$G = \alpha_1 \cdot G_1 + \alpha_2 \cdot G_2 \tag{16}$$

$$\begin{cases} G_1 = \left( V^* - V_{DC2}^N(k+2) \right)^2 \\ G_2 = \left( I_{DC2}^N - I_0 \right)^2 \end{cases} \tag{17}$$

where $\alpha_1$ and $\alpha_2$ are the voltage and current gains, $V^*$ is the compensate reference, $V_{DC2}^N$ is the $N = 3$ possibilities of output voltage value predicted, $I_{DC2}^N$ is the $N = 3$ possibilities of output current value predicted, and $I_0$ is the measure output current.

The first term, $G_1$, is responsible for regulating the voltage. When the output voltage is far from the target value, this term becomes dominant. As the voltage approaches the reference value, the second term, $G_2$, assumes a dominant role, preventing resonance in the output voltage. This oscillation occurs when the control variable ($\delta$, $\tau_1$ and $\tau_2$) deviates from the desired operating point. The adjustment of $\alpha_1$ and $\alpha_2$ is carried out similarly to the adjustment presented in [29] and is defined as $\alpha_1 = 1$ and $\alpha_2 = 2$.

In addition, a penalty is imposed on the cost function if the voltage on the input capacitors drops below 80% of the nominal value. This penalty prevents the control from selecting the $\delta_{old}$ and $\left( \delta_{old} + \delta_{step} \right)$ phase shift presented in (12), ensuring that the converter is able to operate without drastic reduction in the voltage on the input capacitors and preventing the MOSFET's trigger driver protection from activating due to control action.

### 3.4. Flow Chart of AMPC

Figure 9 shows an overview of the AMPC for output voltage, where $\delta_{opt}$ is the optimal phase shift and $\tau_1(\delta_{opt})$ and $\tau_2(\delta_{opt})$ are the inner phase shifts calculated as functions of $\delta_{opt}$. The ANPC-DAB converter switches are triggered by the modulation control block based on the phase shift and inner phase shift delivered by the cost function. The construction of the modulation control block is described in detail in [10].

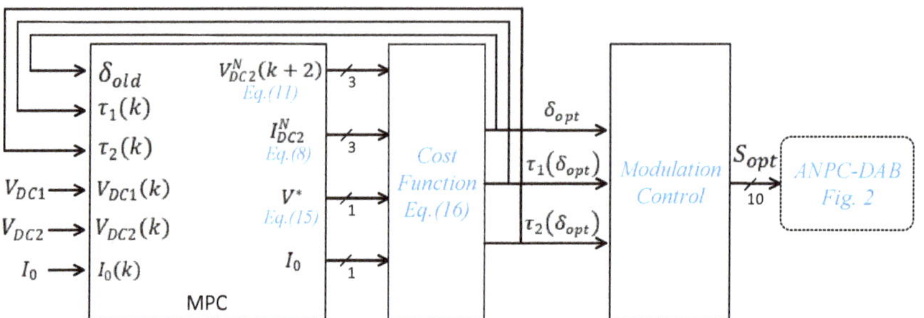


**Figure 9.** AMPC control strategy.

The flow chart of the proposed AMPC is presented in Figure 10, which illustrates the calculation of AMPC from control period $k$ to $(k+1)$.

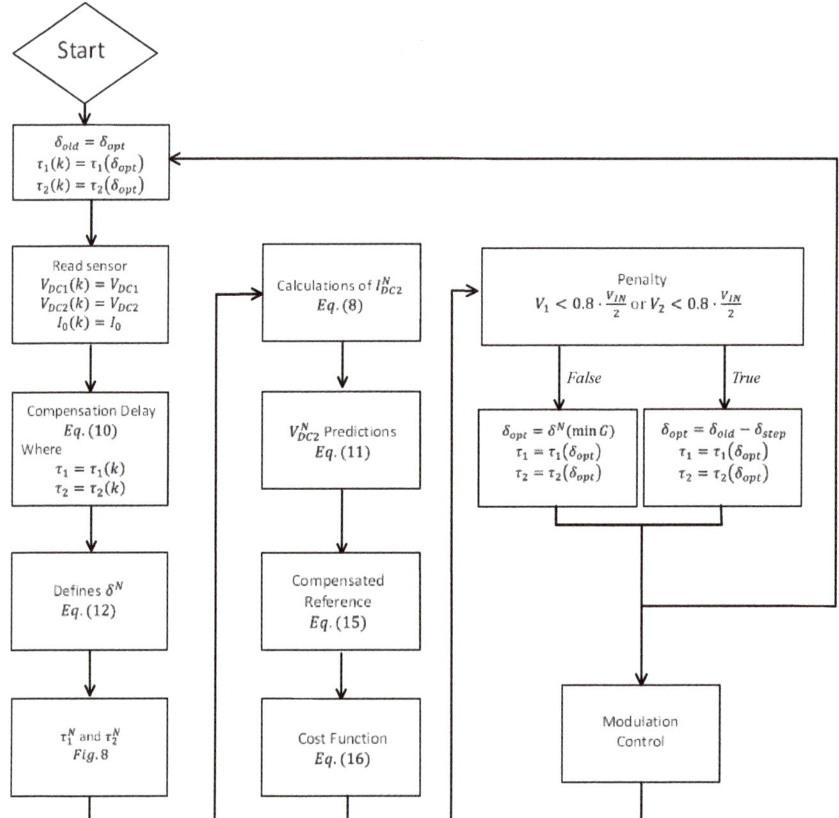


**Figure 10.** Flow chart of AMPC.

The process starts by reading the sensors and saving the optimal phase shift applied at instant $k$. The next step is to apply the compensation delay. Next, the $\delta_{step}$ is calculated and the three phase shifts contained in $\delta^N$ are defined. Once $\delta^N$ is defined, the values of $\tau_1^N$ and $\tau_2^N$ are calculated to define the modulation for each phase shift. The output voltage is then predicted for $N = 3$ possibilities. The next step is to apply the reference compensation and calculate the cost function for $N = 3$ possibilities. If the voltage on the ANPC input capacitors is below 80% of the $V_{in}/2$ value, the value of $\delta_{opt}$ is adjusted to $\delta_{opt} = (\delta_{old} - \delta_{step})$ with their respective $\tau_1(\delta_{opt})$ and $\tau_2(\delta_{opt})$, causing the power transmitted by the DAB to be reduced. Otherwise, $\delta_{opt}$ is adjusted so that the cost function is minimized with their respective $\tau_1(\delta_{opt})$ and $\tau_2(\delta_{opt})$.

## 4. Performance Analysis

This section details a performance analysis of the proposed AMPC control. Real-time simulation results using OPAL OP5700 were used to evaluate the proposed control. The specifications of the ANPC-DAB converter under evaluation are listed in Table 2.

**Table 2.** DAB converter parameters.

| Variable | Symbol | Value |
|---|---|---|
| Switch Frequency | $f_{sw}$ | 20 kHz |
| Voltage Input DAB | $V_{IN}$ | 800 $V$ |
| Voltage Output DAB | $V_{DC2}$ | 400 $V$ |
| Rated Phase Shift | $\Phi$ | 45° |
| Rated Power Transformer | $S$ | 15 kVA |
| HFT Transformer Ratio | $n = N_1/N_2$ | 1.2 |
| Leakage Inductor | $L$ | 32 $\mu$H |
| Input Capacitor | $C_1 = C_2$ | 120 $\mu$F |
| Output Capacitor | $C_3$ | 160 $\mu$F |
| Lowest Phase-Shift Angle | $\delta_{min}$ | 0.05° |
| Gain | $\alpha$ | 1 rad/V |
| Maximum Value for $Vadp$ | $Vm$ | 10 $V$ |

The ANPC-DAB converter may be affected by power-supply noise and load disturbances in real-world scenarios. Therefore, this section examines the load disturbance rejection (LDR) and source voltage disturbance rejection (SVDR), which assess the performance of the AMPC control in real-world conditions.

Another common situation in the real world involves variation in the converter parameters due to temperature. Consequently, this section evaluates the AMPC's response to variations in the inductance and capacitance of the ANPC-DAB converter. Additionally, it examines the significance of item $G_2$ in the cost function.

The real-time simulation data were plotted using a graphical tool for better presentation. The original values were preserved, and no manipulations were carried out.

### 4.1. Source Voltage Disturbance Rejection (SVDR)

Figure 11 illustrates the circuit used in the simulation to assess the SVDR. The input voltage of the ANPC-DAB converter was decomposed into a DC component ($V_{DC}$) and an AC component ($v_{AC}$). The DC component represents the nominal input voltage, while $v_{AC}$ simulates source voltage disturbances. The experiment involved introducing a sinusoidal voltage with a peak value of 80 $V$ in series with a DC source set at 800 $V$.

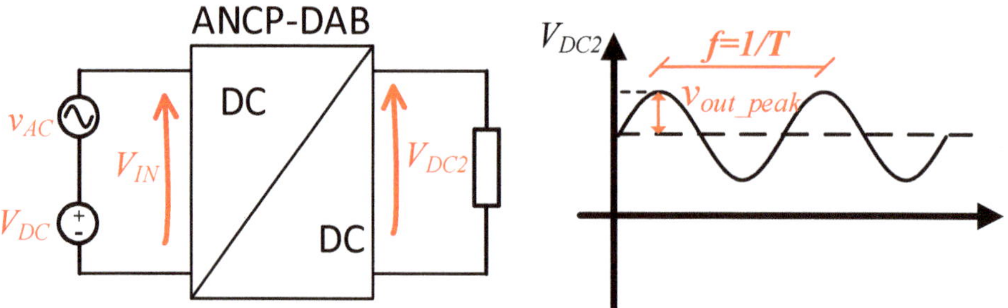



**Figure 11.** Real-time simulation to evaluate SVDR.

SVDR analysis involves Equation (18), where the output voltage $V_{DC2}$ is measured for each injected frequency.

$$G_{svdr} = 20 \cdot \log \frac{v_{out\_peak}(f)}{v_{AC}(f)} \tag{18}$$

where $v_{out\_peak}(f)$ is the ripple value of $V_{DC2}$ at frequency $f$ and $v_{AC}(f) = 80\ V$ for all analyzed frequencies. Figure 12 shows the frequency response for SVDR analysis.

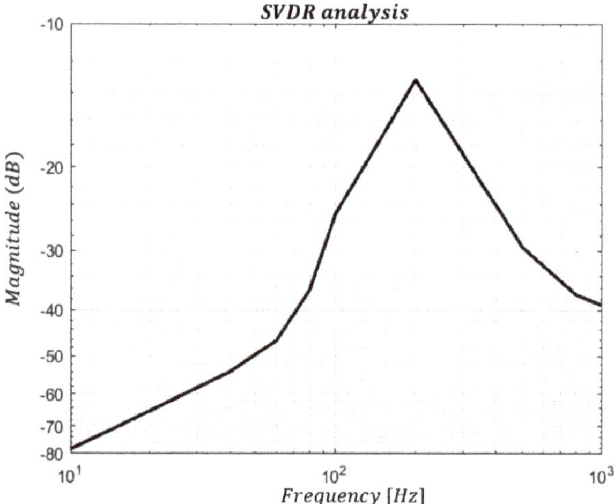



**Figure 12.** AMPC frequency response: SVRD analysis.

The AMPC control demonstrates exceptional performance in rejecting source disturbances. In the low-frequency band, the AMPC has a lower $G_{svdr}$ amplitude, with a value of $-80$ dB at 10 Hz. The most significant degradation occurs at 200 Hz, with a higher $G_{svdr}$ value of $-13$ dB.

### 4.2. Load Disturbance Rejection (LDR)

When the DAB converter is applied in an SST, it can power a variety of switched loads, such as three-phase inverters, electric vehicle charging stations, and other direct current loads. Therefore, it is important to analyze the LDR of the ANPC-DAB converter. The metric used in this study was the output impedance analysis.

To illustrate this point, the converter was represented by a $V_{out}$ source connected in series with the $Z_{out}$ output impedance, as depicted in Figure 13.

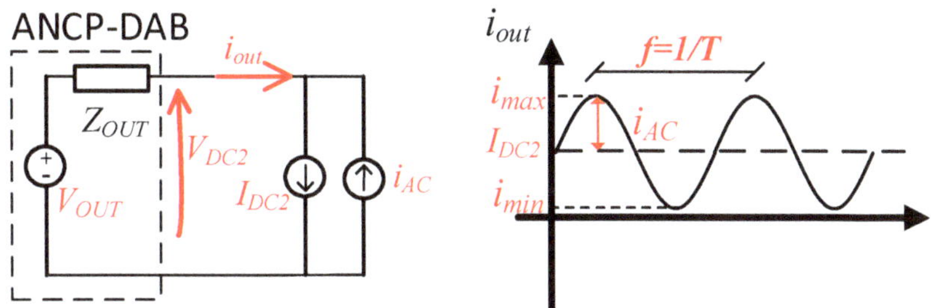



**Figure 13.** Real-time simulation to evaluate LDR.

During the analysis, the $I_{DC2}$ current was set to 8 $A$ and the current $i_{AC}$ was added as a disturbance to simulate the load disturbance. The $i_{AC}$ value was chosen to induce a change of modulation. The output impedance was then calculated using Equation (19).

$$Z_{out} = \frac{v_{out\_peak}(f)}{i_{AC}(f)} \tag{19}$$

where $v_{out\_peak}$ is the ripple value of $V_{DC2}$ at frequency $f$ and $i_{AC}(f) = 7/\sqrt{2}A$ for all analyzed frequencies. Figure 14 shows the frequency response for LDR analysis.

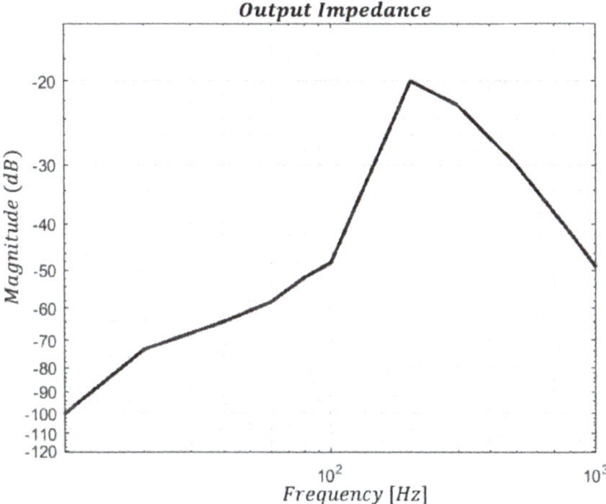



**Figure 14.** AMPC Frequency Response: Real-time simulation results of LDR analysis of the ANPC-DAB converter.

The AMPC control exhibits excellent performance in load disturbance rejection. The output impedance ($Z_{out}$) increases from $-100$ dB to $-20$ dB in the frequency range of 10 Hz to 200 Hz.

### 4.3. Variations of Circuit Parameters

This section evaluates the robustness of the AMPC algorithm against variations in the ANPC-DAB converter parameters. The analysis is particularly significant because the converter parameters can change in real-world scenarios [7]. For this experiment, the voltage $V_{DC2}$ was varied from 380 $V$ to 400 $V$, considering a variation of $\pm50\%$ in the values of $L$ and $C_3$. Figure 15a,b depicts the behavior of the output voltage in response to changes in capacitance $C_3$ and inductance $L$, respectively.

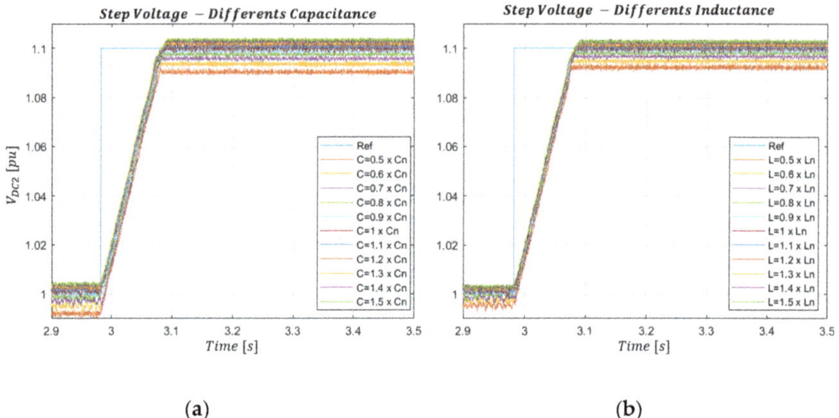



(**a**)          (**b**)

**Figure 15.** Behavior of the output voltage for variation parameters (**a**) inductance, $L$ (**b**) capacitance, $C_3$.

Upon analysis of Figure 15, it is apparent that variations in capacitance and inductance have minimal impacts when the circuit is in a steady state. Table 3 presents the primary quantitative results of the ANPC-DAB converter for all the analyzed scenarios. The mean absolute error (MAE) of the $V_{DC2}$ voltages for each scenario is calculated using Equation (20), where $M$ denotes the total number of collected samples, and $y_k^*$ and $y_k$ represent the reference and measured signals, respectively, at instant $k$.

$$MAE = \frac{\sum_{k=1}^{M} |y_k^* - y_k|}{M} \tag{20}$$

**Table 3.** Quantitative analysis of output voltage.

| Inductor | MAE ($V_{DC2}$) | Capacitor | MAE ($V_{DC2}$) |
|---|---|---|---|
| $L_r = 0.5 \cdot L$ | 6.10 V (1.52%) | $C_r = 0.5 \cdot C_3$ | 6.96 V (1.74%) |
| $L_r = 0.6 \cdot L$ | 5.19 V (1.29%) | $C_r = 0.6 \cdot C_3$ | 5.75 V (1.44%) |
| $L_r = 0.7 \cdot L$ | 4.56 V (1.14%) | $C_r = 0.7 \cdot C_3$ | 4.95 V (1.23%) |
| $L_r = 0.8 \cdot L$ | 4.05 V (1.01%) | $C_r = 0.8 \cdot C_3$ | 4.27 V (1.06%) |
| $L_r = 0.9 \cdot L$ | 3.69 V (0.92%) | $C_r = 0.9 \cdot C_3$ | 3.80 V (0.95%) |
| $L_r = 0.1 \cdot L$ | 3.52 V (0.88%) | $C_r = 0.1 \cdot C_3$ | 3.53 V (0.88%) |
| $L_r = 1.1 \cdot L$ | 3.57 V (0.89%) | $C_r = 1.1 \cdot C_3$ | 3.61 V (0.90%) |
| $L_r = 1.2 \cdot L$ | 3.71 V (0.92%) | $C_r = 1.2 \cdot C_3$ | 3.78 V (0.94%) |
| $L_r = 1.3 \cdot L$ | 3.86 V (0.97%) | $C_r = 1.3 \cdot C_3$ | 3.94 V (0.98%) |
| $L_r = 1.4 \cdot L$ | 3.99 V (0.99%) | $C_r = 1.4 \cdot C_3$ | 4.12 V (1.03%) |
| $L_r = 1.5 \cdot L$ | 4.09 V (1.02%) | $C_r = 1.5 \cdot C_3$ | 4.25 V (1.06%) |

The results show that the control is more robust to inductor variation, yielding an MAE of 1.52%. For capacitor variations, the MAE was 1.74%.

### 4.4. Cost-Function Analysis

As mentioned earlier, the term $G_2$ in the cost function is intended to reduce the resonance at the output voltage $V_{DC2}$. To demonstrate its effectiveness, the $G_2$ term was negated by setting $\alpha_2 = 0$ during a real-time simulation, and the measured resonance frequency in this case was 1 kHz.

Figure 16 shows the impact of the $G_2$ term on the behavior of the output voltage of the DAB converter, and it is clear that steady-state resonance is attenuated. This resonance reduction results in a decrease in acoustic noise emitted by the transformer, as well as a reduction in transformer losses.

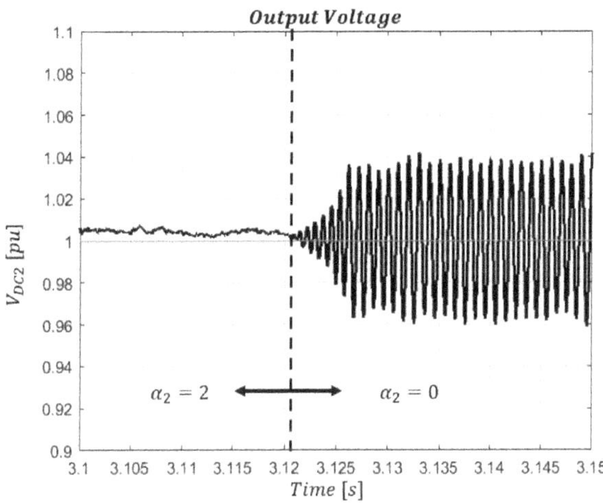



**Figure 16.** Verification of the effectiveness of $G_2$.

## 5. Experimental Results

The AMPC control strategy was validated in a downscaled prototype in an ANPC-DAB converter operating at 8 kW 20 kHz 800 $V$/400 $V$. The experimental setup is shown in Figure 17. The platform consists of the ANPC-DAB converter, control system, load simulator and DC source.

The ANPC-DAB converter employs isolated voltage sensors, LEM DVC 1000-P (two for the ANPC input capacitors and one for the output voltage), and a current sensor, LEM LA 55-P, to monitor the output current. The ANPC and H-bridge were built using CREE CAS120M12BM2 modules, with three modules for ANPC and two modules for the H-bridge. Capacitor M119550731 77 G, 10 µF, 1.3 kV, forms one of the input capacitor banks (two blocks of 12 capacitors—$C_1 = C_2 = 120$ µF) and output capacitor banks (16 capacitors—$C_3 = 160$ µF). The high-frequency transformer (HFT) has a turns ratio of $n = 1.2$ and a power of 15 kVA.

The ANPC-DAB output load was simulated using a Cinergia GE&EL-50, a power electronics device that is capable of simulating both AC and DC electrical networks. In this study, the device was configured to simulate resistive loads. To power the converter, two ePower model SPS 15 kW programmable DC sources were used.

The control system is based on the OPAL OP5700-RCP/HIL Virtex7 FPGA-based Real-Time Simulator, which is equipped with an Intel Xeon E5 processor with 8 cores, 3.2 GHz frequency, and 20 MB cache. The test bench was developed at Laboratory of Electronic Engineering Applied to Renewable Energies, University of Alcalá de Henares, Madrid, Spain. The four analog signals from the DAB sensors are received and processed by OPAL, which generates the pulses to trigger the MOSFETs. The trigger signal is transmitted over fiber optics between OPAL and ANPC-DAB.

This study includes steady-state and transient-behavior analysis, as well as assessment of the computational burden. The steady-state performance of the ANPC-DAB converter was evaluated in terms of output voltage, AC voltages, inductor current, and global losses. Global losses were calculated by measuring the input and output power when the converter operates in steady state.

The transient behavior for $d > 1$ using AMPC control was analyzed observing the converter output voltage. In addition, all experiments were performed with traditional MPC, referred to here as MPC, for comparisons with AMPC.

The goal was to simulate the behavior of the DC-DC stage of an SST in which one side operates at 800 V and the other at 400 V, with the ANPC-DAB converter stabilizing the voltage on the 400 V side.

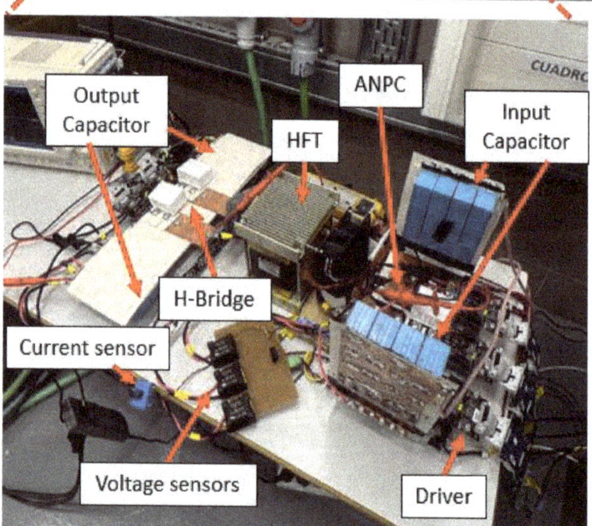



**Figure 17.** Experimental setup.

### 5.1. Computational Burden

The computation time of the proposed AMPC was measured with OPAL resources. The AMPC controller takes 9.5 μs to run, while the time to run MPC was 8.7 μs. The AMPC does not use complex optimization techniques, instead requires almost the same time as the traditional MPC. When a 20 kHz switching frequency is utilized, 50 μs is available in a single sampling period. Thus, there is enough time to implement other functionalities, such as protections, MODBUS communication etc.

### 5.2. Steady-State Analysis

The steady-state analysis aims to compare the overall performance of the converter using AMPC to that using MPC. In the first test, the load was set to 7.36 kW. At this point, the AMPC operates with triangular modulation. Waveforms of $v_{AC1}(t)$, $v_{AC2}(t)$ and $I_L(t)$ using AMPC and MPC are shown in Figures 18a and 18b, respectively.

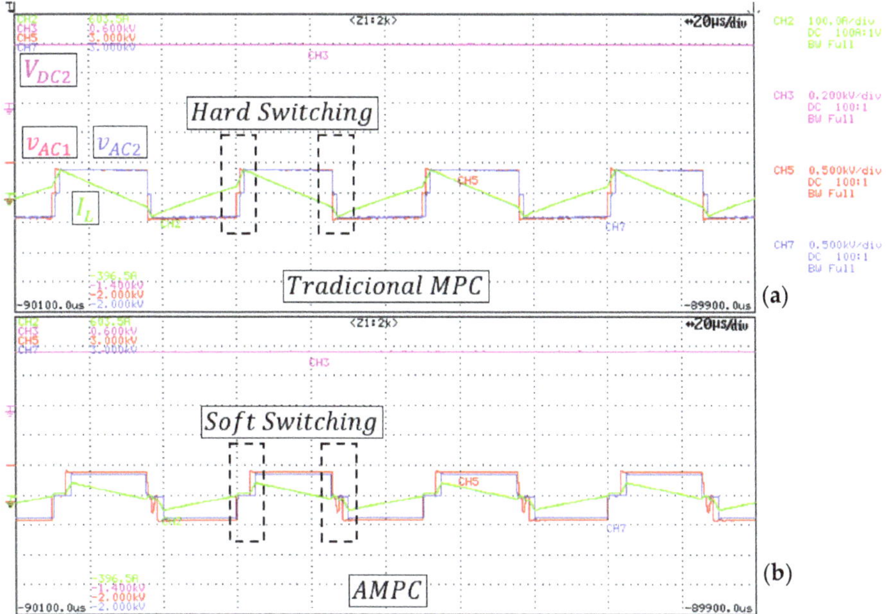



**Figure 18.** Waveforms of $v_{AC1}(t)$, $v_{AC2}(t)$ and $I_L(t)$. (**a**) MPC Control (**b**) AMPC Control.

The second test load was set to 8.4 kW. At this point, the AMPC operates with trapezoidal modulation. Waveforms of $v_{AC1}(t)$, $v_{AC2}(t)$ and $I_L(t)$ using MPC and AMPC are shown in Figures 19a and 19b, respectively.

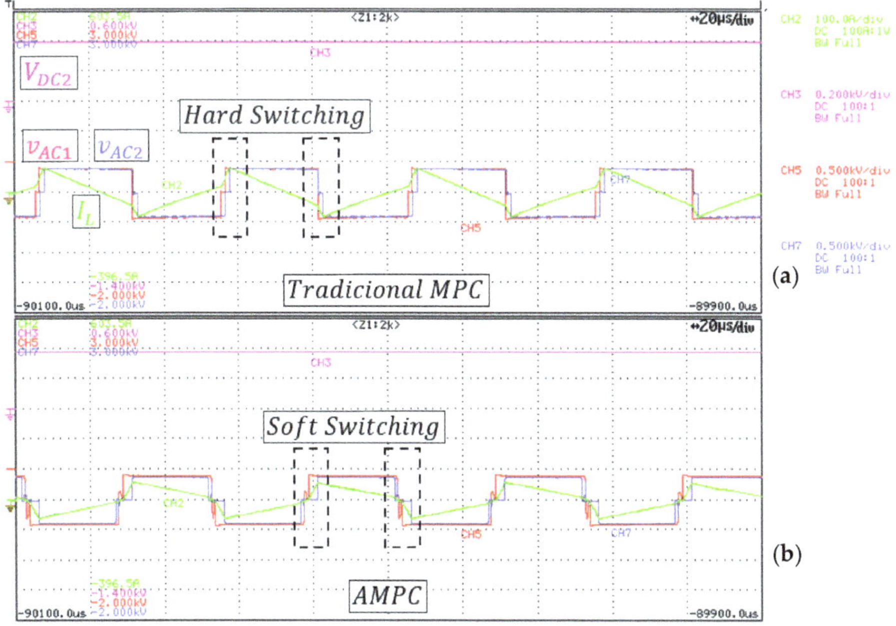



**Figure 19.** Waveforms of $v_{AC1}(t)$, $v_{AC2}(t)$, and $I_L(t)$. (**a**) MPC control (**b**) AMPC control.

Qualitatively, Figures 18b and 19b show that the converter operating with the MPC has hard switching in state changes ($P$, $Z$, and $N$), whereas this change occurs in the presence of current, dissipating more power in the switches. In contrast, the AMPC control causes the converter to operate with soft switching, i.e., six soft switchings out of the eight commutations in one cycle when the control chooses triangular modulation and four soft switchings out of the eight commutations in one cycle when the control chooses trapezoidal modulation.

Table 4 presents values of input voltage, output voltage, input current, and output current for the two operating points mentioned above with MPC and AMPC control.

**Table 4.** Global losses in ANPC-DAB for AMPC and MPC control.

| MPC | AMPC |
| --- | --- |
| $Load = 7.36$ kW | |
| $V_{IN} = 800.8\ V$ | $V_{IN} = 800.6\ V$ |
| $I_{IN} = 9.75\ A$ | $I_{IN} = 9.53\ A$ |
| $V_{DC2} = 396.3\ V$ | $V_{DC2} = 394.5\ V$ |
| $I_{DC2} = 18.23\ A$ | $I_{DC2} = 18.13\ A$ |
| $\eta = 92.53\%$ | $\eta = 93.77\%$ |
| $Load = 8.4$ kW | |
| $V_{IN} = 800.6\ V$ | $V_{IN} = 800.5\ V$ |
| $I_{IN} = 11.02\ A$ | $I_{IN} = 10.83\ A$ |
| $V_{DC2} = 396.2\ V$ | $V_{DC2} = 394.8\ V$ |
| $I_{DC2} = 20.8\ A$ | $I_{DC2} = 20.71\ A$ |
| $\eta = 93.37\%$ | $\eta = 94.33\%$ |

The proposed AMPC control demonstrated accurate response, with less than 1.4% error on the output voltage mean value. This control increased the converter's global efficiency without changing the hardware or increasing computational efforts. In microgrid applications, where the SST operates at low loads for extended periods, high efficiency is essential. The AMPC control strategy can save approximately 600 kWh of energy in one year of operation (8.4 kW, 18 h of low load per day), equivalent to the monthly consumption of three houses with a 200 kWh usage, simply through use of the AMPC control.

*5.3. Transient Analysis*

The transient behavior of the AMPC control was analyzed for load and voltage step on the converter. The MPC control was subjected to the same analyses for comparison purposes.

In the first experiment, the load was changed from 7.36 kW to 8.40 kW to evaluate the response of the AMPC to a modulation change, which is a critical situation. In order to visualize the chosen modulation on the scope within other DAB variables, an analog output was used, where 3 $V$ indicates triangular modulation, 4 $V$ indicates trapezoidal modulation, and 5 $V$ indicates SPS modulation (used in MPC).

Figures 20 and 21 show the behavior of the output voltage for AMPC and the MPC controls, respectively, where the purple, blue, and yellow lines are the $V_{DC2}$ voltage, the $I_{DC2}$ current, and the modulation used by the control, respectively.

The AMPC control can keep the output voltage stable and choose between triangular and trapezoidal modulations according to the operating point, ensuring better converter efficiency. The response time was less than 282 ms with an overshoot of less than 6.86%. Comparing traditional MPC with AMPC, the differences in transient behavior are small. The summarized results of the experiment are shown in Table 5.

In the second experiment, the $V_{ref}$ voltage was changed from 380 $V$ to 400 $V$. Figures 22 and 23 show the behavior of the AMPC and the MPC controls, respectively, where the purple, blue, yellow, and red lines are the $V_{DC2}$ voltage, the $I_{DC2}$ current, the modulation used by the control, and the desired value for the output voltage $V_{ref}$, respectively.

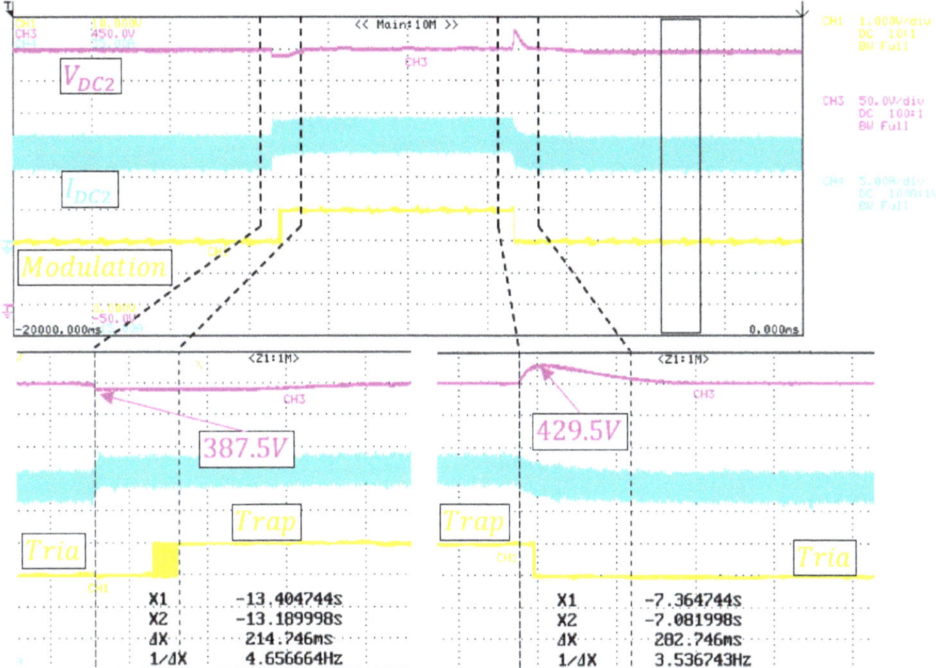

Figure 20. AMPC performance under a load step.

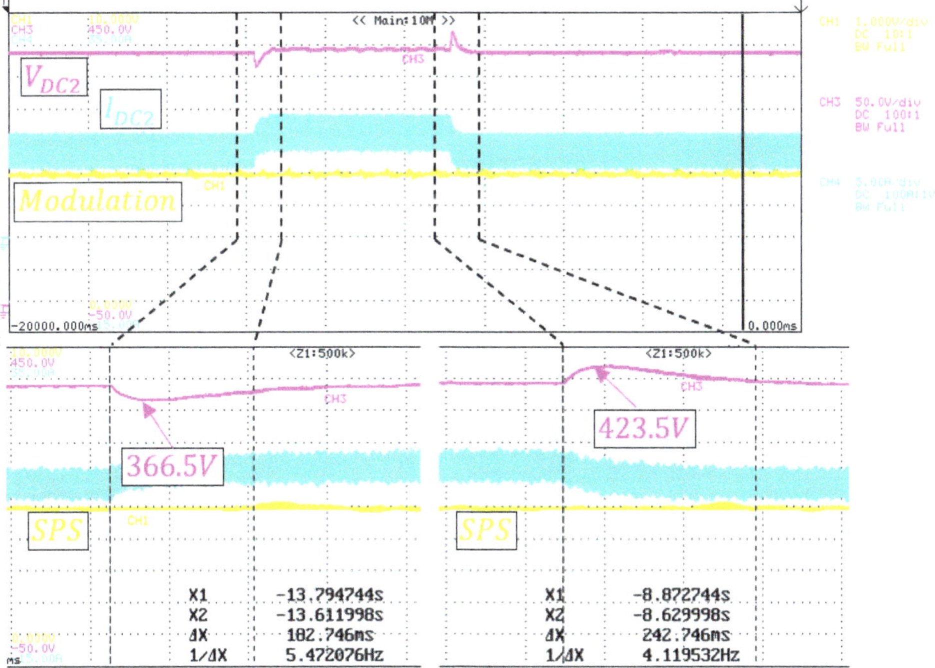

Figure 21. MPC performance under a load step.

**Table 5.** Load-step results: comparison of AMPC × MPC.

| Transition | AMPC Settling Time | MPC Settling Time | AMPC Overshoot | MPC Overshoot |
|---|---|---|---|---|
| 7.36 kW $\rightarrow$ 8.40 kW | 214 ms | 182 ms | $\frac{(400-387.5)}{400} = 3.1\%$ | $\frac{(400-366.5)}{400} = 8.4\%$ |
| 8.40 kW $\rightarrow$ 7.36 kW | 282 ms | 242 ms | $\frac{(400-429.5)}{400} = -7.4\%$ | $\frac{(400-423.5)}{400} = -5.9\%$ |

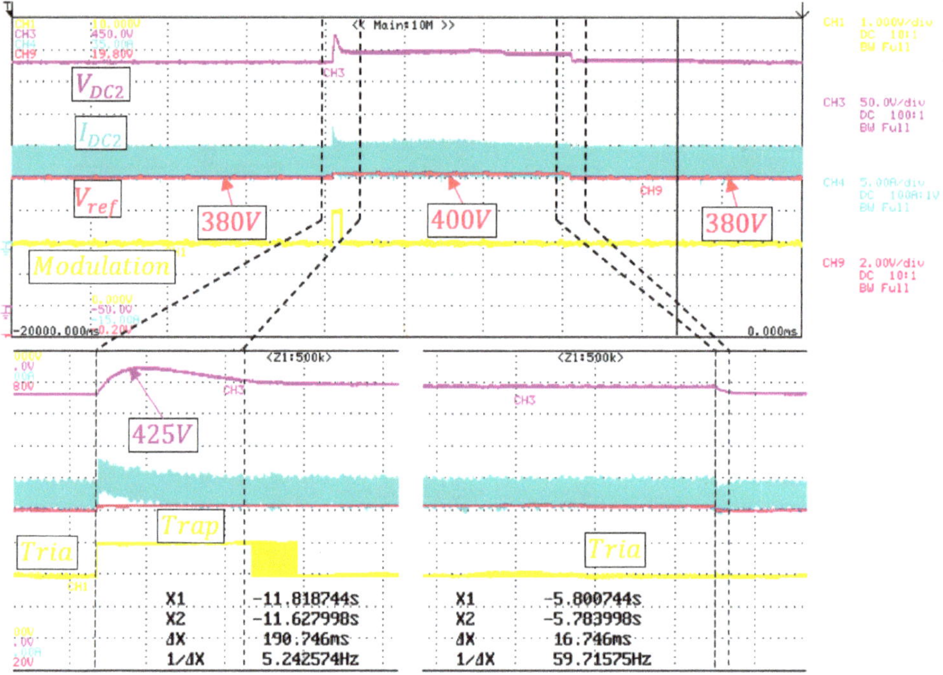


**Figure 22.** AMPC performance under a $V_{ref}$ step.

In this scenario, the AMPC strategy dynamically changes the phase shift and inner phase shift in order to charge the output capacitor so that the $V_{DC2}$ voltage reach the new requested reference value.

AMPC and MPC showed similar performance. However, in the transition $V_{ref} = 400V \rightarrow 380V$, there was no undervoltage and the settling time was much shorter than in the MPC. The summarized results of the experiment are shown in Table 6.

**Table 6.** Voltage step results: comparison of AMPC and MPC.

| Transition | AMPC Settling Time | MPC Settling Time | AMPC Overshoot | MPC Overshoot |
|---|---|---|---|---|
| 380 V $\rightarrow$ 400 V | 190 ms | 178 ms | $\frac{(400-425)}{400} = -6.3\%$ | $\frac{(400-440)}{400} = -10\%$ |
| 400 V $\rightarrow$ 380 V | 16 ms | 358 ms | $\frac{(380-380)}{380} = 0.00\%$ | $\frac{(380-346)}{380} = 8.95\%$ |

Unlike a simulated environment, the converter's transient power transfer is limited, which delays the response time in relation to other topologies. The insertion of the penalty in the cost function prevents the ANPC input capacitor from discharging to levels that cause the MOSFET's protection trigger drivers to activate.

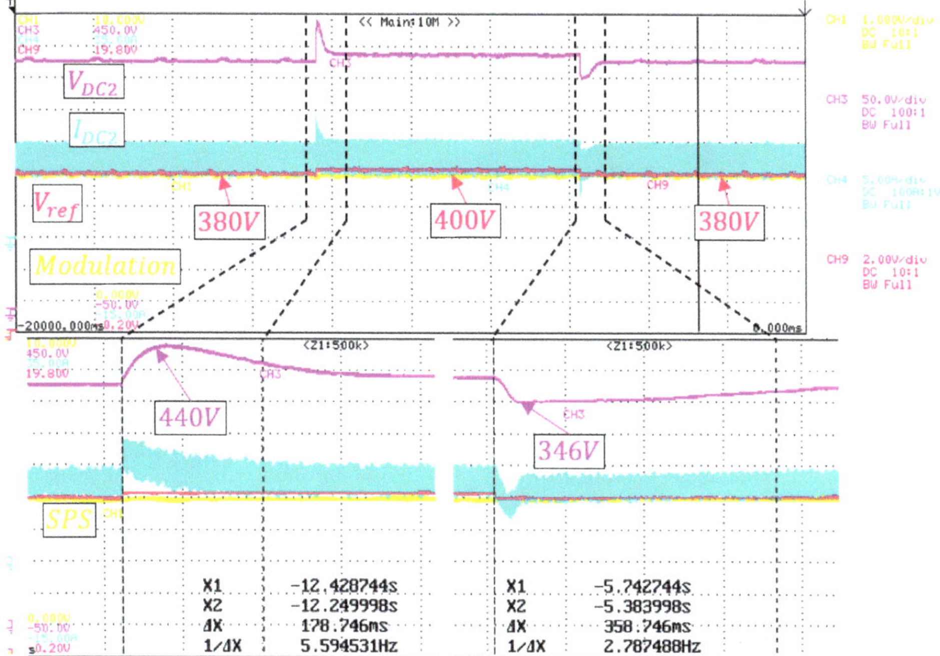



**Figure 23.** MPC performance under a $V_{ref}$ step.

## 6. Discussion

The DAB converter has many advantages for use in SSTs, including galvanic isolation, bidirectional power flow, high power density, and the ability to operate as a buck or boost with ZVS over a wide range. However, low loads and high voltages can pose challenges to its efficiency and operation capacity. To address this difficulty, multilevel topologies such as the ANPC-DAB can be used. In this topology, ANPC is applied on the high-voltage side and an H-bridge is used on the low-voltage side. Experiments were conducted on an 8 kW 20 kHz 800 $V$/400 $V$ prototype.

The traditional control method used in the DAB converter controls only the phase shift, limiting its efficiency at low loads. To improve efficiency at low loads, different modulations can be used to enable soft switching throughout the operating range. However, these solutions require the selection of voltage values for the converter's active bridges, which involves complex optimization methods that can increase computational costs and hinder practical application.

MPC is an advanced technique used in many electrical systems. Unlike conventional control, it predicts the system's future behavior using mathematical modeling and calculates an optimal control signal to maintain the desired output. To enhance efficiency at low loads, AMPC control, which employs triangular and trapezoidal modulations, is used. The natural transition between these modulations results in a gradual change in the inductor's current waveform from triangular to trapezoidal or vice versa. This method is used to obtain the highest efficiency when the DAB converter operates at low loads.

Real-time simulation was used to evaluate the frequency response and robustness of the AMPC applied to the ANPC-DAB topology. The converter attenuates voltage-source noise by −80 dB at low frequencies, with the worst case at 200 Hz, at which it attenuates the disturbance by −13 dB. Additionally, AMPC offers low output impedance, allowing for high-intensity output current without significant voltage drops, which is essential for maintaining output current stability in applications with load variations.

Robustness tests were conducted by varying the power transfer inductance and output capacitance. There are several approaches to designing predictive controllers that are robust to converter parameter variations. In this work, the reference is adjusted to ensure lower error when the converter operates in steady state. Mean absolute error (MAE) values of 1.54% and 1.74% were obtained for a 50% variation in the nominal values for the inductor and capacitor, respectively.

The impact of the $G_2$ term in the cost function on the output voltage was demonstrated. When $\alpha_2 = 0$, a resonance occurs in the output voltage, even in steady state. Resonance can cause system instability, low performance, electrical noise, and electromagnetic interference, which can affect nearby electronic systems. Therefore, the importance of the $G_2$ term in reducing oscillations in the system is evident.

Experimental results from the prototype indicate a significant improvement in converter efficiency with the AMPC control, achieving 94.33% at 8.4 kW and 93.77% at 7.36 kW in steady state, compared to 93.37% and 92.53%, respectively, with the MPC. While the dynamic response of the AMPC control did not differ significantly from that of the MPC, the MPC was faster and the AMPC showed smaller overshoots in most of the tests.

Finally, an important advantage of AMPC is that it eliminates the need for complex optimization processes to choose the phase shift and inner phase shift, while also avoiding the need for intense computational efforts. This study shows that the processing time difference between MPC and AMPC is only 0.7 μs.

For future work, the control methodology discussed in this study can be applied to input series output parallel (ISOP) structures to verify the energy efficiency of the structure and evaluate the control behavior for transient situations and steady-state operation.

## 7. Conclusions

The present article proposes the application of the advanced AMPC control technique in the ANPC-DAB topology. This topology enables the converter to operate at high voltages, providing an alternative to ISOP structures, which operate in series and parallel.

Several real-time simulated analyses have shown that the AMPC control has high capacity for voltage-source noise rejection and exhibits low output impedance. Additionally, dynamic reference adjustment has improved the control's robustness to variations in converter parameters.

The experimental results from the prototype showed excellent dynamic and steady-state performance. The AMPC exhibited dynamic response similar to that of the MPC. However, in steady state, the efficiency of the converter was significantly improved with use of the AMPC, reaching 94.33% at 8.4 kW and 93.77% at 7.36 kW in steady state, compared to 93.37% and 92.53%, respectively, without the AMPC. Additionally, the control achieved error rates below 1.4% in relation to the desired value.

Despite the benefits of using AMPC, the computational cost was not significantly increased compared to MPC. The processing time was measured using OPAL resources, which showed that the MPC processing time was 8.7 μs, while that of the AMPC was 9.5 μs. Considering that the processor has 50 μs available ($T = 1/20$ kHz), applying AMPC leaves enough time for other applications, such as protections and data communication.

**Author Contributions:** Conceptualization, A.N. and R.F.; Formal analysis, A.N. and R.F.; Investigation, D.M.; Methodology, A.N.; Software, D.M. and E.B.; Supervision, E.B., L.E. and W.S.; Validation, A.N. and D.M.; Writing—original draft, A.N. and A.A.; Writing—review & editing, A.A., L.E. and W.S. All authors have read and agreed to the published version of the manuscript.

**Funding:** This research was funded by Espírito Santo Research and Innovation Support Foundation—FAPES (grant number 514/2021, 382/2021, 1024/2022, 1068/2022 and 336/2023), National Council for Scientific and Technological Development—CNPq (grant number 311848/2021-4) and Ministerio de Ciencia e Innovación (grant numbers TED2021-130610B-C21 and PID2021-125628OB-C22).

**Data Availability Statement:** Data are contained within the article.

**Acknowledgments:** Thanks to the Alcalá University and Federal University of Espírito Santo for providing equipment, physical structure, support, and assistance from professors; and to the Federal Institute of Espírito Santo for allowing me to use my working hours to conduct research with other educational institutions.

**Conflicts of Interest:** The authors declare no conflict of interest.

# References


1. Ruiz, F.; Perez, M.A.; Espinosa, J.R.; Gajowik, T.; Stynski, S.; Malinowski, M. Surveying Solid-State Transformer Structures and Controls: Providing Highly Efficient and Controllable Power Flow in Distribution Grids. *IEEE Ind. Electron. Mag.* **2020**, *14*, 56–70. [CrossRef]
2. She, X.; Huang, A.Q.; Burgos, R. Review of Solid-State Transformer Technologies and Their Application in Power Distribution Systems. *IEEE J. Emerg. Sel. Top. Power Electron.* **2013**, *1*, 186–198. [CrossRef]
3. Huber, J.E.; Kolar, J.W. Applicability of Solid-State Transformers in Today's and Future Distribution Grids. *IEEE Trans. Smart Grid* **2017**, *10*, 317–326. [CrossRef]
4. Gu, Q.; Yuan, L.; Yi, S.; Nie, J.; Zhao, Z. Active Selection of Current Commutation Loop for Hybrid Three-Level Dual Active Bridge DC-DC Converter with TPS Control. In Proceedings of the 2019 IEEE 10th International Symposium on Power Electronics for Distributed Generation Systems (PEDG), Xi'an, China, 3–6 June 2019; IEEE: Piscataway, NJ, USA, 2019; pp. 155–161.
5. Nabae, A.; Takahashi, I.; Akagi, H. A new neutral-point-clamped PWM inverter. *IEEE Trans. Ind. Appl.* **1981**, *5*, 518–523. [CrossRef]
6. Bruckner, T.; Bernet, S.; Guldner, H. The active NPC converter and its loss-balancing control. *IEEE Trans. Ind. Electron.* **2005**, *52*, 855–868. [CrossRef]
7. Bruckner, T.; Bemet, S. Loss Balancing in Three-Level Voltage Source Inverters Applying Active NPC Switches. In Proceedings of the 2001 IEEE 32nd Annual Power Electronics Specialists Conference (IEEE Cat. No. 01CH37230), Vancouver, BC, Canada, 17–21 June 2001; IEEE: Piscataway, NJ, USA, 2021; Volume 2.
8. De Doncker, R.W.; Divan, D.M.; Kheraluwala, M.H. A three-phase soft-switched high-power-density DC/DC converter for high-power applications. *IEEE Trans. Ind. Appl.* **1991**, *27*, 63–73. [CrossRef]
9. Schibli, N. Symmetrical Multilevel Converters with Two Quadrant DC-DC Feeding. Thesis, École Polytechnique Fédérale de Lausanne, Lausanne, Switzerland, 2000.
10. Nardoto, A.; Amorim, A.; Santana, N.; Bueno, E.; Encarnação, L.; Santos, W. Adaptive Model Predictive Control for DAB Converter Switching Losses Reduction. *Energies* **2022**, *15*, 6628. [CrossRef]
11. Hu, J.; Cui, S.; De Doncker, R.W. Natural boundary transition and inherent dynamic control of a hybrid-mode-modulated dual-active-bridge converter. *IEEE Trans. Power Electron.* **2021**, *37*, 3865–3877. [CrossRef]
12. Rauf, U.; Schütt, M.; Eckel, H.G. Model Predictive Control for Space Vector Modulation of a Three-Level ANPC Inverter for Efficient Loss Distribution and Neutral Point Balancing. In Proceedings of the 2019 21st European Conference on Power Electronics and Applications (EPE'19 ECCE Europe), Genova, Italy, 3–5 September 2019; IEEE: Piscataway, NJ, USA, 2019.
13. Deng, Y.; Li, J.; Shin, K.H.; Viitanen, T.; Saeedifard, M.; Harley, R.G. Improved modulation scheme for loss balancing of three-level active NPC converters. *IEEE Trans. Power Electron.* **2016**, *32*, 2521–2532. [CrossRef]
14. Guan, Q.X.; Zhang, Y.; Zhao, H.B.; Kang, Y. Optimized Switching Strategy for ANPC–DAB Converter Through Multiple Zero States. *IEEE Trans. Power Electron.* **2021**, *37*, 2885–2898. [CrossRef]
15. Kumar, S.; Akin, B.; Gohil, G. EMI Performance of Active Neutral Point Clamped Phase Leg for Dual Active Bridge DC–DC Converter. *IEEE Trans. Ind. Appl.* **2021**, *57*, 6093–6104. [CrossRef]
16. dos Santos, W.M.; Martins, D.C. Introdução ao conversor DAB monofásico. *Eletrônica Potência* **2014**, *19*, 36–46.
17. Shi, H.; Wen, H.; Chen, J.; Hu, Y.; Jiang, L.; Chen, G.; Ma, J. Minimum-backflow-power scheme of DAB-based solid-state transformer with extended-phase-shift control. *IEEE Trans. Ind. Appl.* **2018**, *54*, 3483–3496. [CrossRef]
18. Shao, S.; Jiang, M.; Ye, W.; Li, Y.; Zhang, J.; Sheng, K. Optimal phase-shift control to minimize reactive power for a dual active bridge DC–DC converter. *IEEE Trans. Power Electron.* **2019**, *34*, 10193–10205. [CrossRef]
19. Gu, Q.; Yuan, L.; Nie, J.; Sun, J.; Zhao, Z. Current stress minimization of dual-active-bridge DC–DC converter within the whole operating range. *IEEE J. Emerg. Sel. Top. Power Electron.* **2018**, *7*, 129–142. [CrossRef]
20. Kheraluwala, M.N.; Gascoigne, R.W.; Divan, D.M.; Baumann, E.D. Performance characterization of a high-power dual active bridge DC-to-DC converter. *IEEE Trans. Ind. Appl.* **1992**, *28*, 1294–1301. [CrossRef]
21. Chen, L.; Shao, S.; Xiao, Q.; Tarisciotti, L.; Wheeler, P.W.; Dragičević, T. Model predictive control for dual-active-bridge converters supplying pulsed power loads in naval DC micro-grids. *IEEE Trans. Power Electron.* **2019**, *35*, 1957–1966. [CrossRef]
22. Eremeeva, A.M.; Ilyushin, Y.V. Automation of the control system for drying grain crops of the technological process for obtaining biodiesel fuels. *Sci. Rep.* **2023**, *13*, 14956. [CrossRef]
23. Ilyushin, Y.V. Development of a process control system for the production of high-paraffin oil. *Energies* **2022**, *15*, 6462. [CrossRef]
24. Dutta, S.; Hazra, S.; Bhattacharya, S. A digital predictive current-mode controller for a single-phase high-frequency transformer-isolated dual-active bridge DC-to-DC converter. *IEEE Trans. Power Electron.* **2016**, *63*, 5943–5952. [CrossRef]

25. Shen, K.; Feng, J.; Zhang, J. Finite control set model predictive control with feedback correction for power converters. *CES Trans. Electr. Mach. Syst.* **2018**, *2*, 312–319. [CrossRef]
26. Aguilera, R.P.; Lezana, P.; Quevedo, D.E. Finite-control-set model predictive control with improved steady-state performance. *IEEE Trans. Ind. Inform.* **2012**, *9*, 658–667. [CrossRef]
27. Du, G.; Li, J.; Liu, Z. The Improved Model Predictive Control Based on Novel Error Correction between Reference and Predicted Current. In Proceedings of the 2018 IEEE Applied Power Electronics Conference and Exposition (APEC), San Antonio, TX, USA, 4–8 March 2018; IEEE: Piscataway, NJ, USA, 2018.
28. Quevedo, D.E.; Aguilera, R.P.; Pérez, M.A.; Cortés, P. Finite Control Set MPC of an AFE Rectifier with Dynamic References. In Proceedings of the 2010 IEEE International Conference on Industrial Technology, Via del Mar, Chile, 14–17 March 2010; IEEE: Piscataway, NJ, USA, 2010.
29. Dragicevic, T.; Novak, M. Weighting Factor Design in Model Predictive Control of Power Electronic Converters: An Artificial Neural Network Approach. *IEEE Trans. Ind. Electron.* **2018**, *66*, 8870–8880. [CrossRef]

*Review*

# A Comprehensive Review on Voltage Stability in Wind-Integrated Power Systems

Farhan Hameed Malik [1], Muhammad Waseem Khan [2], Tauheed Ur Rahman [2], Muhammad Ehtisham [2], Muhammad Faheem [2], Zunaib Maqsood Haider [3,*] and Matti Lehtonen [4,*]

[1]  Department of Electromechanical Engineering, Abu Dhabi Polytechnic, Abu Dhabi 13232, United Arab Emirates; farhan.malik@adpoly.ac.ae
[2]  Department of Electrical Engineering, Bahauddin Zakariya University, Multan 60800, Pakistan
[3]  Department of Electrical Engineering, The Islamia University of Bahawalpur, Bahawalpur 63100, Pakistan
[4]  Department of Electrical Engineering and Automation, Aalto University, 2150 Espoo, Finland
*  Correspondence: zunaib.haider@iub.edu.pk (Z.M.H.); matti.lehtonen@aalto.fi (M.L.)

**Citation:** Malik, F.H.; Khan, M.W.; Rahman, T.U.; Ehtisham, M.; Faheem, M.; Haider, Z.M.; Lehtonen, M. A Comprehensive Review on Voltage Stability in Wind-Integrated Power Systems. *Energies* **2024**, *17*, 644. https://doi.org/10.3390/en17030644

Academic Editors: Tek Tjing Lie, Saeed Sepasi and Quynh Thi Tu Tran

Received: 9 August 2023
Revised: 3 January 2024
Accepted: 24 January 2024
Published: 29 January 2024

**Abstract:** The fast growth of the world's energy demand in the modernized world has stirred many countries around the globe to focus on power generation by abundantly available renewable energy resources. Among them, wind energy has attained significant attention owing to its environment-friendly nature along with other fabulous advantages. However, wind-integrated power systems experience numerous voltage instability complexities due to the sporadic nature of wind. This paper comprehensively reviews the problems of voltage instability in wind-integrated power systems, its causes, consequences, improvement techniques, and implication of grid codes to keep the operation of the network secure. Thorough understanding of the underlying issues related to voltage instability is necessary for the development of effective mitigation techniques in order to facilitate wind integration into power systems. Therefore, this review delves into the origin and consequences of voltage instability, emphasizing its adverse impacts on the performance and reliability of power systems. Moreover, it sheds light on the challenges of integrating wind energy with existing grids. This manuscript provides a comprehensive overview of the essential features required for critical analysis through a detailed examination of Voltage Stability Indices (VSIs). To address voltage stability issues in wind-integrated power systems, this review examines diverse techniques proposed by researchers, encompassing the tools utilized for assessment and mitigation. Therefore, in the field of power system operation and renewable energy integration, this manuscript serves as a valuable resource for researchers by comprehensively addressing the complexities and challenges associated with voltage instability in wind-integrated power systems.

**Keywords:** voltage stability; wind farm integration; wind penetration level; low voltage ride-through (LVRT); high voltage ride-through (HVRT); FACTS; voltage stability limit; voltage stability indices; weakest bus; weak grid

## 1. Introduction

The never-ending potential of renewable energy resources has made them a viable solution to get rid of costly fossil fuel-dependent energy generation portfolio and to cope with the ever-increasing demand of energy due to fast population growth and intense industrialization. Economic as well as eco-friendly attributes have nominated wind energy as an excellent resource over various other renewable energy resources. Government policies of many countries around the globe seem to be more inclined towards switching their power generation pathways to wind energy [1]. Power electronics have revamped the wind generation technology and altered the generation paradigm utterly to make it as one of the most attractive energy origins [2]. The heavily loaded systems and the increasing penetration of wind farms in the conventional power system have given rise to the complexity in power systems. Because of this, power systems are operating at the verge of the stability limits that impose severe threats on the reliability of the system [3]. Many countries have published grid codes that are significant drivers for wind farm integration to ensure system stability, power quality, protection of equipment, and demand for interconnections and generation to the grid [4].

Multifarious factors like grid codes, Low Voltage Ride-Through (LVRT), High Voltage Ride-Through (HVRT), Doubly Fed Induction Generator (DFIG) role, and permissible penetration level of wind power need to be analyzed before proper wind power integration to avoid a voltage instability aftermath. Thus, multiple solutions are presented by different researchers to improve the power system stability that is mostly related to the voltage stability of wind turbines' generation and distribution systems. Many publications have proposed different methodologies to improve voltage stability by using DFIG, but most of them did not fully explore the crowbar circuit, which is a protecting circuit against high voltages during power supply malfunction or power surge operation [5]. In [6], the crowbar operation and its role during voltage dips are discussed; however, some of the disadvantages of crowbar like increasing mechanical stress, power oscillations, and consumption of more reactive power from the grid are not highlighted.

Reference [7] addressed a bunch of plus points about DFIG-based wind turbine systems, and the structure of DFIG with the interconnection between Rotor Side Converter (RSC) and Grid Side Converter (GSC) in terms of reactive power support was also discussed; however, they did not talk about the reactive power limitation by these generators. In the last few decades, most of the research work have highlighted and presented various methods and techniques to improve the performance of wind-integrated power systems, but some of its problems need more research. For example, extensive research has been performed on LVRT. However, the researchers did not fully illustrate the emerging HVRT behavior of Doubly Fed Induction Generators (DFIGs) [8–10].

High-Wind Speed Shut Down is also an essential consideration in wind farm integration, which few researchers have taken into account [11]. For voltage stability improvement, most of the researchers have just focused on FACTS devices [12,13], even though other tools like D-SMES (Distributed Superconducting Magnetic Energy Storage) for voltage stability improvement exist, as has been discussed by [14]. Monitoring and measurement of the power system are generally performed by SCADA, which is a sophisticated supervisory system, but real-time tracking with any system changes makes the WAMS technology a remarkable choice. There is vast research present on the SCADA system, while researchers should focus on the WAMS technology [15–17].

In wind-integrated power systems, one of the major reasons for voltage instability is the reduction in system inertia due to the reliance on energy conversion from wind, unlike the rotational inertia of the conventional synchronous generators. Therefore, during faults, the power grid is more susceptible to voltage and frequency fluctuations. Furthermore, reactive power deficit and weak grid connections are also major concerns to the maintenance of voltage stability. Wind turbines might not be able to provide sufficient reactive power support owing to the technology employed and the limited capacity of the grid to transmit power, leading to voltage instability. In addition, the intermittent nature of wind power and the limited fault response also contribute to voltage and system instability. The reduced contribution of wind turbines in fault current compared to conventional generators makes it difficult to clear faults, potentially causing system instability and cascading effects. To address these challenges, researchers have proposed different approaches to enhance voltage and system stability, which includes upgradation of transmission and distribution infrastructure and deployment of FACTS devices. Moreover, some studies have suggested boosting the grid resilience with energy storage technology and demand side management techniques. However, stress on the utility grid can be further relived by integrating advanced forecasting tools, efficient monitoring systems, and hybrid power systems.

This paper is compiled in two sections (2 and 3). Section 2 highlights issues related to voltage stability of a power system and the techniques for their prevention. Section 3 covers the wind farms' evolution and the integration of wind farms with conventional grids. Lastly, penetration levels and the optimal location for wind farm integration are also studied.

## 2. Overview of Voltage Stability

Voltage stability has become the most vital concern for today's complex electrical networks. It refers to the ability of a power system to maintain constant voltages at all buses in the system after being subjected to a disturbance from a given initial operating condition [18]. Multifarious causes are at the back-end that lay the foundation for voltage instability. Some of the top-listed reasons include intensive use of capacitor banks for reactive power compensation, voltage-insensitive loads, low system voltages, and over-stressing the system with heavy loads [19,20]. M Johnson in [21] prescribed the slow voltage instability phenomenon, and through mathematical computation, he manifested that there is a fundamental relationship between system voltage (V) and reactive power (Q) and thereby confirmed that limitations of the reactive power transfer across the transmission lines are the prime cause of voltage instability that in turn gives rise to voltage collapse.

The load characteristic is one of the significant factors responsible for dynamic voltage instability [22]. P W Sauer et al. explained the relationship between loading characteristics and voltage stability [23]. Bradley R. Williams et al. in [24] presented that stalled air conditioner compressors delay the voltage recovery following the fault. High Voltage Direct Current (HVDC) link, induction motor load, and air conditioner loads affect the short-term voltage stability, while long-term voltage stability is affected by thermostatic loads, distributed voltage regulation, and tap changing of transformers [25,26]. Voltage stability may also stem from rotor angle or frequency instability. In [27], Thierry Van Cutsem illustrated two aspects of voltage instability. Firstly, voltage stability is distressed by the tripping of generation or transmission equipment and secondly by a maximum loading power attempt beyond its limits. The greatest threat to a power system is "blackout" that is the complete outage of power. Deplorable voltage profile of any power system causes blackouts that typically arise due to voltage collapse [28,29]. In case if any line trips, the rest of the lines go on carrying the power, resulting in more reactive power consumption, and thus, the voltage profile decays further [30]. Weak grids, climatic effects, abrupt load changes, heavily loaded systems, and human errors are the factors because of which a system encounters blackout [31–33]. Because of this, people suffer from communication and traffic interruptions, and the countries have to face a considerable decline in the economy and stock markets [32].

A deficiency of reactive power results in blackouts each year all around the world. A glance over the globe in terms of significant outages over the past few decades highlights the Tokyo blackout in 1987 [34], the S/SE Brazilian system blackout in 1997 [35], the US and Canadian blackout in 2003, the Athen's blackout in 2004, the Pakistan blackout in 2006, which are among the significant blackouts. According to engineers and operators, the root of these blackouts was voltage instability that originated because of inadequate reactive power supply between generation and load points [30,36]. While discussing voltage stability of a power system, voltage stability limit is considered to be a worthy mentioning aspect that renders an insight about the range of load increment on a particular bus of a power system before it undergoes voltage instability. Load increment necessitates the determination of the voltage stability limit to restrict a power system operation within the permissible limits and stability margin [37–39]. Many researchers have shown their interest in developing different methods to estimate the voltage stability limit in an efficient, precise, and easy way. Some of the commonly utilized methods have been addressed here.

- PV and QV curves are commonly used to find out the voltage stability limit, which is not an effective method because one has to generate a family of the curves using simulation tools to determine the ceiling [37,38,40].
- Voltage stability index can be found using the static voltage stability index in terms of the minimum singular value of the power flow Jacobian matrix [38,41,42].

- Conventional load flow solution method and non-linear optimization technique can also be utilized to find the voltage stability limit [43–45].

When a power system is equipped with Static Var Compensator (SVC), the voltage stability limit of the system can be efficiently determined. M.H. Haque has verified it through tests on a two-bus system and IEEE 14-bus system [37]. An Energy Storage System (ESS) application is addressed in [46] for voltage stability enhancement, and the smallest eigenvalue is exploited to evaluate the lower voltage stability limit. With a view on the voltage stability limit, Chaisit Wannoi et al. suggested a repeated power flow-based technique for better installing positions of renewable power generation units that are integrated into a power system [47]. To cope with the need of maximizing the voltage stability limit, M. Tripathy et al. exploited the Bacteria Foraging Algorithm (BFA) in [39,48] for the proper allocation of transformer taps and the injection of reactive power at specific buses. Furthermore, the authors proclaimed the superiority of the BFA through MATLAB simulation based on the Interior-Point Successive Linearization Program (IPSLP) technique. Simulation-based comparison between the Real-Coded Genetic Algorithm (RCGA) and the Firefly Algorithm (FA) testifies the better efficiency of the latter to uplift the voltage stability limit [49]. The load flow equations concerning the minimum modulus eigenvalue have convergent behavior; therefore, a new approach is discussed in [50] for determining the static voltage limit and identifying the weak bus. The On-load Tap Changer (OLTC) having dynamic characteristics influences the system voltage stability, and therefore, it describes the static voltage limit and the maximum transmission power by adjusting its variable ratio properly [51].

During contingency and the standard operating scenario, an optimization-based strategy is used for controlling the reactive power. From the solutions of mixed integer programming problems and the backward–forward search approach with linear complexity, magnitude and location of reactive power control can be determined [52]. The physical and electrical characteristics like the thermal, voltage, and stability boundary decide the limit for the maximum power transfer of an interconnected power system [53]. For the calculation of the maximum power transfer limit, the static security constraints are considered because the transient and dynamic constraints are very hard for online applications [54,55]. A predictor–corrector framework is presented in [56] to obtain a fast estimation of the maximum power transfer limit for the load increase model with a common scaling factor and to derive a particular outline of the impedance matching state for the maximum power transfer.

In the last few decades, researchers have been trying to develop various techniques to maximize the power transfer capability of a power system. In 1989, C.S. Indulkar et al. determined the maximum power transfer capability of transmission systems of 220–500 kV within stability limits by conducting tests on five different compensation schemes and concluded that by increasing the degree of series compensation, the extent of the maximum power transfer at the verge of stability could be enhanced. Moreover, the maximum power transfer limit is highest in the case of leading power factor loads [57]. The maximum power transfer expression bounded by voltage stability indicates that the point of the maximum power transfer is independent of the connected load type and changes only with the power factor of the load [51]. For different power factors, the PV and QV curves are used to estimate the parameters like the maximum power transfer value and reactive power transfer [58]. Maximum PowerPoint Tracking is highly essential in modern wind farms because it determines the operating point where maximum power can be extracted from wind streams. To achieve this economic benefit, four control strategies have been summarized in [59], namely, Optimal Torque (OT), Control Speed Ratio (TSR) Control, Perturbation and Observation (P&O) Control, and Power Signal Feedback (PSF) Control.

Xuan Wei and Joe H. Chow describe that, in the presence of Unified Power Flow Converters (UPFC), maximum power can be transferred at its conventional voltage within a rated capacity operation. FACTS devices improve the transfer capability of the power system, and they must operate at their rated limits (mega Volt-Ampere, voltage, current) to achieve maximum power transfer [60]. Static Synchronous Compensators (STATCOM) are shunt controllers that improve voltage stability by providing reactive power, thus improving power transfer [61]. Reference [62] formulated the direct relationship between impedance matching (IM) and power injection sensitivities, and through this work, the Sensitivity Impedance Matching (SIM) tool is proposed to figure out the following: maximum power transfer point, maximum load ability point, and instability conditions. Reference [63] deals with the various loading characteristics on the maximum power transfer limit for the critical values of series and shunt compensators by examining these two different schemes. To quantify the amount of reactive power that could be added to the load end of an AC transmission line system for maximum power transfer, the controlled compensation scheme is addressed by [64] for different load models by neglecting the effect of line resistance and capacitance.

### 2.1. Types of Voltage Stability

Based on the magnitude of disturbance, voltage stability is mainly categorized as large disturbance and small disturbance. Each of these is further classified into short-term and long-term depending upon the period.

### 2.1.1. Large Disturbance Voltage Stability

Enormous disturbance voltage stability refers to the system's ability to maintain steady voltages following significant disturbances such as system faults, loss of generation, or circuit contingencies [18]. Large disturbance voltage instability usually arises due to long transmission lines or tripping of generators [65]. The voltages at the several bus nodes tend to decline by increasing the reactive power requirement during a massive disturbance voltage instability, thereby the operating point and steady-state operating points get apart from each other [66]. To shield the system from a failure, the large disturbance voltage instability needs to be uprooted as swiftly as possible. Phasor Measurement Units (PMUs) are very handy to deal with these circumstances [67]. It can lead the system to voltage instability or rotor angle stability depending on the magnitude of the load attached. For large loads, the system undergoes a voltage collapse point, while for lighter loads, the rotor angle instability becomes prominent [68].

Using simulations, R.B.L. Guedes et al. argues in [68] that the Extended Lyapunov Function can be utilized to determine the clearing time when a light load power system undergoes voltage instability due to a significant disturbance. Tang Yong et al. believes that load dynamics' characteristics have a significant impact on short-term substantial disturbance voltage stability and presented solutions for overcoming the short-term massive disturbance voltage instability by either of the following: appropriate load shedding in the under-voltage condition or running the system at high voltage and incorporating reactive power injection devices [69]. Based on the induction motor's electromechanical transients, Sun Huadong et al. established an improved criterion involving less simulative calculations for judging the short-term considerable disturbance voltage stability [70]. For the improvement of the substantial disturbance voltage stability, G Naveen Kumar et al. have suggested the appropriate use of Thyristor Controlled Series Capacitor (TCSC) and SVC FACTS devices due to their robust nature against alternating climatic conditions [71]. Traditional load models cannot precisely describe the characteristics of load shedding under low voltage. Xiao-yu Zheng et al. overcame this problem by presenting an improved and novel load model in [72] with several advantages.

### 2.1.2. Small Disturbance Voltage Stability

Small disturbance voltage stability is the ability of the power system, for a given initial operating condition, to maintain a steady voltage when subjected to a small disturbance such as an incremental change in load demand [18,73,74]. By making use of sensitivity information, factors affecting the small disturbance voltage stability can be determined. This needs to linearize the power system under an equilibrium state [75]. Load characteristics, discrete control, and continuous control at a given time-instant influences the small disturbance voltage stability [76]. For the determination of small disturbance voltage stability, Liancheng Wang et al. proposed a novel Q angle index [77]. Using PMU with AR model adaptation, Chao Xu proposed a new technique for real-time small disturbance voltage stability analysis [78].

To assess small disturbance voltage stability, quite a few methods have been employed that use either local or global measurements. Although these methods are fast, but being less robust during the transient period, they cannot determine the precise equivalent parameters [79]. Keeping in view the fact that small disturbance voltage stability is independent of the nature of the disturbance, Robin Preece et al. proposed the risk-based probabilistic small disturbance security analysis (PSSA) to compute small disturbance stability risks [80]. Unlike large disturbance voltage stability, the small disturbance voltage stability is concerned with small incremental variations in the system load. The processes undergoing small disturbance voltage instability have a steady-state nature. The higher the reactive power injection to the system, the higher the voltage rise. Hence, this benchmark can be used to identify the small disturbance voltage stability by considering the performance of each bus in the system. Q-V sensitivity is the yardstick to check the stability. If Q-V sensitivity is positive for every bus, the system voltage is stable and vice versa [81,82]. Under various operating conditions, [73] carried out detailed analysis on grid-connected Wound Rotor Induction Generator (WRIG)-based wind turbine system to study the small disturbance voltage stability and concluded that the rotor resistance might reduce the voltage stability margin.

### 2.1.3. Long-Term Voltage Stability

Long-term voltage stability involves slower acting equipment such as tap-changing transformers, thermostatically controlled loads, and generator current limiters [18]. Caused by a large disturbance, long-term voltage stability is an interlinked local phenomenon that leads to power blackout [83]. The period of interest varies from 0.5 to several minutes [84]. Regardless of the use of power recovery components having load recovery characteristic and the extensive use of On-load Tap Changer (OLTC), voltage collapse course may still elongate, which is referred to as long-term voltage stability [85]. Inadequate supply of reactive power to some buses results in long-term voltage stability. Many key contributors like stressed power systems, maloperation of relays, load characteristics, improper fast reactive power resources, and tap-changer response leading to long-term voltage collapse have been reported in [86].

Based on the timescale, T.M.L. Assis et al. also presented midterm voltage stability [87] and proposed a novel technique for voltage security assessment. This methodology is a blend of the simulation techniques of Quasi Steady State (QSS) and Artificial Neural Networks (ANNs). Non-linear time domain dynamic simulations show the impact of induction generators on long-term voltage stability and also indicate that their existence squeezes the voltage stability margin in distribution networks [88]. OLTC with Line Drop Compensator (LDC) and Automatic Voltage Controller (AVC) relay has been used as a prevalent voltage control method. For preventing excessive tap changing, OLTC maintains an appropriate voltage at its secondary with a time delay of 30 to 60 s [89]. Model Predictive Control (MPC) is a magnificent technology because of its numerous benefits in long-term voltage stability improvement by performing trajectory sensitivity-based load shedding [90]. Angel Perez et al. briefly discussed various methods to evaluate long-term voltage stability [91]. Wenjie Zheng et al. [92] worked for the enhancement of long-term voltage stability through direct dynamic optimization.

### 2.1.4. Short-Term Voltage Stability

During the transient region, after being subjected to large disturbances, short-term voltage stability is one of the major concerns for a power system [93]. Short-term voltage stability refers to the dynamics of the faster components of load like HVDC converters, induction motors, and electronically controlled loads [18]. Various researchers have figured out that the extensive use of capacitor banks, voltage-intensive loads, air conditioners, and compressor motors having low inertia made the short-term voltage stability a consequential phenomenon [94–97]. During short-term voltage instability, the motor loads frequently compensated by shunt capacitor banks draw very high currents [19]. The occurrence of voltage drop due to a fault leads to dynamic load to restore the consumed power. This eventually results in short-term voltage instability. Long-term instability may lead to short-term instability. It causes a synchronization loss of generators and delays for motors [98].

SVC and STATCOM are dynamic VAR compensators with appropriate control strategies that can efficiently improve the short-term voltage stability by supporting fast dynamic reactive power [99]. The methods to improve the short-term voltage stability problem can be classified in two ways as an emergency and preventive control [97,100]. Voltage instability problems with their solutions, including proper placement of dynamic VAR, are presented by Ashutosh Tiwari et al. in [95].

An appropriate solution of differential equations is required in the analysis of short-term voltage stability. The period for the short-term instability may vary from few to many seconds [18]. With the help of Quasi Steady State (QSS) time domain simulation, we can calculate short-term voltage stability considering short-term voltage stability points [85]. Based on the Neural Networks with Random Weights (NNRWs) algorithm and ensemble learning strategy, Yan Xu et al. proposed a hierarchical intelligence system (IS) in [101], and K. Kawabe et al. presented a PV plane stability boundary line indicating the deceleration or acceleration of the induction motor [102] for the assessment of short-term voltage instability.

### 2.2. Voltage Stability Assessment

Voltage stability assessment is imperative for taking precautionary measures to mitigate the complications leading to voltage collapse. Conventionally, PV and QV curves are used to estimate the voltage stability margin. Owing to the assiduous time-consuming repetitive power flow simulations under various loading conditions, researchers are more inclined toward Voltage Stability Indices (VSIs) [103]. Voltage Stability Indices are powerful tools for detecting the proximity to the voltage collapse point and providing valuable information regarding critical lines [104], system loading, and weak nodes of a loaded power system [105]. The weakest lines and buses are the ones that have a voltage stability index closest to a critical value. Therefore, it is essential to evaluate VSIs at each node of a system, and the value of VSI varies between 0 and 1 [106]. Voltage Stability Indices can be broadly categorized into two types, namely [3,107]:

1. Jacobian matrix and System variable-based Voltage Stability Indices.
2. Bus, line, and overall Voltage Stability Indices.

A system-based VSIs make use of system parameters such as line power, bus voltage, and admittance matrix and are advantageous for online monitoring, while 'Jacobian matrix-based VSIs' are suitable for the accurate calculation of the voltage stability margin, which is difficult to calculate by the former [108]. Moreover, system variable-based indices require much less computing time and a weak bus/line can be identified precisely, while the Jacobian-based indices are very effective in the discovery of the voltage collapse point and maximum load ability, but a very high computational time results in an inappropriate online assessment, which makes them less preferable [3,109]. Offline VSIs are used in planning while online VSIs determine the voltage collapse proximity for controlling voltage instability automatically or manually by the operator to prevent blackout [108]. Voltage Stability Indices are very helpful to observe the online changes in the system parameters, and they are scalar magnitudes, which quantify the distance of the voltage collapse point from the operating point, and they are derived from the properties of voltage collapse [110,111]. Table 1 summarizes different VSIs in terms of their characteristics, critical values, and their formulations.

**Table 1.** List of VSIs.

| Sr. | Index | Proposed by | Expression/Formula | Critical Value | Salient Features | Input variables (of Formula) | References |
|---|---|---|---|---|---|---|---|
| 1 | Fast Voltage Stability Index (*FVSI*) | Ismail Musiril et al. | $FVSI = (4Z2Q_j)/V_i^2 X_{ij}$ | Critical value $\geq$ 1 | Used in determining the weakest bus, line criticalness, and maximum load-ability | $Z$ = Line impedance<br>$X_{ij}$ = Line reactance<br>$Q_j$ = Receiving end reactive power<br>$V_i$ = Sending end voltage | [106] |
| 2 | Line Stability Index ($L_{mn}$) | M. Moghavvemi | $L_{mn} = \dfrac{4xQ_r}{[V_s \sin (\theta - \delta)]^2}$ | If *Lmn* = 1, system is near to voltage collapse; *Lmn* < 1 means system is stable; *Lmn* = 0 means system has no load | Based on "Power flow through a single line." Uses little information like the specific local measurements | $V_s$ = Sending bus voltage<br>$\theta$ = Line impedance angle<br>$\delta$ = Angle difference between the supply voltage and receiving end voltage<br>$Q_r$ = Reactive power flow of receiving bus<br>$4X$ = Line reactance | [112–114] |
| 3 | Line Voltage Stability Index ($L_i$) | Arya et al. | $L_i = AvCos(\delta - \alpha)$<br>$\delta = \delta_k - \delta_m$ | $L_i < 0.5$ | This index halves at voltage collapse | $A = \dfrac{V_R}{V_S}$<br>$V_R$ = Receiving end voltage<br>$V_S$ = Sending end voltage<br>Phase angle across the line | [103] |
| 4 | Equivalent Node Voltage Collapse Index (*ENVCI*) | Yang Wang et al. | $ENVCI = 2E_k V_n \cos \theta_{kn} - E_k^2$ | For strongest bus, *ENVCI* = 1; for weakest bus, *ENVCI* = 0 | Useful for online assessment. It is quickly determined by using the voltage phasors | $V_n$ = Voltage at Nth node<br>$E_k$ = Voltage of the external system<br>$\theta_{kn} = \theta_k - \theta_n$ | [115–117] |
| 5 | Novel Line Stability Index (*NLSI*) | A. Yazdanpanah-Goharrizi et al. | $NLSI = \dfrac{PR + QX}{0.25V_1^2 \cos^2\delta}$ | Value close to 1 indicates voltage collapse point | Consider both active and reactive power; accurate results | $P$ = Active power<br>$Q$ = Reactive power<br>$R$ = Resistance of line<br>$X$ = Reactance of line<br>$V_1$ = Sending end voltage<br>$\delta$ = Angle difference between sending and receiving buses | [118–120] |
| 6 | Voltage Collapse Point Indicators (*VCPIs*) | M. Moghavvemi and O. Faruque | $VCPI(1) = \dfrac{P_r}{P_r(max)}$<br>$VCP(2) = \dfrac{Q_r}{Q_r(max)}$<br>$VCPI(3) = \dfrac{P_l}{P_l(max)}$<br>$P_l$ = Real power loss in the line<br>$P_l(max)$ Maximum possible real power loss<br>$VCPI(4) = \dfrac{Q_l}{Q_l(max)}$ | Lmn = 1 means system is near to voltage collapse; Lmn < 1 means system is stable; Lmn = 0 means system has no load | Based on "Power flow through a single line." This is based on the concept of maximum power transferred through the lines of the network. A higher degree of reliability; higher accuracy; simple; easy to calculate; We do not need the information on other lines in the system for calculation. We need specific local measurements | $P_r$ = Real power transferred to the receiving end<br>$P_r(max)$ = Maximum real power that can be transferred<br>$Q_r$ = Reactive power transferred to the receiving end<br>$Q_r(max)$ = Maximum reactive power that can be transferred<br>$P_l$ = Real power loss in the line<br>$P_l(max)$ = Maximum possible real power loss<br>$Q_r$ = Reactive power loss in the line<br>$Q_r(max)$ = Maximum possible reactive power loss | [3,121,122] |
| 7 | LQP index | Mohammad et al. | $LQP = 4(X/V_s^2)(X/V_s^2(P_s^2 + Q_R))$<br>$LQP = 4(\dfrac{X}{V_i^2})(Q_r + \dfrac{XP_i^2}{V_i^2})$ | *LQP* = 1 means the system is near to voltage collapse | Based on line stability factors; fast computation capability; helps to find the cause of voltage collapse | $X$ = Reactance<br>$R$ = Resistance<br>$V_i$ = Input Voltage<br>$P_i$ = Input power<br>$Q_r$ = Reactive power at receiving end | [123,124] |
| 8 | *VSI_1* | Y. Gong et al. | $VSI\_1 = (\dfrac{P_{margin}}{P_{max}}, \dfrac{Q_{margin}}{Q_{max}}, \dfrac{S_{margin}}{S_{max}})$ | *VSI_1* = 0 System collapse | Provides voltage stability margin of each bus. Also diversifies the most critical load bus | $P_{margin} = P_{max} - P$<br>$Q_{margin} = Q_{max} - Q$<br>$S_{margin} = S_{max} - S$<br>$P_{margin}$ = Active load margin<br>$Q_{margin}$ = Reactive load margin<br>$S_{margin}$ = Apparent load margin | [125,126] |

**Table 1.** *Cont.*

| Sr. | Index | Proposed by | Expression/Formula | Critical Value | Salient Features | Input variables (of Formula) | References |
|---|---|---|---|---|---|---|---|
| 9 | Power Transfer Stability Index ($PTSI$) | M. Nizam et al. | $PTSI = \dfrac{2S_L Z_{Thev}(1+\cos(\beta-\alpha))}{E_{Thev}^2}$ | 1 indicates voltage collapse point | Helps in finding dynamic voltage collapse | $E_{Thev}$ = Thevenin voltage $Z_{Thev}$ = Thevenin impedance $S_L$ = Maximum load apparent power $\beta,\ \alpha$ = Load phase angles | [127] |
| 10 | Line Index ($L_{ij}$) | C.Subramani et al. | $L_{ij} = \dfrac{4Z^2 Q_j}{V_i^2 (R\sin\delta + X\cos\delta)^2}$ | $L_{ij}$ = 0 line is close to voltage collapse | Easily identifies the weak areas of a nearly stressed system | $Z$ = Line impedance $X$ = Line reactance $V_i$ = Sending end voltage $Q_j$ = Receiving end reactive power | [128,129] |
| 11 | $L_{SR}$ | Mario A. Albuquerque and Carlos A. Castro | $L_{SR} = \dfrac{S_r}{S_{r\ max}}$ | $L_{sr}$ = 1 voltage stability indication | Easily used in real time; exactly recognizes the critical contingencies | $S_r$ = Apparent power flow $S_{r\ max}$ = Maximum power transfer | [130] |
| 12 | New Voltage Stability Index ($NVSI$) | R. Kanimozhi et al. | $NVSI = \dfrac{2X\sqrt{P_j^2 + Q_j^2}}{2Q_j X - V_i^2}$ | 1 indicates critical line | This index relates real and reactive powers; high standard voltage instability prediction; relates real and reactive power; exact voltage stability prediction | $X$ = Line reactance $P$ = Real Power $V_i$ = Sending end voltage $Q_j$ = Reactive power at receiving end | [131,132] |
| 13 | Voltage reactive power index ($VQI_{Line}$) | F.A. Althowibi et al. | $VQI_{Line} = \dfrac{4Q_{km}}{\left\|B_{km}\right\|\cdot\left\|V_k\right\|^2}$ | 1 is the point at which voltage collapses | Easily tells the distance from the point of voltage collapse of any line; high speed and accuracy | $Q_m$ = Receiving reactive power at bus $V_k$ = Sending voltage at system bus $Y_{km}$ = Line admittance between bus k and m $B_{km} = Y_{km}\sin\theta$ | [133] |

## 2.3. Voltage Stability Improvement by FACTS Devices

Originated by a reactive power imbalance, voltage instability has always been a serious attention-seeking problem in a power system. Fast control features of FACTS devices make them a great candidate to control and regulate both active and reactive powers, thereby restricting the voltage magnitudes within the stability margins [134,135]. FACTS are the power-electronic devices used to escalate reactive power generation and hence can improve voltage stability [136]. Suitable site selection for the allocation of FACTS devices is a must to exploit them efficiently. Incorporating these devices at the weakest bus yields in boosting up the voltage profile of a power system [13,137,138]. The evolutionary algorithm-based method concerning cost function is presented in [139] and the Genetic Algorithm (GA)-based novel technique is presented in [12,140,141] to dilate the voltage stability margin and diminish system losses by the optimal allocation of Multi-type FACTS devices. Introduced by Eberhart and Kennedy, the Particle Swarm Optimization (PSO) algorithm can also be employed to maximize system load-ability regarding a minimum cost of the installation of the devices [142]. However, the Group Search Optimizer with Multiple Producer (GSOMP) algorithm has been used in [143] for improving voltage and lessening the real power loss by multiple FACTS devices. It is effective for finding the multiple FACTS device parameters, because it gives better results in average and minimum form for power loss optimization as compare to Particle Swarm Optimizer (PSO).

Dr. J. Amarnath et al. suggested the optimal location of Static VAR Compensator at the particular bus that has the most negative sensitivity index for the voltage stability enhancement, whereas the optimal location for Thyristor Controlled Series Compensator (TCSC) is the bus with the most positive loss sensitivity index [144]. Reference [135] examined the effectiveness of TCSC, SVC, and UPFC to study voltage stability in various scenarios by implementing them in a common Newton Raphson power flow. The authors concluded that UPFC amalgamates the benefits of both TCSC and SVC, namely reduction in active power loss and improvement in voltage profile. The authors of [145] proposed a general methodology for the improvement of steady-state voltage stability using FACTS devices in an IEEE 39-bus test system environment. A. Rathi et al. offered a different Distributed Flexible AC Transmission System (D-FACTS), for controlling system voltage by examining sensitivities of voltage magnitudes with respect to variations in line impedances [146]. Figure 1 demonstrates types of FACTS devices, and Table 2 summarizes some important FACTS devices used for voltage stability improvement with special findings asserted by several researchers.

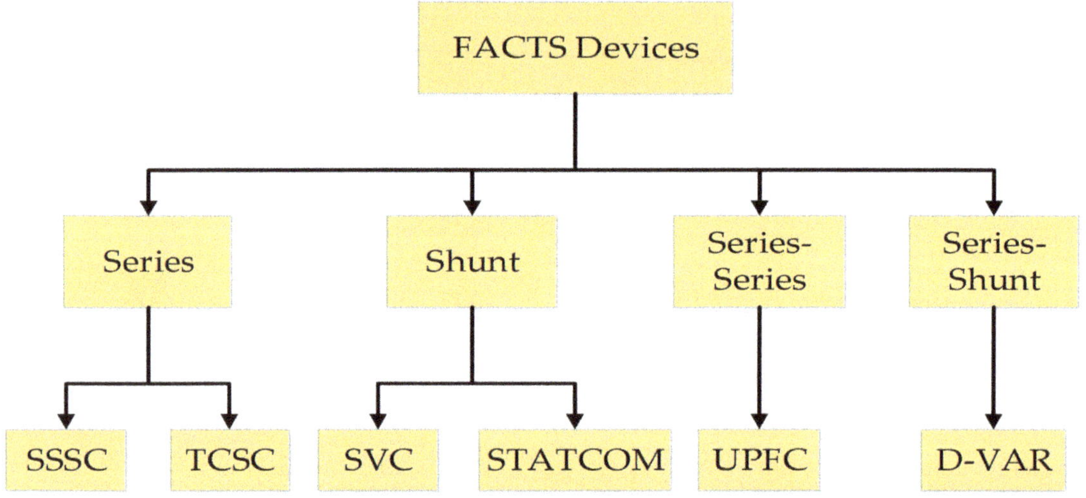


**Figure 1.** Classification of FACTS devices.

**Table 2.** List of FACTS devices.

| Sr | FACT Device | Category | Salient Features | Special Findings | Refs. |
|---|---|---|---|---|---|
| 1 | SSSC | Series | • To elude voltage unbalance, SSSC uses the voltage source inverter (VSI), which controls the rotor current, which in turn controls the reactive and active power fed from the wind farm to the connected grid<br>• Provides a better series of compensation<br>• Provides the active DFIG | • C. Udhaya Shankar et al. conducted tests at a specific bus of two machine power systems and concluded via simulations that voltage stability had been improved along with the damped power oscillations using SSSC.<br>• SSSC has been called the best controller for voltage stability improvement by Hridya.K.R et al. based on their simulations in PSAT. They made use of line index voltage stability $L_{mn}$ to conduct the test to simulate two disturbances to get the desired outcomes. | [147–150] |
| 2 | TCSC | | • It is located at the line that held the most positive 'loss sensitivity index' for reactive power load reduction<br>• Used at the line having the most negative 'sensitivity index' for PI method<br>• Can be operated in both capacitive and inductive regions and hence acts as a controllable series impedance for wind voltage instability | • F. Z. GHERBI and ABDSALLEM analyzed the impact of integrating wind generation and FACTS on the voltage collapse and active losses of IEEE 30 bus system, and they concluded that TCSC is the best choice for the active power control.<br>• Joshi and Mohan used the TCSC for the interconnection of wind turbines to grid in terms of voltage stability issues. In their literature, they performed modeling of TCSC connected to a wind turbine along with an induction generator that resulted in its effectiveness for voltage unbalance compensation and fault current limitation. | [151–154] |

**Table 2.** *Cont.*

| Sr | FACT Device | Category | Salient Features | Special Findings | Refs. |
|---|---|---|---|---|---|
| 3 | SVC | Shunt | • SVC is based on thyristor-controlled rectifier (TCR)<br>• SVC has to be located at a bus with the greatest negative 'Cij sensitivity index'<br>• Controls the equivalent resistance of a bus to control its voltage and SVC is like a shunt reactor<br>• SVC improves the bus voltage more than that of a TCSC present in a line, so SVC is better<br>• Based on annual operating cost expectation, location of SVC has been proposed | • By employing SVC in a wind farm with PSCAD/EMTDC program, Zakieldeen Elhassan et al. presented that terminal voltage ($V_T$) has improved from 0.9078 pu to 1.0 pu. The voltage at the point of common coupling ($V_{PCC}$) is enhanced from 0.9395 pu to 1.003 pu.<br>• Tang Baofeng et al. presented a substation in Liutun connecting three 220 kV fault buses of Hancun, Cangxi, and Yuzhuang. The simulation result showed enrichment in voltage profile by SVC usage from 0.004 pu to 0.009 pu at Liutun substation. | [144,155–158] |
| 4 | STATCOM | | • Gives fast and smooth steady state and moreover the transient voltage control at the points of voltage dips in the system<br>• The response is faster in STATCOM as compared to SVC<br>• As compared to SVC, compensating current of STATCOM is independent of common point voltage. So, it does not change with a voltage dip<br>• Operates at full capacity even during low voltages | • Keping Zhu et al. employed cascade STATCOM with a grid-connected wind form to enhance voltage stability. Through simulations, they showed that cascade STATCOM could govern the voltage profile quickly and smoothly by providing reactive power to the grid-connected wind farm.<br>• Ahvand Jalali et al. showed how to upsurge the voltage stability of power networks with embedded wind farms by combining STATCOM with Energy Storage Systems (ESSs) and also highlighted the placements of these devices to boost up VSM. | [159–162] |
| 5 | UPFC | Combined series shunt | • Controls active and reactive power in both ways either simultaneously or selectively<br>• Gives direct bus and line control<br>• Good series compensator<br>• Phase shifter<br>• Controls power flow for only one transmission line<br>• UPFC arrangement with the wind form sustains the voltage imbalance during integration with the grid | • Yasser M. Alharbi et al. applied UPFC to improve the LVRT capability of DFIG-based wind turbine systems. Via simulations, they showed that the proposed controller for UPFC can appreciably improve the LVRT ability of the wind turbines and assure the continuity of power delivery during small voltage dips.<br>• A. Papantoniou et al. simulated a wind farm model whose voltage profile was 10% less than the nominal voltage and stabilized to nominal value after insertion of UPFC. | [163–165] |
| 6 | D-Var | | • A type of STATCOM<br>• Effectively solve the problems of wind energy generation<br>• Controls the overall voltage profile at the collector bus<br>• Additionally, it controls the nearby switching of capacitor devices to improve voltage stability<br>• Diminishes the transients of capacitor-switching | • A. Kehrli et al. presented an 8 MVar D-Var system controlling nine banks to correct power factor. Between the utility and turbine, 1 MVA D-Var incorporated injections of about 2.3 MVar at point of common coupling to diminish the voltage dip and bring the wind farm to the normal condition after the voltage dips. | [159] |

## 2.4. Identification of Weakest Bus

Power system stability is at stack when the voltage of one of the buses approaches its collapse point, and ultimately, the system leads to a blackout. To ensure the operation of the power system is within the security constraints, it is utmost important to figure out the weakest bus in the power system under various operating conditions [166,167]. The weakest bus is the one that is heavily loaded, has the most sensitive voltage, and first reaches the minimum/maximum limit after which system security gets affected. The most influential factors that determine the strength of a bus are the R/X ratio of interconnections, load directions, load models, the different network configuration, installation of compensators, and generators. Moreover, the lack of local and remote VARs can have the worst impact on the critical bus [168–171]. A power system can be analyzed in two aspects: one is static and the other is dynamic. The former approach is utilized to signify the weakest bus. However, this approach can cause the loss of dynamic stability of the system [172]. The purpose of weakest bus identification is to improve the voltage profile on these buses for increasing the power transfer capability and maximum loading limit [173]. Secondly, it furnishes the exact information about the optimal reactive power planning, and it directs better sites for the installation of reactive power sources [174]. As it is vital to determine the weakest bus in the system, different researchers used various computational tools and methods to spot such a bus. For example, reference [175] deals with the voltage collapse proximity indicator (VCPI) method; the Singular Value Decomposition (SVD) method is examined in [176]; in [177], the authors used the real-coded simple Security Constraint Genetic Algorithm (SCGA) to compute the maximum loading limit for different IEEE standard bus systems; and [178] makes use of the nodal voltage security assessment tool. M. K. Jalboub et al. made use of Singular Value Decomposition, multivariable control, and modal analysis to study voltage stability and determine the weakest bus in the system [179].

In references [171,180], various Voltage Stability Indices have been explained to evaluate the strength of a bus irrespective of the load type. These are:

❖ voltage sensitivity factor;
❖ reactive power margin;
❖ L-Index.

The estimated value of L-Index varies between 0 and 1 [181]. In a nutshell, determination of the weakest bus in a power system is inevitable to restrict the system within permissible limits.

## 3. Wind-Integrated Power Systems

Due to the intensive demand growth for electric power in every sector of life, the appropriate financial and environmentally friendly resources are getting a great deal of attention. The wind energy market has been a far-reaching subject of interest, especially in European, North American, and Asian countries like China in the several past decades. A glance over some statistics about wind power capacity in the near past indicates that the wind capacity has increased from 6000 MW (1996) to 197,000 MW (December 2010) [177]. Up until 2006, the countries with the highest total installed capacity were Germany (20,622 MW), Spain (11,615 MW), USA (11,603 MW), India (6270 MW), and Denmark (3136 MW), and this kept on increasing year after year [182]. In 2015, the increase in wind generation was equal to almost half of the global electricity growth [183]. Globally, the year 2015 crossed the 60 GW mark for the first time in history, and 63 GW of online power capacity was injected to the system. In Asia, China sustained its premiership position and installed 30.8 GW of wind energy at the end of 2015. India remained unbeatable at the second position in Asia and installed 2623 MW for the total of 25,088 MW. The US market, which is the second single largest market after China in terms of total installed capacity, added 4000 new turbines for the total capacity of 8598 MW in 2014 and got a 77% increase over the year 2014 [183]. At the end of 2016, the new installed capacities for different regions around the globe include Africa and Middle East (418 MW), Asia (27,680 MW), Europe (13,926 MW), Latin America and Caribbean (3079 MW), North America (9359 MW), and the Pacific Region (54,600 MW) [178]. The study on the rate and pace of wind energy growth reflects that 12% of all the electricity generation will come from wind resources by 2020 [184,185]. Figure 2 graphically shows the gradual increase in the wind market over the last two decades, whereas the regional installed capacity (MW) by the end of 2016 is shown in Figure 3 [186,187].

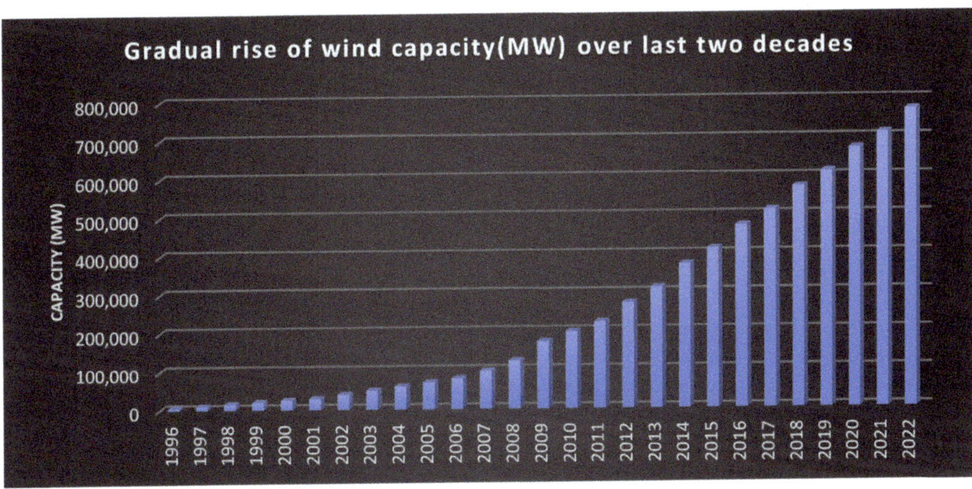


**Figure 2.** Gradual rise of wind capacity (MW) over last two decades.

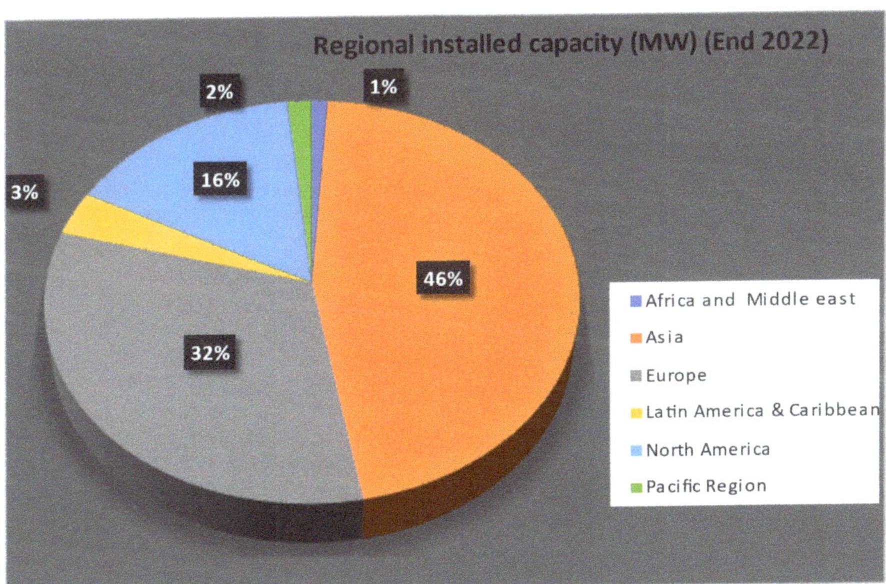


**Figure 3.** Regional installed capacity (MW).

Pakistan, a developing country, has been going through energy crises for the past several decades. Currently, the amounts of generating and demanding power are not getting closer. Hence, researchers and engineers are trying to explore more resources of power generation to get rid of the energy crises. Among renewable energy resources, wind energy is also being paid heed in many areas of Pakistan. The availability of wind speed needed to operate wind turbines is low in Pakistan [188]. A few areas have sufficient wind speed like the Jhimpir Wind Power Plant that is located at Jhimpir in Thatta, District of Sindh province in Pakistan. This power plant comprises 33 wind turbines, injecting a total capacity of 56.4 MW of power into the system. At present, 45 wind power projects are working in Pakistan, with 3200 MW capacity [189].

### 3.1. Grid Codes

Nature has gifted us numerous efficient wind sites around the globe, but unfortunately, most of them exist far away from the load centers and are weakly interlinked via long transmission lines. Under such conditions, the steady state voltage level gets affected by pouring a large amount of power into the system. To make the system secure, stable, and reliable, we have to consider grid codes [190]. Grid codes are formulated to ensure steadfast operating conditions and to correlate the on-working generating unit disturbances. A grid code presents the requirements for the proper interconnections of wind power plants with the grid [191]. At the point of interconnection, grid codes primarily circumscribe the grid specifications and requirements. Grid codes differ substantially on several bases such as the nature of country/region or the type of transmission and distribution system [192]. The main causes of varying requirements of grid codes correlate with the network structure, short circuit capacity, the technology used in the wind turbine, total installed capacity, and the amount of the penetration level [193]. There are plenty of mandatory requirements in grid codes, but from the wind farm outlook, some are as follows [192]:

- active power control;
- frequency control;
- voltage regulation;
- frequency variations;
- overvoltage conditions;
- fault ride-through;
- voltage regulation and power factor;
- power quality.

Andreas Armenakis et al. proposed some additional parameters such as requirements for reactive power, short-circuit current, and provision of ancillary services on significant grid users for the small wind farms connected to small distribution grid in Cyprus [194]. The reactive power output control for the wind farm is essentially regulated within the reactive power range to reach this set point [195]. E. Fagan et al. proposed the Irish grid code design and examined the supplementary requirements like provision of signals, control, and communications added on by the System Dispatch Codes. To employ the voltage and frequency requirements, the signals and control commands are needed for transmission system operators (TSOs) [196]. Another requirement for the Irish wind grid code is that the wind farm should be adequate to supervise the ramp rate of its active power output with a maximum MW per minute ramp rate that is adjusted by the transmission system operators (TSOs) [193]. Germany (EON) spells out both the maximum permissible active power change and the minimum required active power reduction capability. While the grid codes of China (SGCC), Britain (NGC), and Ireland (ESB) only set the maximum ramp rate [193].

### 3.2. Wind Farm Integration

To overcome the deficiency of non-renewable energy sources, wind energy has turned out to be the most popular energy generation method, and wind farms are being integrated with the conventional grid at a high rate and pace in many developed countries [196]. In this process, one of the serious problems that we have to deal with is the voltage instability, which arises due to the fluctuating nature of wind power [197,198]. That is why, now, power system utilities' interests are more inclined towards stability issues rather than power quality [199,200]. Generators used in wind turbines can be broadly categorized into two classes, which are fixed speed and variable speed. Squirrel Cage Induction Generator (SCIG) falls in the first category as its output is directly coupled with the grid. SCIGs can only absorb reactive power, whereas variable rotational speed DFIGs can produce or consume reactive power. This makes DFIGs a better choice in wind farm integration [196,199,200]. The installation of Automatic Voltage Regulators (AVRs) in all the generators can be helpful to avoid the small fluctuations in voltage and restore grid stability after the integration of wind farm [201].

Without sufficient reactive power support, the wind farm connected to a weak node may be tripped due to AC under-voltage protection if the node voltages stay too low, eventually leading the system towards voltage collapse [202]. The increased wind power capacity decreases the voltage stability margin even in the presence of the crowbar; the voltage stability of DFIG-based systems remain vulnerable at the point of fault clearance as long as the fault persists [203]. Different types of wind turbines can be integrated with the conventional grid, and the impact of some of the turbines on voltage stability has been reported in [202]. The impact on voltage stability with respect to the type of wind turbine, for e.g., DFIG vs. full-converter, depends on various factors. One of the major factors is reactive power control capability. Full-converter wind turbines offer high flexibility in reactive power control. Regardless of the amount of active power generation, they can absorb and inject reactive power to provide voltage regulation and grid stability. On the other hand, reactive power control capabilities of doubly-fed induction generators (DFIGs) are very limited. In order to provide adequate reactive power support, they usually require additional equipment such as static var compensators (SVCs). Consequently, under certain conditions, DFIG-based wind farms are more prone to voltage dips. Another important factor is the fault ride-through (FRT) capability. During temporary faults or voltage dips, wind turbines with strong FRT capabilities can remain connected to the grid. Therefore, the impact on grid stability will be minimized, cascading outages can be prevented, and faster recovery of the grid is possible. However, during faults, wind turbines with weaker FRT capabilities may detach from the power grid, posing a risk to system stability and further exacerbating voltage dips. Furthermore, wind turbines associated with weak grids (high line impedance) or operating at low power may not be able to provide sufficient reactive power support during faults and therefore are more likely to contribute to voltage drops. Conversely, necessary reactive power support can be provided by the wind turbine attached to strong grids or operating at high power to enhance voltage stability. Therefore, owing to strong FRT capabilities and flexible reactive power control, full-converter wind turbines are preferred for voltage stability. However, with careful planning and an appropriate grid support, DFIG wind turbines can also be installed. Operating conditions and the entire system configuration play a vital role in any wind turbine type's response to the faults. Consequently, in order to analyze the impact of wind integration, detailed studies and simulations must be carried out on a case-by-case basis. Based on simulation results, reference [204] concluded that the increase in the penetration level of wind farm results in system instability that imposes severe effects on the voltage profile of the system.

Hierarchical Security Constrained Optimal Power Flow (SCOPF) is a cost-saving method used for large wind-integrated system that makes the calculations easy by separating the AC system from the DC system [205]. The effect of wind farm integration on the system's voltage stability is dependent on the load characteristics. Based on load types, a comparative study is conducted in [206], with and without considering load voltage dependency, concerning the location of the synchronous compensator and penetration of wind farm generation.

### 3.3. Weak Grids

Integration of wind farm with the conventional grid possesses serious threats on the stability, security, and reliability of the normal operation of a power system. The bulk amount of wind energy injection via long transmission lines with a limited short circuit ratio (SCR) result in low short circuit current and hence diminishes the system strength. When a wind farm is being integrated with a weak grid, it leads the system stability to a more precarious state. Keeping in mind the severity of the fact, many of the researchers have paid due attention to this troublesome phenomenon [207–209]. Considering the static and dynamic performance, IEEE standard 1204 [210] defines the weak alternating grid:

1.  the AC system impedance may be high relative to DC power at the point of common coupling (PCC);

2. AC system mechanical inertia which may be inadequate relative to the DC power infeed [210,211].

Among various gauges mentioned in the literature, Low Short Circuit Ratio (LSCR) is the main indicator of the weakness of a grid [207]. In [209], Short Circuit Ratio (SCR) is described as:

$$\text{SCR} = S''_{K\_PCC} / S_{\text{wind\_power\_plant}} \tag{1}$$

where $S''_{K\_PCC}$ = short circuit power at the wind power plant point of common coupling and $S_{\text{Wind\_Power\_Plant}}$ = wind power plant capacity.

Refs. [211–213] designated the grid as strong when the SCR is between 20 and 25, and for weakness, the value lies below 6–10. A grid is recognized as weak if the value of the effective SCR is less than 2.5 [214]. Low effective DC inertia constant also acts as an indication of a weak grid [215] if its value is below two [214,216,217], which addressed different problems associated with the weak grid itself and when it is connected with the wind farm and provided some useful solutions for the issues. Among innumerable complications of the weak grid, voltage instability is a drastic one, because the occurrence of fault at the point of common connection (PCC) will cause a voltage dip that results in tripping of several turbines if not controlled well in time [218]. To meet the challenge for voltage control, the adequate amount of reactive power compatibility is ensured. Under a weak grid, the $dV/dQ$ sensitivity is high. According to [219], a grid having the value of its SCR less than three is considered as feeble. Consequently, an SCR less than its threshold value leads to noncompliance action of the turbine, thus generating voltage oscillations. Voltage stability analysis is performed that is based on PV and QV curve method with the different short-circuit ratios and $X/R$ of the external grid [217].

The impedance could increase further because of several faults, load dynamics, and line tripping [220]. Furthermore, the electromagnetic stability is associated with the mechanical design. The Permanent Magnet Synchronous Generator (PMSG) and DFIG may cause electromagnetic stability problems in a weak grid. DFIG output variation causes the worst voltage fluctuation in a weak grid [221]. Electromagnetic stability can be improved by modifying specific mechanical design features, which includes elimination of the gearbox. This helps to improve the overall system efficiency and reduce the potential of mechanical failures. Furthermore, a faster reactive power response can be achieved to enhance stability. To improve the power output smoothness and minimize the torque fluctuations, an optimal aerodynamic design of the blades can be developed. It will also improve voltage and electromagnetic stability and reduce stress on the electrical system. Additionally, stability can be improved by using a composite material such as carbon fiber to create lighter blades. The overall system stability can be enhanced by utilizing damped or tuned mass damper systems in the tower design, contributing to mitigate oscillations and vibrations caused by the wind gusts. To mitigate the voltage fluctuations due to DFIG wind turbines in weak grids, control algorithms to dynamically adjust the reactive power output according to the grid need can be implemented. Furthermore, grid support technologies such as SVCs, battery energy storage systems (BESSs), and FACTS devices can be employed to improve the overall grid stability and mitigate DFIG-induced fluctuations. To minimize localized voltage disturbances, multiple DFIG wind farms can be coordinated effectively for balanced reactive power contributions. Moreover, upgrading the existing transmission infrastructure and deploying smaller, distributed generation sources closer to wind farms can help mitigate the DFIG-induced fluctuations on longer transmission lines and support local voltage profiles.

*3.4. Low Voltage Ride-through*

In any power system, the stability of the grid functioning and the security of the power system are prominent aspects. For this purpose, the power generating plants, including turbines and generators, must have some suitable methodology and control capabilities. To achieve these objectives, the term Low Voltage Ride-Through (LVRT) is presented, which can be put into words as the requisite to persist in the operation of the generating plants through short periods of low grid voltages [222]. Even if the voltage in any phase becomes less than 80% of its nominal value, then the generators should remain in operation and connected to the power system in an LVRT-capable power system [223]. LVRT is the ability to keep the wind turbines connected with the grid, even under faulty conditions [224]. For wind farms, LVRT is marked as an integral part of grid codes by the regulatory bodies. Popular DFIGs having partial converters do not have the capability of Low Voltage Ride-Through. Non-linear Control-based Modified Bridge-type Fault Current Limiter (NC MBFCL) application has been proposed in [225] for enhancing LVRT capability in DFIG-based wind farms under symmetrical and non-symmetrical faults. One of the top requirements of the grid codes is to keep the DFIGs connected with the load during LVRT.

LVRT capability of a power system is being improved through different superconducting power devices. Lei Chen et al. carried out MATLAB simulations to prove the superiority of the Superconducting Fault Current Limiter (SFCL) on Dynamic Voltage Restorer (DVR) in terms of economic and technical perspectives for improving LVRT of the micro-grid [226]. Huihui Song et al. proposed a new strategy for limiting the adverse impact of voltage drop that integrates the attenuation of L2 disturbance with energy-based control of DFIG. This improved strategy is preferable to conventional energy-based control methods in terms of internal and external faults of the grid [227]. Asymmetrical fault ride-through capability of the DFIG-based WECSs is presented in [228] to study the dynamic behavior of DFIG via the MATLAB/Simulink R2010b environment with the help of space vector control concept.

Large loads and different faults (short circuit or lightning) are considered to be some of the dominant causes of short-term voltage dips. Every grid connection requires a certification test, which is measured by its LVRT characteristics with the help of different simulation techniques for voltage dips at several scenarios [221]. The viable function of LVRT is to compare the characteristic with the terminal voltage. Use of LVRT results in the system's expensiveness because it suggests additional apparatus and modulation in the control design and approach [229].

Various testing methodologies exist to check the LVRT ability of wind turbines, which includes devices such as series connected Voltage Source Converter (VSC), autotransformers, synchronous generators, etc. Among these, impedance-based Voltage Source Generator (VSG) is mostly used for confirming the LVRT capability of wind turbines. By using the PSCAD/EMTDC time domain simulation, [230] demonstrated the VSC-based novel LVRT test technique, which can help in wind turbine voltage control.

The need for on-site LVRT testing cannot be over-emphasized because of the following reasons [231]:

1. it gives more realistic results as compared to the ones found in the laboratory;
2. to ensure the LVRT capability of all the installed wind turbines;
3. to counter-check the capability of wind turbines after major events in a system's maintenance.

To fulfil these requirements, the authors of [231] presented a novel technique for the on-site assessment of LVRT for Full Power Converter (FPC) wind turbines and showed via MATLAB-based simulations that without any hardware modifications, the on-site LVRT tests could be performed by proper modifications in the control algorithm. Reference [232] pointed out some of the latest methods of improving LVRT. It can exclusively be performed either by using external devices or by controller improvement methods. These have been summarized in Figure 4.

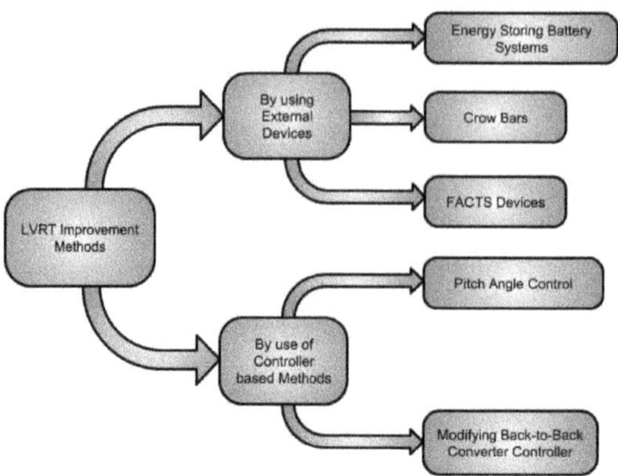


**Figure 4.** LVRT improvement techniques.

### 3.5. High Voltage Ride-through

Numerous grid code requirements must be accomplished to sustain the uninterrupted overall stability and reliability of any wind turbine system. Among these various requirements, the implementation of High Voltage Ride-Through (HVRT) capability is integral. Contrary to LVRT, the HVRT capability of the wind turbine is the ability to keep itself connected with the grid throughout the voltage rise circumstances up to a certain level. Australia is credited as a pioneer country in developing the criterion for HVRT [233]. Switching OFF large loads, single line to ground fault, unbalance faults, load shedding, or switching ON the capacitor banks can result in voltage swells, so HVRT becomes a critical event in terms of the system stability [234–236]. Since the use of DFIG in the wind industry is promulgating, a main concern with the DFIG is that its stator side is directly connected to the grid, and it makes DFIG more sensitive towards disturbances like voltage swells and voltage sags [237].

Few researchers have emphasized the control strategies for HVRT resulting from voltage swell [238]. In reference [239], the authors proposed a hybrid current controller scheme to enhance the Low and High Voltage Ride-Through capabilities of DFIG-based wind turbine systems. Under normal operating conditions, they used a standard proportional integral (PI) and current controllers to regulate the output currents of the rotor-side and grid-side converters. FACTS devices are also incredibly supportive in increasing the HVRT capabilities of DFIGs. For the improvement of the HVRT capabilities of DFIG systems, STATCOM is investigated in [240]. Using the MATLAB/SIMULINK environment, a suitable algorithm is developed by considering 30% voltage rise for HVRT, and a novel strategy is introduced for enhancing the ride-through capabilities of DFIG [241]. A hybrid current controller is proposed in [239] for HVRT as well as the LVRT improvement of wind farms. Reference [242] proposed a strategy for controlling HVRT based on GSC, while a cost-saving strategy requiring less hardware with little software algorithm changes is reported in [243].

Electric power can be generated under constrained conditions via wind turbines. The minimum speed at which a wind turbine starts generating power is referred to as cut-in wind speed and typically varies between 2 and 3 m/s, whereas it produces rated power in the range of 11–15 m/s [244]. However, there are certain circumstances when the wind speed overshoots these limits and approaches 25 m/s for a predefined spell or a 30–35 m/s sudden puff. Under such deteriorative conditions, it is necessarily important to implement a security feature termed as High Wind Speed Shutdown (HWSS). This technique saves the turbine from severe damage to its various parts, specifically turbine motor and blades, but results in an unpredicted loss of power generation [244,245].

### 3.6. Role of DFIG in Voltage Stability Improvements

The exploitation of fixed and variable speed induction generator is communal in the current era of wind industry [246]. Almost 78% of the total wind installed capacity has employed DFIGs for the generation of electricity [4]. Distribution network-connected wind farms often experience a higher-level voltage disturbance of about 2%. Such disturbances are coped with the use of DFIG [247]. DFIG is essentially a wound rotor induction generator, having its rotor coupled with a voltage source converter, frequently known as the Rotor Side Converter (RSC) [248,249]. Recognition of DFIG as the prevalent wind turbine system is based upon the following pluses [157,232,249–256]:

1.  independent control capability for active and reactive powers;
2.  constant frequency operation at variable speed for robust four quadrants' reactive power control;
3.  handles 20–30% of total system power that eventually lowers power losses;
4.  operate at a higher range of wind speed
5.  lower mechanical stress and acoustical noise;
6.  high energy efficiency;
7.  lower cost due to low rating power electronic converters.

Another overwhelming plus point determined is the penetration level that is 60% more in case of DFIG as compared to its counterpart SCIG [246]. The most imperative responsibility of DFIG is to provide LVRT capability, so that wind farms remain connected with the grid as per grid code requirements when the voltage dip occurs for a shorter time interval [248]. Such low voltage scenarios impose serious threats on the system equipment, especially on the RSC [257]. A widely used protection technique is connecting the crowbar to the rotor circuit that will bypass RSC. Nevertheless, this technique reduces DFIG control competency, increases mechanical stress and power oscillations, and devours more reactive power from the grid, eventually leading to exaggerated fault conditions [251,257,258]. Figure 5 shows the schematic diagram of DFIG-based WT integrated with the conventional grid through GSC and RSC voltage source converters and a common DC bus. The main components of a DFIG-based wind turbine system are drive train, pitch angle control system, DFIG system control, and protection system (crowbar). Drive train helps to reduce the electrical torque during grid fault.

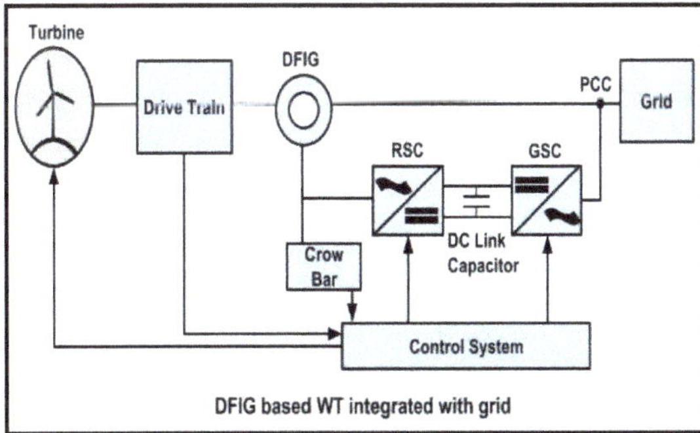


**Figure 5.** DFIG-based WT integrated with grid.

An apt balance in the demand and supply of reactive power of any system controls the voltage stability issues, and DFIG is remarkable in this regard due to its special design [259]. This capability of providing reactive power alters according to the power electronic converters' rating [260]. Jinho Kim et al. proposed a hierarchal voltage control scheme employing a wind generator (WG) controller and wind power plant (WPP) control [261]. Another control strategy employing both the GSC and RSC is presented in [262] to acquire voltage stability. RSC and GSC both control the active and reactive power injected by the generator into the grid, while the GSC maintains the DC link voltage constant independent to the direction of rotor power flow. But GSC has the edge over RSC because the reactive power is primarily produced by GSC [7]. In a doubly fed induction generator (DFIG), RSC adjusts the amplitude and frequency of the voltage applied to the rotor winding to control the active and reactive power generation of the wind turbine. It establishes the connection between the DC link and the rotor windings with the help of IGBTs or GTOs. To maximize the energy generated from fluctuating wind speed, RSC enables the wind turbine to operate at different speeds, and both active and reactive power output control is possible to improve voltage stability and regulation. Furthermore, its cost is lower compared to full-converter wind turbines. However, reactive power control range is limited compared to that of full-converter wind turbines. Consequently, additional support such as SVCs are required in weak grids for full reactive power support. Furthermore, it can experience torque oscillations and stress under fault conditions. GSC controls the power flow between the grid and the wind turbine and maintains a constant DC link voltage. It establishes the connection between the DC link and the grid at the frequency and AC voltage that the grid requires. It maintains a constant DC link voltage and establishes a smooth connection between the wind turbine and the grid. Moreover, in order to improve grid stability, it can inject or absorb reactive power according to the requirement and provides independent control of the active and reactive power flow from the wind turbine. However, in order to maintain DC link voltage stability and grid synchronization, robust control systems need to be employed in GSC. Additionally, its hardware design is very complex, and advanced filtering techniques are required for harmonic generation. Consequently, the cost of GSC is much higher compared to simple grid-connected asynchronous generator. RSC can be used for many purposes like torque controller, speed controller, or active power controller to regulate the DFIG output [263]. However, a few challenges should be taken into consideration in the installation, like voltage sags and short circuits [264]. Owing to the voltage sag behavior of DFIG, early researchers have identified two main issues, namely overcurrent in the rotor and overvoltage in the DC link capacitor [239,255]. The DFIG response towards grid disturbances is more sensitive and requires additional attention for power converter protection as prescribed earlier [255,256,264]. If the imbalance in the voltage is not taken into consideration, the rotor and stator current can be highly unstable because of the negative impedance of DFIG [264]. One more issue with DFIG is the unavoidable use of brush gear and slip rings, and this put a question mark on the DFIG reliability; so, a brushless DFIG can be a better candidate [6].

### 3.7. Penetration Level of Wind Energy

A number of factors affect the permissible wind energy penetration level; some of those are system inertia, headroom, frequency sensitivity index, and the amount of governor responsive generation. Frequency sensitivity has a major impact on the penetration limit [265]. The power system must be stable enough to withstand the frequency fluctuations during penetration. During wind energy penetration, the whole cascaded system may lead to collapse if it is not strong enough to withstand the frequency variations within the reasonable limit [266]. To estimate how much variation within frequency limits is bearable, the following specified parameters are considered:

- frequency deviation range = $\pm 0.2$ Hz;
- maximum predicted load disturbance = 0.04 pu.

Based on the simulation with the parameters given above, [265] concluded that in a stand-alone control area, the penetration level of wind power should not be more than the level required to create 30% reduction in inertia [267]. By appropriate placement of wind turbines, we can maximize the power generation. The maximum loading parameter decreases with the increase in the penetration level, and system moves quickly towards voltage instability. More active power generation and wind farm integration decrease the stability of the power system, as shown by the IEEE 14-bus test system [268]. Output power factor influences the maximum power injection in a system, which is specially controlled by reactive power and simulation results, to show that the maximum power can be injected at a certain power factor [269].

### 3.8. The Optimal Location for Wind Farm

Driven by air puffs, the interconnected wind turbines impart electrical power to the system. To boost up power production level within effective cost criteria, it is preliminary to consider the determination of the most suitable wind farm installation site [270]. The intermittent nature of wind offers a bunch of troubles to designate appropriate sites to install wind farms. The elements like wind speed, disturbances to electrical substations, and the air density are taken into account to select an optimal wind farm location [1] amply. Generally, the selection of wind farm location is based on the following prominent characteristics:

- gusts-free availability of wind;
- regular smooth winds with least variations;
- strong air currents due to thermal gradients;
- distance from the power system;
- land cost and roads to approach the site.

Locations having such attributes are most probably hilltops, valley bottoms, vast plains, forests, and offshores significantly away from common mass [270,271]. The most substantial and significant factor in this regard is the accessibility of stiff wind at the point of installation. [272] presented the performance index PI to investigate the probability of wind availability by setting the criteria given below.

If $\mu pi$ < PI, the site is not recognized as good for wind farm installation.

The power output of the turbine is defined by the power curve or p-v characteristic that varies with the wind speed. This power curve has three regions that need to be examined. Initially, to start the wind turbine, the wind speed must be greater than the cut-in speed, while the third region of the p-v curve is unsafe for turbine blades since the wind speed is at a higher level. To ensure safe turbine operation, it is made to operate between the 'cut in' and 'cut off' during the second region [273,274]. Various methodologies have been proposed by researchers and engineers to facilitate the diagnoses of the proper sites for wind farm plantation. This is a Multi-Criteria Decision Making (MCDM) problem. GIS (Graphic Information System) has been a useful tool for research and gathering information in this regard [1]. It is most often combined with MCDM, and this combination gives a concept termed as GIS-MCDM that is being frequently used in recent years for the purpose being discussed [275,276]. Geovanna Villacreses et al. [1] used OCRA, VIKOR, TOPSIS, GIS-MCDM, and AHP-based nine various methodologies for evaluating the location of wind farms in Ecuador, a South American country. Tim Höfer et al. [277], in their literature review, comprehensively enlisted the different techniques and approaches used by other researchers.

### 3.9. Wide Area Monitoring (Measurement) Systems

High rate and speedy advancement in the information technology sector and the deficiency of SCADA to respond to rapid system changes have stirred the researchers to make use of online tools for the quick assessment of security parameters [278]. To fulfill these necessities, a real-time monitoring system like WAMS is indispensable to assist the system operators with the real-time know-how about different instability occasions and enables them to bring about fast actions to mitigate these issues [279]. Wide Area Monitoring System (WAMS) is a modern concept, which is employed for the prior visualization and conception of transmission capacity and thermal state of the transmission line [280]. WAMS also monitors the oscillations in frequency or voltage to preclude the voltage collapse [281]. A remarkable plus point of WAMS is that the data collected are long-lasting and highly abundant [282]. A vast range of WAMS applications includes island operation and online stability assessment to trace the voltage security and stability against various interruptions in a power system [283,284].

Usually, WAMS is based on a reliable communication system, which connects the power system with the control centers. WAMS is frequently equipped with Phasor Measurement Unit (PMU) located at specific positions in a system to evaluate the phasor voltage of a station and the line current of the adjacent station [285]. The chief responsibility of PMU is that it samples and uploads the time-synchronized real-time data through that communication system [279]. A neuro-fuzzy based novel technique for the accurate online assessment of dynamic voltage stability is proposed in [286] that makes use of synchro-phasors determined by PMUs in a wide area monitoring system. Seyed Sina Mousavi-Seyedi et al. used an algorithm for online monitoring of series-compensated transmission lines. This technique used a few PMUs instead of installing PMUs on both ends of a line [287]. Figure 6 describes the simple flow-chart based on the basic operating scheme of WAMS. Several PMUs are located at different sites, and the signals from these sites are fetched by the Phasor Data Concentrator (PDC) via fast communication channels like fiber optics, and, in turn, the voltage and angle stabilities of the power system are monitored.

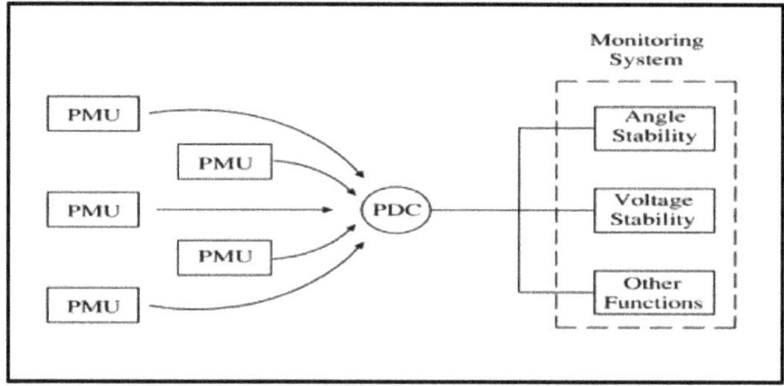


**Figure 6.** Flow chart of operating schemes of WAMS.

With the help of the synchro-phasor system, feasibility for the assessment of real-time transient stability has been increased. Due to fast and reliable real-time synchronized measurements by WAMS, the operator can perform the control actions by accurately monitoring the recently recovered system falling towards instability once again [288,289]. Many researchers have suggested various load shedding schemes based on WAMS, considering only the static characteristics of the induction motor to control the voltage within the permissible limits, but [290] implemented a novel load shedding control strategy based on the dynamic behavior of induction motor, incorporating WAMS with R-index, and proved its authentication on the Power System Analysis Software Package (PSASP) version 34. The authors in [291] derived WAMS-based voltage stability L-Q and L-P sensitivities, making use of the L-index, which determine the reactive power compensation and emergency load shedding for voltage control. In their case study, Pavel Hering et al. [292] presented the two key applications of PMU-based WAMS system in the distribution system. The first case study is about on-line ampacity monitoring by using WAMS, while the second application is about frequency oscillation monitoring. By using the WAMS/PMU technique, Chun Wang et al. [292] investigated the fault location model, which uses synchronized fault voltages.

In the real-time power system, paper [293] discussed a WAMS-based voltage stability indicator that takes into account reactive margin, generator capacity limit, etc., to evaluate system security and voltage collapse proximity. In [294], ZHAO Jinli et al. developed a method for the online assessment of voltage stability based on WAMS. Critical regions of voltage instability were determined by using the Recursive Spectral Partitioning algorithm.

To put into a nutshell, WAMS network has increased the system ability to respond to fast changes arising in system conditions. Even though the SCADA system is mature, the WAMS system has great accuracy with a higher speed of convergence [278].

## 4. Discussion

Despite significant efforts by the researchers, the voltage stability issue in the wind-integrated power systems still confronts many challenges. One of the major problems is the grid integration and intermittency of wind farms. It is very difficult to predict and manage the fluctuations in voltage caused by the unpredictable and intermittent nature of wind power, therefore resulting in grid instability. Furthermore, in developing countries, the majority of the existing power grids are not flexible and capable enough to integrate large wind farms, and stability concerns can be exacerbated owing to long transmission lines with limited current carrying capacity that are thereby unable to transfer reactive power from wind farms for voltage support. Moreover, reactive power management requires coordination among grid operators, wind farms, and distributed energy resources such as the energy storage system (ESS). Therefore, communication infrastructure and advanced control systems must be implemented for efficient coordination, which can be complex and costly. Tailored and flexible control strategies for each type of wind turbines are required as converter technologies in different wind turbines (e.g., DFIG vs. full-converter) have varying capabilities for reactive power control.

In order to effectively analyze the impact of wind integration on voltage stability in a complex grid, accurate modeling of the dynamic behavior of wind farms is still a challenge, especially with the increase in wind penetration levels. Furthermore, for accurate forecasting and control of the wind farms, data-driven approaches such as machine learning need to be incorporated. However, their integration in the existing grid requires further research and development. To deal with the forecasting issue owing to the inherent uncertainty associated with wind power, robust uncertainty management techniques must be employed to avoid a voltage stability problem. Another major challenge in the wind-integrated power systems is the cost involved for upgradation of existing grids in order to integrate grid support technologies such as FACTS devices or distributed generation (DG). Moreover, attractive regulatory frameworks must be developed for adequate incentivization of the investments in reactive power support and adaptable control strategies for voltage stability management. Establishment of protocols and clear standards for control and communication among grid operators and wind farms is still a major concern for the improvement of system-wide stability and to facilitate efficient integration of FACTS devices and DGs.

The integration of other renewable energy resources such as solar power in the wind-powered electrical systems may enhance the overall flexibility and reactive power support to improve the voltage stability. However, it poses further challenges owing to the intermittency and variability of solar power that need to be addressed in conjunction with wind power integration. Furthermore, in order to ensure grid stability and maximize the benefits of renewable energy integration, the optimal and coordinated control techniques for wind farms, solar plants, and other renewable energy resources (RES) still remain an ongoing challenge.

Continuous research efforts, technological advancements, and collaborations between researchers, policymakers, and industry stakeholders are required to address the aforementioned challenges. Therefore, we can pave the way for a reliable and efficient integration of wind power into power systems by devising innovative strategies and surmounting these hurdles.

## 5. Conclusions

Voltage stability in wind-integrated power systems is one of the major concerns to deal with for a secure and reliable grid. Therefore, a comprehensive analysis focusing on the complexities associated with voltage instability and its implications for wind power integration with the power system is provided in this manuscript. To ensure the stable and reliable integration of wind energy, the development and implementation of grid codes are necessitated, owing to the intermittent nature of wind power generation. In developed countries, the increasing integration of wind farms with the existing power network have resulted in voltage instability issues due to the fluctuation in wind power. Consequently, the focus of power system utilities has been shifted towards stability concerns over power quality. To address the challenges of voltage instability, a range of techniques with a specific focus on FACTS devices and WAMS have been discussed. The significance of real-time monitoring systems such as WAMS is emphasized to alleviate potential issues by taking swift actions against instability events due to the availability of timely information. Moreover, to identify weak buses in the power system and assess overall voltage stability, a comprehensive list of Voltage Stability Indices is presented in the manuscript. It also serves as a practical tool for researchers in the evaluation and enhancement of voltage stability in wind-integrated power systems.

To maximize the benefits and the smooth transition of wind power, advanced forecasting and prediction models along with flexible and adaptive control systems and energy storage solutions should be employed. Furthermore, developments in material and blade design as well as an improved drivetrain and efficient generator design should be developed to optimize wind turbine technology. Moreover, to analyze the impacts of integrating other renewable energy resources with wind farms, further detailed studies are required.

**Author Contributions:** Conceptualization, F.H.M., Z.M.H. and M.L.; methodology, F.H.M., T.U.R., M.E. and M.F.; software, F.H.M., M.W.K. and Z.M.H.; validation, T.U.R., Z.M.H. and M.L.; formal analysis, F.H.M., M.E. and M.F.; investigation, M.W.K., T.U.R. and Z.M.H.; resources, T.U.R. and M.L.; data curation, M.W.K. and M.F.; writing—F.H.M. and Z.M.H.; writing—review and editing, T.U.R. and M.L.; visualization, M.W.K., T.U.R. and M.F.; supervision, Z.M.H. and M.L. All authors have read and agreed to the published version of the manuscript.

**Funding:** This research received no external funding.

**Data Availability Statement:** Not applicable.

**Acknowledgments:** We are very grateful to the Office of Research, Innovation and Commercialization (ORIC), The Islamia University of Bahawalpur, Pakistan (No. 3900/ORIC/IUB/2021) for their support in this research.

**Conflicts of Interest:** The authors declare no conflicts of interest.

# References

1. Villacreses, G.; Gaona, G.; Martínez-Gómez, J.; Jijón, D.J. Wind farms suitability location using geographical information system (GIS), based on multi-criteria decision making (MCDM) methods: The case of continental Ecuador. *Renew. Energy* **2017**, *109*, 275–286. [CrossRef]
2. Meegahapola, L.; Littler, T.; Perera, S. Capability curve based enhanced reactive power control strategy for stability enhancement and network voltage management. *Int. J. Electr. Power Energy Syst.* **2013**, *52*, 96–106. [CrossRef]
3. Goh, H.H.; Chua, Q.S.; Lee, S.W.; Kok, B.C.; Goh, K.C.; Teo, K.T.K. Evaluation for Voltage Stability Indices in Power System Using Artificial Neural Network. *Procedia Eng.* **2015**, *118*, 1127–1136. [CrossRef]
4. Ahmed, M.; EL-Shimy, M.; Badr, M.A. Advanced modeling and analysis of the loading capability li mits of doubly-fed induction generators. *Sustain. Energy Technol. Assess.* **2014**, *7*, 79–90.
5. Khezri, R.; Bevrani, H. Voltage performance enhancement of DFIG-based wind farms integrated in large-scale power systems: Coordinated AVR and PSS. *Int. J. Electr. Power Energy Syst.* **2015**, *73*, 400–410. [CrossRef]
6. Tohidi, S.; Tavner, P.; McMahon, R.; Oraee, H.; Zolghadri, M.R.; Shao, S.; Abdi, E. Low voltage ride-through of DFIG and brushless DFIG: Similarities and differences. *Electr. Power Syst. Res.* **2014**, *110*, 64–72. [CrossRef]
7. Rahimi, M. Coordinated control of rotor and grid sides converters in DFIG based wind turbines for providing optimal reactive power support and voltage regulation. *Sustain. Energy Technol. Assess.* **2017**, *20*, 47–57. [CrossRef]
8. Liu, X. Quality of Optical Channels in Wireless SCADA for Offshore Wind Farms. *IEEE Trans. Smart Grid* **2012**, *3*, 225–232. [CrossRef]
9. Hosseini, S.H.; Tang, C.Y.; Jiang, J.N. Calibration of a Wind Farm Wind Speed Model with Incomplete Wind Data. *IEEE Trans. Sustain. Energy* **2014**, *5*, 343–350. [CrossRef]
10. Ericsson, G.N. Cyber Security and Power System Communication—Essential Parts of a Smart Grid Infrastructure. *IEEE Trans. Power Deliv.* **2010**, *25*, 1501–1507. [CrossRef]
11. Macdonald, H.; Hawker, G.; Bell, K. Analysis of wide-area availability of wind generators during storm events. In Proceedings of the 2014 International Conference on Probabilistic Methods Applied to Power Systems (PMAPS), Durham, UK, 7–10 July 2014; pp. 1–6.

12. Baghaee, H.R.; Jannati, M.; Vahidi, B.; Hosseinian, S.H.; Rastegar, H. Improvement of voltage stability and reduce power system losses by optimal GA-based allocation of multi-type FACTS devices. In Proceedings of the 2008 11th International Conference on Optimization of Electrical and Electronic Equipment, Brasov, Romania, 22–24 May 2008; pp. 209–214.
13. Deb, G.; Chakraborty, K.; Deb, S. Voltage stability analysis using reactive power loading as indicator and its improvement by FACTS device. In Proceedings of the 2016 IEEE 1st International Conference on Power Electronics, Intelligent Control and Energy Systems (ICPEICES), Delhi, India, 4–6 July 2016; pp. 1–5.
14. Ross, M.; Borodulin, M.; Kazachkov, Y. Using D-SMES devices to improve the voltage stability of a transmission system. In Proceedings of the 2001 IEEE/PES Transmission and Distribution Conference and Exposition. Developing New Perspectives (Cat. No.01CH37294), Atlanta, GA, USA, 2 November 2001; Volume 2, pp. 1144–1148.
15. Ruan, J.Y.; Lu, Z.X.; Qiao, Y.; Min, Y. Analysis on Applicability Problems of the Aggregation-Based Representation of Wind Farms Considering DFIGs' LVRT Behaviors. *IEEE Trans. Power Syst.* **2016**, *31*, 4953–4965. [CrossRef]
16. Wen, G.; Chen, Y.; Zhong, Z.; Kang, Y. Dynamic Voltage and Current Assignment Strategies of Nine-Switch-Converter-Based DFIG Wind Power System for Low-Voltage Ride-Through (LVRT) under Symmetrical Grid Voltage Dip. *IEEE Trans. Ind. Appl.* **2016**, *52*, 3422–3434. [CrossRef]
17. Rashid, G.; Ali, M.H. Nonlinear Control-Based Modified BFCL for LVRT Capacity Enhancement of DFIG-Based Wind Farm. *IEEE Trans. Energy Convers.* **2017**, *32*, 284–295. [CrossRef]
18. Kundur, P.; Paserba, J.; Ajjarapu, V.; Andersson, G.; Bose, A.; Canizares, C.; Hatziargyriou, N.; Hill, D.; Stankovic, A.; Taylor, C.; et al. Definition and classification of power system stability IEEE/CIGRE joint task force on stability terms and definitions. *IEEE Trans. Power Syst.* **2004**, *19*, 1387–1401.
19. de Leon, J.A.D.; Taylor, C.W. Understanding and solving short-term voltage stability problems. In Proceedings of the IEEE Power Engineering Society Summer Meeting, Chicago, IL, USA, 21–25 July 2002; Volume 2, pp. 745–752.
20. Nikolaidis, V.C. Emergency Zone 3 Modification as a Local Response-Driven Protection Measure Against System Collapse. *IEEE Trans. Power Deliv.* **2016**, *31*, 2114–2122. [CrossRef]
21. Johnson, M.; Vaziri, M.; Vadhva, S. Understanding slow voltage instability. In Proceedings of the 2012 IEEE 13th International Conference on Information Reuse & Integration (IRI), Las Vegas, NV, USA, 8–10 August 2012; pp. 502–508.
22. Duan, J.; Huang, J. The Mechanism of Voltage Instability Analysis Considering Load Characteristic. *Energy Power Eng.* **2013**, *5*, 1497–1502. [CrossRef]
23. Sauer, P.W.; Lesieutre, B.C.; Pai, M.A. Maximum loadability and voltage stability in power systems. *Int. J. Electr. Power Energy Syst.* **1993**, *15*, 145–153. [CrossRef]
24. Williams, B.R.; Schmus, W.R.; Dawson, D.C. Transmission voltage recovery delayed by stalled air conditioner compressors. *IEEE Trans. Power Syst.* **1992**, *7*, 1173–1181. [CrossRef]
25. Tirtashi, M.R.S.; Samuelsson, O.; Svensson, J. Dynamic and static analysis of the shunt capacitors control effect on the long-term voltage instability. In Proceedings of the 2016 IEEE Power and Energy Society General Meeting (PESGM), Boston, MA, USA, 17–21 July 2016; pp. 1–5.
26. de Mello, F.P.; Feltes, J.W. Voltage oscillatory instability caused by induction motor loads. *IEEE Trans. Power Syst.* **1996**, *11*, 1279–1285. [CrossRef]
27. Van Cutsem, T. Voltage instability: Phenomena, countermeasures, and analysis methods. *Proc. IEEE* **2000**, *88*, 208–227. [CrossRef]
28. Augugliaro, A.; Dusonchet, L.; Mangione, S. Voltage collapse proximity indicators for radial distribution networks. In Proceedings of the 2007 9th International Conference on Electrical Power Quality and Utilisation, Barcelona, Spain, 9–11 October 2007; pp. 1–6.
29. Amjady, N.; Ansari, M.R. Small disturbance voltage stability assessment of power systems by modal analysis and dynamic simulation. *Energy Convers. Manag.* **2008**, *49*, 2629–2641. [CrossRef]
30. Atputharajah, A.; Saha, T.K. Power system blackouts—Literature review. In Proceedings of the 2009 International Conference on Industrial and Information Systems (ICIIS), Peradeniya, Sri Lanka, 28–31 December 2009; pp. 460–465.
31. Brayley, H.; Redfern, M.A.; Bo, Z.Q. The Public Perception of Power Blackouts. In Proceedings of the 2005 IEEE/PES Transmission & Distribution Conference & Exposition: Asia and Pacific, Dalian, China, 15–18 August 2005; pp. 1–5.
32. Kamali, S.; Amraee, T. Blackout prediction in interconnected electric energy systems considering generation re-dispatch and energy curtailment. *Appl. Energy* **2017**, *187*, 50–61. [CrossRef]
33. Wang, P.; Qin, W.; Han, X.; Ding, Y.; Du, X. Reliability assessment of power systems considering reactive power sources. In Proceedings of the 2009 IEEE Power & Energy Society General Meeting, Calgary, AB, Canada, 26–30 July 2009; pp. 1–7.
34. Xu, C.; Li, P.; Li, X.; Chen, D.; Zhang, Y.; Lei, B. Small disturbance voltage stability considering thermostatically controlled load. In Proceedings of the 2011 International Conference on Advanced Power System Automation and Protection, Beijing, China, 16–20 October 2011; pp. 862–866.
35. Chen, H.; Chen, J.; Shi, D.; Duan, X. Power flow study and voltage stability analysis for distribution systems with distributed generation. In Proceedings of the 2006 IEEE Power Engineering Society General Meeting, Montreal, QC, Canada, 18–22 June 2006; p. 8.
36. Veloza, O.P.; Santamaria, F. Analysis of major blackouts from 2003 to 2015: Classification of incidents and review of main causes. *Electr. J.* **2016**, *29*, 42–49. [CrossRef]
37. Haque, M.H. Determination of steady-state voltage stability limit of a power system in the presence of SVC. In Proceedings of the 2001 IEEE Porto Power Tech Proceedings (Cat. No.01EX502), Porto, Portugal, 10–13 September 2001; Volume 2, p. 6.

38. Haque, M.H. Use of V–I characteristic as a tool to assess the static voltage stability limit of a power system. *IEE Proc.-Gener. Transm. Distrib.* **2004**, *151*, 1–7. [CrossRef]
39. Tripathy, M.; Mishra, S. Optimizing Voltage Stability Limit and Real Power Loss in a Large Power System using Bacteria Foraging. In Proceedings of the 2006 International Conference on Power Electronics, Drives and Energy Systems, New Delhi, India, 12–15 December 2006; pp. 1–6.
40. Ramalingam, K.; Indulkar, C.S. Determination of Steady State Voltage Stability limit using PQ curves for voltage sensitive loads. In Proceedings of the 2008 Joint International Conference on Power System Technology and IEEE Power India Conference, New Delhi, India, 12–15 December 2008; pp. 1–5.
41. Koessler, R.J. Voltage instability/collapse—An overview. In Proceedings of the IEE Colloquium on Voltage Collapse (Digest No: 1997/101), London, UK, 24 April 1997; pp. 1/1–1/6.
42. Tiranuchit, A.; Thomas, R.J. A posturing strategy against voltage instabilities in electric power systems. *IEEE Trans. Power Syst.* **1988**, *3*, 87–93. [CrossRef]
43. Zhang, Y.; Song, W. Power system static voltage stability limit and the identification of weak bus. In Proceedings of the TENCON '93. IEEE Region 10 International Conference on Computers, Communications and Automation, Beijing, China, 19–21 October 1993; Volume 5, pp. 157–160.
44. Yokoyama, A.; Kumano, T.; Sekine, Y. Static voltage stability index using multiple load-flow solutions. *Elect. Eng. Jpn.* **1991**, *111*, 69–79. [CrossRef]
45. Obadina, O.O.; Berg, G.J. Determination of voltage stability limit in multimachine power systems. *IEEE Trans. Power Syst.* **1988**, *3*, 1545–1554. [CrossRef]
46. Le, H.T.; Santoso, S.; Nguyen, T.Q. Augmenting Wind Power Penetration and Grid Voltage Stability Limits Using ESS: Application Design, Sizing, and a Case Study. *IEEE Trans. Power Syst.* **2012**, *27*, 161–171. [CrossRef]
47. Wannoi, C.; Khumdee, A.; Wannoi, N. An Optimum Technique for Renewable Power Generations Integration to Power System Using Repeated Power Flow Technique Considering Voltage Stability Limit. *Procedia Comput. Sci.* **2016**, *86*, 357–360. [CrossRef]
48. Tripathy, M.; Mishra, S. Bacteria Foraging-Based Solution to Optimize Both Real Power Loss and Voltage Stability Limit. *IEEE Trans. Power Syst.* **2007**, *22*, 240–248. [CrossRef]
49. Balachennaiah, P.; Suryakalavathi, M.; Nagendra, P. Optimizing real power loss and voltage stability limit of a large transmission network using firefly algorithm. *Eng. Sci. Technol. Int. J.* **2016**, *19*, 800–810. [CrossRef]
50. Duan, J.; Zhu, S. The Effect of OLTC on Static Voltage Stability Limit. In Proceedings of the 2011 Asia-Pacific Power and Energy Engineering Conference, Wuhan, China, 25–28 March 2011; pp. 1–4.
51. Raviprakasha, N.S.; Ramar, K. Maximum power transfer in AC transmission lines limited by voltage stability. In Proceedings of the International Conference on Power Electronics, Drives and Energy Systems for Industrial Growth, New Delhi, India, 8–11 January 1996; Volume 2, pp. 776–781.
52. Liu, H.; Jin, L.; McCalley, J.; Kumar, R.; Ajjarapu, V.; Elia, N. Planning reconfigurable reactive control for voltage stability limited power systems. In Proceedings of the 2009 IEEE Power & Energy Society General Meeting, Calgary, AB, Canada, 26–30 July 2009; p. 1.
53. NERC Transmission Transfer Capability Task Force. *Available Transfer Capability Definitions and Determination*; North American Electric Reliability Council: North Princeton, NJ, USA, 1996.
54. Jain, T.; Singh, S.N.; Srivastava, S.C. Dynamic available transfer capability computation using a hybrid approach. *IET Gener. Transm. Distrib.* **2008**, *2*, 775–788. [CrossRef]
55. Shaaban, M.; Li, W.; Liu, H.; Yan, Z.; Ni, Y.; Wu, F. ATC calculation with steady-state security constraints using Benders decomposition. *IEE Proc.-Gener. Transm. Distrib.* **2003**, *150*, 611–615. [CrossRef]
56. Li, W.; Chen, T.; Xu, W. On impedance matching and maximum power transfer. *Electr. Power Syst. Res.* **2010**, *80*, 1082–1088. [CrossRef]
57. Indulkar, C.S.; Viswanathan, B.; Venkata, S.S. Maximum power transfer limited by voltage stability in series and shunt compensated schemes for AC transmission systems. *IEEE Trans. Power Deliv.* **1989**, *4*, 1246–1252. [CrossRef]
58. Basu, K.P. Power transfer capability of transmission line limited by voltage stability: Simple analytical expressions. *IEEE Power Eng. Rev.* **2000**, *20*, 46–47. [CrossRef]
59. Jadhav, H.T.; Roy, R. A comprehensive review on the grid integration of doubly fed induction generator. *Int. J. Electr. Power Energy Syst.* **2013**, *49*, 8–18. [CrossRef]
60. Wei, X.; Chow, J.H.; Fardanesh, B.; Edris, A.A. A dispatch strategy for a unified power-flow controller to maximize voltage-stability-limited power transfer. *IEEE Trans. Power Deliv.* **2005**, *20*, 2022–2029. [CrossRef]
61. Wei, X.; Cjow, J.H.; Edris, A.A.; Fardanesh, B. A dispatch strategy for multiple unified power flow controllers to maximize voltage-stability limited power transfer. In Proceedings of the IEEE Power Engineering Society General Meeting, San Francisco, CA, USA, 12–16 June 2005; Volume 1, pp. 865–870.
62. Tobón; Esteban, J.; Ramirez, J.M.; Gutierrez, R.E.C. Tracking the maximum power transfer and loadability limit from sensitivities-based impedance matching. *Electr. Power Syst. Res.* **2015**, *119*, 355–363. [CrossRef]
63. El-Metwally, M.M.; El-Emary, A.A.; El-Azab, M. Effect of load characteristics on maximum power transfer limit for HV compensated transmission lines. *Int. J. Electr. Power Energy Syst.* **2004**, *26*, 467–472. [CrossRef]

64. Ramar, K.; Raviprakasha, M.S. Design of compensation schemes for long AC transmission lines for maximum power transfer limited by voltage stability. *Int. J. Electr. Power Energy Syst.* **1995**, *17*, 83–89. [CrossRef]
65. Gayathri; Reddy, K.; Maruteswar, G.V. Frequency and Voltage Stability Assessment Applied to Load Shedding. *Int. J. Eng. Trends Technol.* **2015**, *19*, 106–115. [CrossRef]
66. Meegahapola, L.; Littler, T. Characterisation of large disturbance rotor angle and voltage stability in interconnected power networks with distributed wind generation. *IET Renew. Power Gener.* **2015**, *9*, 272–283. [CrossRef]
67. Glavic, M.; Van Cutsem, T. Detecting with PMUs the onset of voltage instability caused by a large disturbance. In Proceedings of the 2008 IEEE Power and Energy Society General Meeting—Conversion and Delivery of Electrical Energy in the 21st Century, Pittsburgh, PA, USA, 20–24 July 2008; pp. 1–8.
68. Guedes, R.B.L.; Silva, F.H.J.R.; Alberto, L.F.C.; Bretas, N.G. Large disturbance voltage stability assessment using extended Lyapunov function and considering voltage-dependent active loads. In Proceedings of the IEEE Power Engineering Society General Meeting, San Francisco, CA, USA, 12–16 June 2005; Volume 2, pp. 1760–1767.
69. Yong, T.; Shiying, M.; Wuzhi, Z. Mechanism Research of Short-Term Large-Disturbance Voltage Stability. In Proceedings of the 2006 International Conference on Power System Technology, Chongqing, China, 22–26 October 2006; pp. 1–5.
70. Huadong, S.; Xiaoxin, Z. A Quick Criterion on Judging Short-Term Large-Disturbance Voltage Stability Considering Dynamic Characteristic of Induction Motor Loads. In Proceedings of the 2006 International Conference on Power System Technology, Chongqing, China, 22–26 October 2006; pp. 1–6.
71. Kumar, G.N.; Kalavathi, M.S. Reactive power compensation for large disturbance voltage stability using FACTS controllers. In Proceedings of the 2011 3rd International Conference on Electronics Computer Technology, Kanyakumari, India, 8–10 April 2011; pp. 164–167.
72. Zheng, X.Y.; He, R.M.; Ma, J. A new load model suitable for transient stability analysis with large voltage disturbances. In Proceedings of the 2010 International Conference on Electrical Machines and Systems, Incheon, Republic of Korea, 10–13 October 2010; pp. 1898–1902.
73. Najafi, H.R.; Robinson, F.V.P.; Dastyar, F.; Samadi, A.A. Small-disturbance voltage stability of distribution systems with wind turbine implemented with WRIG. In Proceedings of the 2009 International Conference on Power Engineering, Energy and Electrical Drives, Lisbon, Portugal, 18–20 March 2009; pp. 191–195.
74. Wang, W.; Ni, J.; Ngan, H.W. Small-Disturbance Voltage Stability Study on Shaanxi Power System. In Proceedings of the 2005 IEEE/PES Transmission & Distribution Conference & Exposition: Asia and Pacific, Dalian, China, 15–18 August 2005; pp. 1–5.
75. Amjady, N.; Ansari, M.R. Small disturbance voltage stability evaluation of power systems. In Proceedings of the 2008 IEEE/PES Transmission and Distribution Conference and Exposition, Chicago, IL, USA, 21–24 April 2008; pp. 1–9.
76. Rumiantsev, V.V.; Oziraner, A.S. *Stability and Stabilization of Motion with Respect to Part of the Variables*; Moscow Izdatel Nauka: Moscow, Russia, 1987.
77. Wang, L.; Girgis, A.A. On-line detection of power system small disturbance voltage instability. *IEEE Trans. Power Syst.* **1996**, *11*, 1304–1313. [CrossRef]
78. Xu, C.; Liang, J.; Yun, Z.; Zhang, L. The Small-disturbance Voltage Stability Analysis through Adaptive AR Model Based on PMU. In Proceedings of the 2005 IEEE/PES Transmission & Distribution Conference & Exposition: Asia and Pacific, Dalian, China, 15–18 August 2005; pp. 1–5.
79. Khoshkhoo, H.; Shahrtash, S.M. On-line small disturbance voltage stability assessment in power systems based on wide area measurement. In Proceedings of the 2011 10th International Conference on Environment and Electrical Engineering, Rome, Italy, 8–11 May 2011; pp. 1–5.
80. Preece, R.; Milanović, J.V. Risk-Based Small-Disturbance Security Assessment of Power Systems. *IEEE Trans. Power Deliv.* **2015**, *30*, 590–598. [CrossRef]
81. Kundur, P. *Power System Stability and Control*, 1st ed.; McGraw Hill: New York, NY, USA, 1994.
82. Farmer, R.G. Power System Dynamics and Stability. In *The Electric Power Engineering Handbook*; Grigsby, L.L., Ed.; CRC Press LLC: Boca Raton, FL, USA, 2001.
83. Londero, R.R.; Affonso, C.M.; Vieira, J.P.A.; Bezerra, U.H. Impact of different DFIG wind turbines control modes on long-term voltage stability. In Proceedings of the 2012 3rd IEEE PES Innovative Smart Grid Technologies Europe (ISGT Europe), Berlin, Germany, 14–17 October 2012; pp. 1–7.
84. Liu, J.H.; Chu, C.C. Long-Term Voltage Instability Detections of Multiple Fixed-Speed Induction Generators in Distribution Networks Using Synchrophasors. *IEEE Trans. Smart Grid* **2015**, *6*, 2069–2079. [CrossRef]
85. Wang, Y.; Chan, K.W.; Mei, S.; Zhang, Y. A novel criterion on judging long-term voltage stability. In Proceedings of the 2008 Third International Conference on Electric Utility Deregulation and Restructuring and Power Technologies, Nanjing, China, 6–9 April 2008; pp. 1542–1547.
86. Löf, P.-A.; Hill, D.J.; Arnborg, S.; Andersson, G. On the analysis of long-term voltage stability. *Int. J. Electr. Power Energy Syst.* **1993**, *15*, 229–237. [CrossRef]
87. Assis, T.M.L.; Nunes, A.R.; Falcao, D.M. Mid and Long-Term Voltage Stability Assessment using Neural Networks and Quasi-Steady-State Simulation. In Proceedings of the 2007 Large Engineering Systems Conference on Power Engineering, Montreal, QC, Canada, 10–12 October 2007; pp. 213–217.

88. Freitas, W.; Vieira, J.C.M.; da Suva, L.C.P.; Affonso, C.M.; Morelato, A. Long-term voltage stability of distribution systems with induction generators. In Proceedings of the IEEE Power Engineering Society General Meeting, San Francisco, CA, USA, 12–16 June 2005; Volume 3, pp. 2910–2913.

89. Xu, T.; Taylor, P.C. Voltage Control Techniques for Electrical Distribution Networks Including Distributed Generation. *IFAC Proc. Vol.* **2008**, *41*, 11967–11971. [CrossRef]

90. Liu, S.; Liu, M.; Xie, M. MPC-based Load Shedding for Long-term Voltage Stability Enhancement Using Trajectory Sensitivities. In Proceedings of the 2010 Asia-Pacific Power and Energy Engineering Conference, Chengdu, China, 28–31 March 2010; pp. 1–5.

91. Perez, A.; Jóhannsson, H.; Vancraeyveld, P.; Östergaard, J. Suitability of voltage stability study methods for real-time assessment. In Proceedings of the IEEE PES ISGT Europe 2013, Lyngby, Denmark, 6–9 October 2013; pp. 1–5.

92. Zheng, W.; Liu, M. Long Term Voltage Stability Enhancement Using Direct Dynamic Optimization Method. In Proceedings of the 2010 Asia-Pacific Power and Energy Engineering Conference, Chengdu, China, 28–31 March 2010; pp. 1–4.

93. Stewart, V.; Camm, E.H. Modeling of stalled motor loads for power system short-term voltage stability analysis. In Proceedings of the IEEE Power Engineering Society General Meeting, San Francisco, CA, USA, 12–16 June 2005; Volume 2, pp. 1887–1892.

94. Dong, Y.; Xie, X.; Zhou, B.; Shi, W.; Jiang, Q. An integrated high side var-voltage control strategy to improve short-term voltage stability of receiving-end power systems. In Proceedings of the 2016 IEEE Power and Energy Society General Meeting (PESGM), Boston, MA, USA, 17–21 July 2016; p. 1.

95. Tiwari, A.; Ajjarapu, V. Addressing short term voltage stability problem—Part I: Challenges and plausible solution directions. In Proceedings of the 2016 IEEE/PES Transmission and Distribution Conference and Exposition (T&D), Dallas, TX, USA, 3–5 May 2016; pp. 1–5.

96. Vournas, C.D.; Potamianakis, E.G. Induction Machine Short-term Voltage Stability and Protection Measures. In Proceedings of the 2006 IEEE PES Power Systems Conference and Exposition, Atlanta, GA, USA, 29 October–1 November 2006; pp. 993–998.

97. Dong, Y.; Xie, X.; Shi, W.; Zhou, B.; Jiang, Q. Demand-Response Based Distributed Preventive Control to Improve Short-term Voltage Stability. *IEEE Trans. Smart Grid* **2017**, *9*, 4785–4795. [CrossRef]

98. Hossain, M.J.; Pota, H.R.; Ugrinovskii, V. Short and Long-Term Dynamic Voltage Instability. *IFAC Proc. Vol.* **2008**, *41*, 9392–9397. [CrossRef]

99. Han, T. Surrogate Modeling Based Multi-objective Dynamic VAR Planning Considering Short-term Voltage Stability and Transient Stability. *IEEE Trans. Power Syst.* **2017**, *33*, 622–633. [CrossRef]

100. Wehenkel, L.; Pavella, M. Preventive vs. emergency control of power systems. In Proceedings of the IEEE PES Power Systems Conference and Exposition, New York, NY, USA, 10–13 October 2004; Volume 3, pp. 1665–1670.

101. Xu, Y.; Zhang, R.; Zhao, J.; Dong, Z.Y.; Wang, D.; Yang, H.; Wong, K.P. Assessing Short-Term Voltage Stability of Electric Power Systems by a Hierarchical Intelligent System. *IEEE Trans. Neural Netw. Learn. Syst.* **2016**, *27*, 1686–1696. [CrossRef]

102. Kawabe, K.; Tanaka, K. Analytical Method for Short-Term Voltage Stability Using the Stability Boundary in the P-V Plane. *IEEE Trans. Power Syst.* **2014**, *29*, 3041–3047. [CrossRef]

103. Arya, L.D.; Choube, S.C.; Shrivastava, M. Technique for voltage stability assessment using newly developed line voltage stability index. *Energy Convers. Manag.* **2008**, *49*, 267–275. [CrossRef]

104. Reis, C.; Andrade, A.; Maciel, F.P. Line stability indices for voltage collapse prediction. In Proceedings of the 2009 International Conference on Power Engineering, Energy and Electrical Drives, Lisbon, Portugal, 18–20 March 2009; pp. 239–243.

105. Ismail, N.A.M.; Zin, A.A.M.; Khairuddin, A.; Khokhar, S. A comparison of voltage stability indices. In Proceedings of the 2014 IEEE 8th International Power Engineering and Optimization Conference (PEOCO2014), Langkawi, Malaysia, 24–25 March 2014; pp. 30–34.

106. Musirin, I.; Rahman, T.K.A. Novel fast voltage stability index (FVSI) for voltage stability analysis in power transmission system. In Proceedings of the Student Conference on Research and Development, Shah Alam, Malaysia, 17 July 2002; pp. 265–268.

107. Modarresi, J.; Gholipour, E.; Khodabakhshian, A. A comprehensive review of the voltage stability indices, Renewable and Sustainable. *Energy Rev.* **2016**, *63*, 1–12.

108. Cupelli, M.; DoigCardet, C.; Monti, A. Voltage stability indices comparison on the IEEE-39 bus system using RTDS. In Proceedings of the 2012 IEEE International Conference on Power System Technology (POWERCON), Auckland, New Zealand, 30 October–2 November 2012; pp. 1–6.

109. Cupelli, M.; DoigCardet, C.; Monti, A. Comparison of line voltage stability indices using dynamic real-time simulation. In Proceedings of the 2012 3rd IEEE PES Innovative Smart Grid Technologies Europe (ISGT Europe), Berlin, Germany, 14–17 October 2012; pp. 1–8.

110. Lof, P.A.; Andersson, G.; Hill, D.J. Voltage stability indices for stressed power systems. *IEEE Trans. Power Syst.* **1993**, *8*, 326–335. [CrossRef] [PubMed]

111. Oukennou, A.; Sandali, A. Assessment and analysis of Voltage Stability Indices in electrical network using PSAT Software. In Proceedings of the 2016 Eighteenth International Middle East Power Systems Conference (MEPCON), Cairo, Egypt, 27–29 December 2016; pp. 705–710.

112. Moghavvemi, M.; Omar, F.M. Technique for contingency monitoring and voltage collapse prediction. *IEE Proc.-Gener. Transm. Distrib.* **1998**, *145*, 634–640. [CrossRef]

113. Balamourougan, V.; Sidhu, T.S.; Sachdev, M.S. Technique for online prediction of voltage collapse. *IEE Proc.-Gener. Transm. Distrib.* **2004**, *151*, 453–460. [CrossRef]

114. Kamel, M.M.M.; Karrar, A.A.; Eltom, A.H. Development and Application of a New Voltage Stability Index for On-Line Monitoring and Shedding. *IEEE Trans. Power Syst.* **2017**, *33*, 1231–1241. [CrossRef]
115. Wang, Y.; Li, W.; Lu, J. A new node voltage stability index based on local voltage phasors. *Electr. Power Syst. Res.* **2009**, *79*, 265–271. [CrossRef]
116. Verbic, G.; Gubina, F. A new concept of voltage-collapse protection based on local phasors. *IEEE Trans. Power Deliv.* **2004**, *19*, 576–581. [CrossRef]
117. Chattopadhyay, T.K.; Banerjee, S.; Chanda, C.K. Impact of distributed generator on voltage stability analysis of distribution networks under critical loading conditions. In Proceedings of the 2014 1st International Conference on Non Conventional Energy (ICONCE 2014), Kalyani, India, 16–17 January 2014; pp. 288–291.
118. Yazdanpanah-Goharrizi, A.; Asghari, R. A novel line stability index (NLSI) for voltage stability assessment of power systems. In Proceedings of the 7th WSEAS International Conference on Power Systems, Beijing, China, 15–17 September 2007; pp. 164–167.
119. Tiwari, R.; Niazi, K.R.; Gupta, V. Line collapse proximity index for prediction of voltage collapse in power systems. *Int. J. Electr. Power Energy Syst.* **2012**, *41*, 105–111. [CrossRef]
120. Eminoglu, U.; Hocaoglu, M.H. A voltage stability index for radial distribution networks. In Proceedings of the 2007 42nd International Universities Power Engineering Conference, Brighton, UK, 4–6 September 2007; pp. 408–413.
121. Feltes, C.; Engelhardt, S.; Kretschmann, J.; Fortmann, J.; Koch, F.; Erlich, I. High voltage ride-through of DFIG-based wind turbines. In Proceedings of the 2008 IEEE Power and Energy Society General Meeting—Conversion and Delivery of Electrical Energy in the 21st Century, Pittsburgh, PA, USA, 20–24 July 2008; pp. 1–8.
122. Moghavvemi, M.; Faruque, O. Real-time contingency evaluation and ranking technique. *IEE Proc.-Gener. Transm. Distrib.* **1998**, *145*, 517–524. [CrossRef]
123. Mohamed, A.; Jasmon, G.B.; Yusoff, S. A static voltage collapse indicator using line stability factors. *J. Ind. Technol.* **1989**, *7*, 73–85.
124. Althowibi, F.A.; Mustafa, M.W. Power system network sensitivity to Voltage collapse. In Proceedings of the 2012 IEEE International Power Engineering and Optimization Conference, Melaka, Malaysia, 6–7 June 2012; pp. 379–383.
125. Gong, Y.; Schulz, N.; Guzmán, A. Synchrophasor-Based Real-Time Voltage Stability Index. In Proceedings of the 2006 IEEE PES Power Systems Conference and Exposition, Atlanta, GA, USA, 29 October–1 November 2006; pp. 1029–1036.
126. Pérez-Londoño, S.; Rodríguez, L.F.; Olivar, G. A Simplified Voltage Stability Index (SVSI). *Int. J. Electr. Power Energy Syst.* **2014**, *63*, 806–813. [CrossRef]
127. Nizam, M.; Mohamed, A.; Hussain, A. Dynamic Voltage Collapse Prediction in Power Systems Using Power Transfer Stability Index. In Proceedings of the 2006 IEEE International Power and Energy Conference, Putra Jaya, Malaysia, 28–29 November 2006; pp. 246–250.
128. Subramani, C.; Dash, S.S.; Bhaskar, M.A.; Jagadeeshkumar, M.; Sureshkumar, K.; Parthipan, R. Line outage contingency screening and ranking for voltage stability assessment. In Proceedings of the 2009 International Conference on Power Systems, Kharagpur, India, 27–29 December 2009; pp. 1–5.
129. Smon, I.; Verbic, G.; Gubina, F. Local voltage-stability index using tellegen's Theorem. In Proceedings of the 2007 IEEE Power Engineering Society General Meeting, Tampa, FL, USA, 24–28 June 2007; p. 1.
130. Albuquerque, M.A.; Castro, C.A. A contingency ranking method for voltage stability in real time operation of power systems. In Proceedings of the 2003 IEEE Bologna Power Tech Conference Proceedings, Bologna, Italy, 23–26 June 2003; Volume 1, p. 5.
131. Kanimozhi, R.; Selvi, K. A novel line stability index for voltage stability analysis and contingency ranking in power system using fuzzy based load flow. *J. Electr. Eng. Technol. (JEET)* **2013**, *8*, 694–703. [CrossRef]
132. Wiszniewski, A. New Criteria of Voltage Stability Margin for the Purpose of Load Shedding. *IEEE Trans. Power Deliv.* **2007**, *22*, 1367–1371. [CrossRef]
133. Althowibi, F.A.; Mustafa, M.W. Line voltage stability calculations in power systems. In Proceedings of the 2010 IEEE International Conference on Power and Energy, Kuala Lumpur, Malaysia, 29 November–1 December 2010; pp. 396–401.
134. Hawisa, K.; Ibsaim, R.; Daeri, A. Voltage instability remedy using FACTS, TCSC compensation first A. In Proceedings of the 2016 17th International Conference on Sciences and Techniques of Automatic Control and Computer Engineering (STA), Sousse, Tunisia, 19–21 December 2016; pp. 104–107.
135. Jena, R.; Swain, S.C.; Panda, P.C.; Roy, A. Analysis of voltage and loss profile using various FACTS devices. In Proceedings of the 2016 International Conference on Signal Processing, Communication, Power and Embedded System (SCOPES), Paralakhemundi, Odisha, 3–5 October 2016; pp. 787–792.
136. Hridya, K.R.; Mini, V.; Visakhan, R.; Kurian, A.A. Analysis of voltage stability enhancement of a grid and loss reduction using series FACTS controllers. In Proceedings of the 2015 International Conference on Power, Instrumentation, Control and Computing (PICC), Thrissur, India, 9–11 December 2015; pp. 1–5.
137. Roselyn, J.P.; Devaraj, D.; Dash, S.S. Multi-Objective Genetic Algorithm for voltage stability enhancement using rescheduling and FACTS devices. *Ain Shams Eng. J.* **2014**, *5*, 789–801. [CrossRef]
138. Wibowo, R.S.; Yorino, N.; Eghbal, M.; Zoka, Y.; Sasaki, Y. FACTS Devices Allocation with Control Coordination Considering Congestion Relief and Voltage Stability. *IEEE Trans. Power Syst.* **2011**, *26*, 2302–2310. [CrossRef]
139. Nascimento, S.D.; Gouvêa, M.M. Voltage Stability Enhancement in Power Systems with Automatic Facts Device Allocation. *Energy Procedia* **2017**, *107*, 60–67. [CrossRef]

140. Tiwari, R.; Niazi, K.R.; Gupta, V. Optimal location of FACTS devices for improving performance of the power systems. In Proceedings of the 2012 IEEE Power and Energy Society General Meeting, San Diego, CA, USA, 22–26 July 2012; pp. 1–8.

141. Tiwari, P.K.; Sood, Y.R. Optimal location of FACTS devices in power system using Genetic Algorithm. In Proceedings of the 2009 World Congress on Nature & Biologically Inspired Computing (NaBIC), Coimbatore, India, 9–11 December 2009; pp. 1034–1040.

142. Saravanan, M.; Slochanal, S.M.R.; Venkatesh, P.; Abraham, P.S. Application of PSO technique for optimal location of FACTS devices considering system loadability and cost of installation. In Proceedings of the 2005 International Power Engineering Conference, Singapore, 29 November–2 December 2005; Volume 2, pp. 716–721.

143. Yu, D.; Li, M.S.; Ji, T.Y.; Wu, Q.H. Optimal voltage control of power systems with uncertain wind power using FACTS devices. In Proceedings of the 2016 IEEE Innovative Smart Grid Technologies—Asia (ISGT-Asia), Melbourne, VIC, Australia, 28 November–1 December 2016; pp. 937–941.

144. Rao, N.S.; Amarnath, J.; Rao, V.P. Effect of FACTS devices on enhancement of Voltage Stability in a deregulated power system. In Proceedings of the 2014 International Conference on Circuits, Power and Computing Technologies [ICCPCT-2014], Nagercoil, India, 20–21 March 2014; pp. 642–650.

145. Lakkireddy, J.; Rastgoufard, R.; Leevongwat, I.; Rastgoufard, P. Steady-state voltage stability enhancement using shunt and series FACTS devices. In Proceedings of the 2015 Clemson University Power Systems Conference (PSC), Clemson, SC, USA, 10–13 March 2015; pp. 1–5.

146. Rathi, A.; Sadda, A.; Nebhnani, L.; Maheshwari, V.M.; Pareek, V.S. Voltage stability assessment in the presence of optimally placed D-FACTS devices. In Proceedings of the 2012 IEEE 5th India International Conference on Power Electronics (IICPE), Delhi, India, 6–9 December 2012; pp. 1–6.

147. Muthu, V.U.; Bhaskar, M.A.; Arthur, A.; Kumar, T.; Surya, S.K. Modeling of wind energy based SSSC for voltage stability enhancement. In Proceedings of the 2014 International Conference on Electronics and Communication Systems (ICECS), Coimbatore, India, 13–14 February 2014; pp. 1–6.

148. Bhatt, P. Short term active power support from DFIG with coordinated control of SSSC and SMES in restructured power system. In Proceedings of the 2014 IEEE PES Asia-Pacific Power and Energy Engineering Conference (APPEEC), Hong Kong, China, 7–10 December 2014; pp. 1–6.

149. Shankar; Udhaya, C.; Thottungal, R.; Mythili, S. Voltage stability improvement and power oscillation damping using Static Synchronous Series Compensator (SSSC). In Proceedings of the 2015 IEEE 9th International Conference on Intelligent Systems and Control (ISCO), Coimbatore, India, 9–10 January 2015; pp. 1–6.

150. Hridya, K.R.; Mini, V.; Visakhan, R.; Kurian, A.A. Comparative study of voltage stability enhancement of a grid and loss reduction using STATCOM and SSSC. In Proceedings of the 2015 International Conference on Power, Instrumentation, Control and Computing (PICC), Thrissur, India, 9–11 December 2015; pp. 1–4.

151. Besharat, H.; Taher, S.A. Congestion management by determining optimal location of TCSC in deregulated power systems. *Int. J. Electr. Power Energy Syst.* **2008**, *30*, 563–568. [CrossRef]

152. Joshi, N.N.; Mohan, N. Application of TCSC in wind farm application. In Proceedings of the International Symposium on Power Electronics, Electrical Drives, Automation and Motion, SPEEDAM 2006, Taormina, Italy, 23–26 May 2006; pp. 1196–1200.

153. Mohammadpour, H.A.; Ghaderi, A.; Mohammadpour, H.; Ali, M.H. Low voltage ride-through enhancement of fixed-speed wind farms using series FACTS controllers. *Sustain. Energy Technol. Assess.* **2015**, *9*, 12–21. [CrossRef]

154. Adjoudj, L.; Lakdja, F.; Gherbi, F.Z.; OuldAbdsallem, D. Synthesis integrating wind generation and FACTS of network. In Proceedings of the 2014 International Conference on Electrical Sciences and Technologies in Maghreb (CISTEM), Tunis, Tunisia, 3–6 November 2014; pp. 1–6.

155. Berrouk; Rachedi, B.A.; Lemzadmi, A.; Bounaya, K.; Zeghache, H. Applications of shunt FACTS controller for voltage stability improvment. In Proceedings of the 2014 International Conference on Electrical Sciences and Technologies in Maghreb (CISTEM), Tunis, Tunisia, 3–6 November 2014; pp. 1–6.

156. Yorino, N.; El-Araby, E.E.; Sasaki, H.; Harada, S. A new formulation for FACTS allocation for security enhancement against voltage collapse. *IEEE Trans. Power Syst.* **2003**, *18*, 3–10. [CrossRef]

157. Elhassan, Z.; Li, Y.; Tang, Y. Simplified voltage control of paralleling doubly fed induction generators connected to the network using SVC. *Int. Trans. Electr. Energy Syst.* **2015**, *25*, 2847–2864. [CrossRef]

158. Tang, B.; Fan, H.; Wang, X.; Yang, X.; Huang, C. The dynamic simulation research on application of SVC in the south Hebei power grid. In Proceedings of the CICED 2010 Proceedings, Nanjing, China, 3–6 September 2010; pp. 1–4.

159. Kehrli, A.; Ross, M. Understanding grid integration issues at wind farms and solutions using voltage source converter FACTS technology. In Proceedings of the 2003 IEEE Power Engineering Society General Meeting (IEEE Cat. No.03CH37491), Toronto, ON, Canada, 13–17 July 2003; Volume 3, p. 1828.

160. SreeLatha, K.; Kumar, M.V. STATCOM for enhancement of voltage stability of a DFIG driven wind turbine. In Proceedings of the 2014 Power and Energy Systems: Towards Sustainable Energy, Bangalore, India, 13–15 March 2014; pp. 1–5.

161. Zhu, K.; Jiang, D.; Lian, X. Cascade STATCOM application in voltage stability of grid-connected wind farm. In Proceedings of the 2011 International Conference on Electronics, Communications and Control (ICECC), Ningbo, China, 9–11 September 2011; pp. 174–177.

162. Jalali, A.; Aldeen, M. Placement and operation of STATCOM-storage for voltage stability enhancement of power systems with embedded wind farms. In Proceedings of the 2016 IEEE Innovative Smart Grid Technologies—Asia (ISGT-Asia), Melbourne, VIC, Australia, 28 November–1 December 2016; pp. 948–953.

163. Rao, R.S.; Rao, V.S. A generalized approach for determination of optimal location and performance analysis of FACTs devices. *Int. J. Electr. Power Energy Syst.* **2015**, *73*, 711–724.

164. Papantoniou, A.; Coonick, A. Simulation of FACTS for wind farm applications. In Proceedings of the IEE Colloquium on Power Electronics for Renewable Energy (Digest No: 1997/170), London, UK, 16 June 1997; pp. 8/1–8/5.

165. Alharbi, Y.M.; Yunus, A.M.S.; Siada, A.A. Application of UPFC to improve the LVRT capability of wind turbine generator. In Proceedings of the 2012 22nd Australasian Universities Power Engineering Conference (AUPEC), Bali, Indonesia, 26–29 September 2012; pp. 1–4.

166. Chen, Y.-L. Weak bus-oriented optimal multi-objective VAr planning. *IEEE Trans. Power Syst.* **1996**, *11*, 1885–1890. [CrossRef]

167. Rong, Y.; Zheng, X.; Niu, H. Power system static voltage stability analysis including large-scale wind farms and system weak nodes identification. In Proceedings of the 2014 International Conference on Power System Technology, Chengdu, China, 20–22 October 2014; pp. 2747–2753.

168. Qin, W.; Zhang, W.; Wang, P.; Han, X. Power system reliability based on voltage weakest bus identification. In Proceedings of the 2011 IEEE Power and Energy Society General Meeting, San Diego, CA, USA, 24–29 July 2011; pp. 1–6.

169. Acharjee, P.; Mallick, S.; Thakur, S.S.; Ghoshal, S.P. Detection of maximum loadability limits and weak buses using Chaotic PSO considering security constraints. *Chaos Solitons Fractals* **2011**, *44*, 600–612. [CrossRef]

170. Chen, Y.-L.; Chang, C.-W.; Liu, C.-C. Efficient methods for identifying weak nodes in electrical power networks. *IEE Proc.-Gener. Transm. Distrib.* **1995**, *142*, 317–322. [CrossRef]

171. Aziz, T.; Saha, T.K.; Mithulananthan, N. Identification of the weakest bus in a distribution system with load uncertainties using reactive power margin. In Proceedings of the 2010 20th Australasian Universities Power Engineering Conference, Christchurch, New Zealand, 5–8 December 2010; pp. 1–6.

172. Canizares, C.A.; De Souza, A.C.Z.; Quintana, V.H. Comparison of performance indices for detection of proximity to voltage collapse. *IEEE Trans. Power Syst.* **1996**, *11*, 1441–1450. [CrossRef]

173. Chen, Y.L. Weak bus oriented reactive power planning for system security. *IEE Proc.-Gener. Transm. Distrib.* **1996**, *143*, 541–545. [CrossRef]

174. Li, H.; Song, Y.H. Identification of weak busbars in large scale power system. In Proceedings of the International Conference on Power System Technology, Kunming, China, 13–17 October 2002; Volume 3, pp. 1700–1704.

175. Acharjee, P. Identification of Maximum Loadability Limit under security constraints using Genetic Algorithm. In Proceedings of the 2011 International Conference on System Science and Engineering, Macao, China, 8–11 June 2011; pp. 234–238.

176. Prada, R.B.; Palomino, E.G.C.; Pilotto, L.A.S.; Bianco, A. Weakest bus, most loaded transmission path and critical branch identification for voltage security reinforcement. *Electr. Power Syst. Res.* **2005**, *73*, 217–226. [CrossRef]

177. Zahedi, A. Current status and future prospects of the wind energy. In Proceedings of the 2012 10th International Power & Energy Conference (IPEC), Ho Chi Minh City, Vietnam, 12–14 December 2012; pp. 54–58.

178. Global Wind Statistics 2016. Available online: https://gwec.net/global-wind-report-2016/ (accessed on 27 July 2023).

179. Jalboub, M.K.; Rajamani, H.S.; Abd-Alhameed, R.A.; Ihbal, A.M. Weakest bus identification based on modal analysis and Singular Value Decomposition techniques. In Proceedings of the 2010 1st International Conference on Energy, Power and Control (EPC-IQ), Basrah, Iraq, 30 November–2 December 2010; pp. 351–356.

180. Chanda, S.; Das, B. Identification of weak buses in a power network using novel voltage stability indicator in radial distribution system. In Proceedings of the India International Conference on Power Electronics 2010 (IICPE2010), New Delhi, India, 28–30 January 2011; pp. 1–4.

181. Divya, B.; Devarapalli, R. Estimation of sensitive node for IEEE-30 bus system by load variation. In Proceedings of the 2014 International Conference on Green Computing Communication and Electrical Engineering (ICGCCEE), Coimbatore, India, 6–8 March 2014; pp. 1–4.

182. Global Wind 2006 Report. Available online: https://gwec.net/wp-content/uploads/2012/06/gwec-2006_final_01.pdf (accessed on 25 July 2023).

183. Global Wind Report 2015. Available online: https://gwec.net/global-wind-report-2015/ (accessed on 27 July 2023).

184. Azam, A.; Rafiq, M.; Arif, I. The Economics of Wind Energy of Pakistan. In Proceedings of the 2016 International Conference on Emerging Technologies (ICET), Islamabad, Pakistan, 18–19 October 2016; pp. 1–6.

185. GWEC. Global Wind Statistics 2012. Available online: https://www.gwec.net/wp-content/uploads/2012/06/Annual_report_20 12_LowRes.pdf (accessed on 28 July 2023).

186. Global Wind 2013. Available online: https://www.gwec.net/wp-content/uploads/2014/04/GWEC-Global-Wind-Report_9-April-2014.pdf (accessed on 28 July 2023).

187. Global Wind Report 2014. Available online: https://www.gwec.net/wp-content/uploads/2015/03/GWEC_Global_Wind_2014 _Report_LR.pdf (accessed on 28 July 2023).

188. Chang, G.W.; Lu, H.J.; Hsu, L.Y.; Chen, Y.Y. A hybrid model for forecasting wind speed and wind power generation. In Proceedings of the 2016 IEEE Power and Energy Society General Meeting (PESGM), Boston, MA, USA, 17–21 July 2016; pp. 1–5.

189. Amber, K.P.; Ashraf, N. Energy outlook in Pakistan. In Proceedings of the 2014 International Conference on Energy Systems and Policies (ICESP), Islamabad, Pakistan, 24–26 November 2014; pp. 1–5.
190. di Marzio, G.; Eek, J.; Tande, J.O.; Fosso, O.B. Implication of Grid Code Requirements on Reactive Power Contribution and Voltage Control Strategies for Wind Power Integration. In Proceedings of the 2007 International Conference on Clean Electrical Power, Capri, Italy, 21–23 May 2007; pp. 154–158.
191. Demirol, M.; Çağlar, R.; Demirol, T.N. Wind farm dynamic analysis in terms of Turkish grid codes. In Proceedings of the 2016 International Conference on Probabilistic Methods Applied to Power Systems (PMAPS), Beijing, China, 16–20 October 2016; pp. 1–9.
192. Sutherland, P.E. Canadian grid codes and wind farm interconnections. In Proceedings of the 2015 IEEE/IAS 51st Industrial & Commercial Power Systems Technical Conference (I&CPS), Calgary, AB, Canada, 5–8 May 2015; pp. 1–7.
193. Zhang, Y.; Duan, Z.; Liu, X. Comparison of Grid Code Requirements with Wind Turbine in China and Europe. In Proceedings of the 2010 Asia-Pacific Power and Energy Engineering Conference, Chengdu, China, 28–31 March 2010; pp. 1–4.
194. Armenakis, A. Grid code compliance test for small wind farms connected to the distribution grid in Cyprus. In Proceedings of the 8th Mediterranean Conference on Power Generation, Transmission, Distribution and Energy Conversion (MEDPOWER 2012), Cagliari, Italy, 1–3 October 2012; pp. 1–6.
195. Fagan, E.; Grimes, S.; McArdle, J.; Smith, P.; Stronge, M. Grid code provisions for wind generators in Ireland. In Proceedings of the IEEE Power Engineering Society General Meeting, San Francisco, CA, USA, 12–16 June 2005; Volume 2, pp. 1241–1247.
196. Naser, S.; Anaya-Lara, O.; Lo, K.L. Study of the impact of wind generation on voltage stability in transmission networks. In Proceedings of the 2011 4th International Conference on Electric Utility Deregulation and Restructuring and Power Technologies (DRPT), Weihai, China, 6–9 July 2011; pp. 39–44.
197. Ding, Q.; Qui, Z.; Yang, H.; Qi, Z. Effect of wind turbines integration on voltage stability. In Proceedings of the 2008 China International Conference on Electricity Distribution, Guangzhou, China, 10–13 December 2008; pp. 1–5. [CrossRef]
198. Hossain, M.J.; Pota, H.R.; Mahmud, M.A.; Ramos, R.A. Impacts of large-scale wind generators penetration on the voltage stability of power systems. In Proceedings of the 2011 IEEE Power and Energy Society General Meeting, San Diego, CA, USA, 24–29 July 2011; pp. 1–8.
199. Chi, Y.; Liu, Y.; Wang, W.; Dai, H. Voltage Stability Analysis of Wind Farm Integration into Transmission Network. In Proceedings of the 2006 International Conference on Power System Technology, Chongqing, China, 22–26 October 2006; pp. 1–7. [CrossRef]
200. Nair, A.Y.; Lie, T.T. Integration of wind power into power systems—Voltage stability studies. In Proceedings of the 8th International Conference on Advances in Power System Control, Operation and Management (APSCOM 2009), Hong Kong, China, 8–11 November 2009; pp. 1–6.
201. Amor, W.O.; Amar, H.B.; Ghariani, M. Grid stability study after the integration of a wind farm. In Proceedings of the 2015 16th International Conference on Sciences and Techniques of Automatic Control and Computer Engineering (STA), Monastir, Tunisia, 21–23 December 2015; pp. 678–683.
202. Zheng, C.; Kezunovic, M. Distribution system voltage stability analysis with wind farms integration. In Proceedings of the North American Power Symposium 2010, Arlington, TX, USA, 26–28 September 2010; pp. 1–6.
203. Zou, Z.; Zhou, K. Voltage stability of wind power grid integration. In Proceedings of the 2011 International Conference on Electrical Machines and Systems, Beijing, China, 20–23 August 2011; pp. 1–5.
204. Nguyen, M.H.; Saha, T.K. Dynamic simulation for wind farm in a large power system. In Proceedings of the 2008 Australasian Universities Power Engineering Conference, Sydney, NSW, Australia, 14–17 December 2008; pp. 1–6.
205. Meng, K.; Zhang, W.; Li, Y.; Dong, Z.Y.; Xu, Z.; Wong, K.P.; Zheng, Y. Hierarchical SCOPF Considering Wind Energy Integration through Multi-Terminal VSC-HVDC Grids. *IEEE Trans. Power Syst.* **2017**, *32*, 4211–4221. [CrossRef]
206. Toma, R.; Gavrilaş, M. Voltage stability assessment for wind farms integration in electricity grids with and without consideration of voltage-dependent loads. In Proceedings of the 2016 International Conference and Exposition on Electrical and Power Engineering (EPE), Iasi, Romania, 14–17 December 2016; pp. 754–759.
207. Sahni, M.; Khoi, V.; Tabrizi, M.; Prakash, N.; Karnik, N.; Bojorquez, W.; Caskey, M. Control Instability Index (CII) based approach for evaluating weak grid integration of wind generation clusters. In Proceedings of the 2015 IEEE Power & Energy Society General Meeting, Denver, CO, USA, 26–30 July 2015; pp. 1–5.
208. Cai, L.J.; Erlich, I.; Karaagac, U.; Mahseredjian, J. Stable operation of doubly-fed induction generator in weak grids. In Proceedings of the 2015 IEEE Power & Energy Society General Meeting, Denver, CO, USA, 26–30 July 2015; pp. 1–5.
209. Cai, L.J.; Erlich, I. Doubly Fed Induction Generator Controller Design for the Stable Operation in Weak Grids. *IEEE Trans. Sustain. Energy* **2015**, *6*, 1078–1084. [CrossRef]
210. *IEEE Std 1204-1997*; IEEE Guide for Planning DC Links Terminating at AC Locations Having Low Short-Circuit Capacities. IEEE: Piscataway, NJ, USA, 1997; pp. 1–216.
211. Etxegarai, A.; Eguia, P.; Torres, E.; Iturregi, A.; Valverde, V. Review of grid connection requirements for generation assets in weak power grids. *Renew. Sustain. Energy Rev.* **2015**, *41*, 1501–1514. [CrossRef]
212. Lorenzen, S.L.; Nielsen, A.B.; Bede, L. Control of a grid-connected converter during weak grid conditions. In Proceedings of the 2016 IEEE 7th International Symposium on Power Electronics for Distributed Generation Systems (PEDG), Vancouver, BC, Canada, 27–30 June 2016; pp. 1–6.

213. Bindner, H. Power Control for Wind Turbines in Weak Grids: Concepts Development. 1999. Available online: https://www.osti.gov/etdeweb/biblio/365955 (accessed on 8 April 2023).
214. Kaur, J.; Yogarathinam, A.; Chaudhuri, N.R. Frequency control for weak AC grid connected to wind farm and LCC-HVDC system: Modeling and stability analysis. In Proceedings of the 2016 IEEE Power and Energy Society General Meeting (PESGM), Boston, MA, USA, 17–21 July 2016; pp. 1–5.
215. Haider, Z.M.; Mehmood, K.K.; Rafique, M.K.; Khan, S.U.; Lee, S.-J.; Kim, C.-H. Water-filling algorithm based approach for management of responsive residential loads. *J. Mod. Power Syst. Clean Energy* **2018**, *6*, 118–131. [CrossRef]
216. Xu, X.; Mu, G.; Shao, G.; Zhang, H.; Hou, K.; Gao, D.; Tao, J.; Ma, X.; Xiao, Y. The problems and solutions for large-scale concentrated integration of wind power to partially weak regional power grid. In Proceedings of the 2009 International Conference on Sustainable Power Generation and Supply, Nanjing, China, 6–7 April 2009; pp. 1–6.
217. Zhou, Y.; Nguyen, D.D.; Kjær, P.C.; Saylors, S. Connecting wind power plant with weak grid—Challenges and solutions. In Proceedings of the 2013 IEEE Power & Energy Society General Meeting, Vancouver, BC, Canada, 21–25 July 2013; pp. 1–7.
218. Chen, X.; Hou, Y. STATCOM control for integration of wind farm to the weak grid. In Proceedings of the 2014 IEEE PES General Meeting | Conference & Exposition, National Harbor, MD, USA, 27–31 July 2014; pp. 1–5.
219. Huang, S.H.; Schmall, J.; Conto, J.; Adams, J.; Zhang, Y.; Carter, C. Voltage control challenges on weak grids with high penetration of wind generation: ERCOT experience. In Proceedings of the 2012 IEEE Power and Energy Society General Meeting, San Diego, CA, USA, 22–26 July 2012; pp. 1–7.
220. Givaki, K.; Xu, L. Stability analysis of large wind farms connected to weak AC networks incorporating PLL dynamics. In Proceedings of the International Conference on Renewable Power Generation (RPG 2015), Beijing, China, 17–18 October 2015; pp. 1–6.
221. Wang, D.; Hou, Y.; Hu, J. Effect of AC voltage control on the stability of weak AC grid connected DFIG system. In Proceedings of the 2016 IEEE PES Asia-Pacific Power and Energy Engineering Conference (APPEEC), Xi'an, China, 25–28 October 2016; pp. 1528–1533.
222. Dirksen, J.; DEWI GmbH. Wilhelmshaven. *Low Voltage Ride-Through (LVRT). DEWI Magazin No 2013, 43*. Available online: https://docplayer.org/19386881-Lvrt-low-voltage-ride-through.html (accessed on 16 August 2023).
223. Dietmannsberger, M.; Grumm, F.; Schulz, D. Simultaneous Implementation of LVRT Capability and Anti-Islanding Detection in Three-Phase Inverters Connected to Low-Voltage Grids. *IEEE Trans. Energy Convers.* **2017**, *32*, 505–515. [CrossRef]
224. Hossain, M.J.; Pota, H.R.; Kumble, C. Decentralized robust static synchronous compensator control for wind farms to augment dynamic transfer capability. *J. Renew. Sustain. Energy* **2010**, *2*, 022701. [CrossRef]
225. Haider, Z.M.; Mehmood, K.K.; Khan, S.U.; Khan, M.O.; Wadood, A.; Rhee, S.-B. Optimal management of a distribution feeder during contingency and overload conditions by harnessing the flexibility of smart loads. *IEEE Access* **2021**, *9*, 40124–40139. [CrossRef]
226. Chen, L.; Chen, H.; Yang, J.; Zhu, L.; Tang, Y.; Koh, L.H.; Xu, Y.; Zhang, C.; Liao, Y.; Ren, L.; et al. Comparison of Superconducting Fault Current Limiter and Dynamic Voltage Restorer for LVRT Improvement of High Penetration Microgrid. *IEEE Trans. Appl. Supercond.* **2017**, *27*, 1–7. [CrossRef]
227. Song, H.; Zhang, Q.; Qu, Y.; Wang, X. An energy-based LVRT control strategy for doubly-fed wind generator. In Proceedings of the 2016 UKACC 11th International Conference on Control (CONTROL), Belfast, UK, 31 August–2 September 2016; pp. 1–6.
228. Shukla, R.D.; Tripathi, R.K. Low voltage ride through (LVRT) ability of DFIG based wind energy conversion system-I. In Proceedings of the 2012 Students Conference on Engineering and Systems, Allahabad, India, 16–18 March 2012; pp. 1–6.
229. Abbey, C.; Joos, G. Effect of low voltage ride through (LVRT) characteristic on voltage stability. In Proceedings of the IEEE Power Engineering Society General Meeting, San Francisco, CA, USA, 12–16 June 2005; Volume 2, pp. 1901–1907.
230. Espinoza, N.; Bongiorno, M.; Carlson, O. Novel LVRT Testing Method for Wind Turbines Using Flexible VSC Technology. *IEEE Trans. Sustain. Energy* **2015**, *6*, 1140–1149. [CrossRef]
231. Xiang, D.; Turu, J.C.; Muratel, S.M.; Wang, T. On-Site LVRT Testing Method for Full-Power Converter Wind Turbines. *IEEE Trans. Sustain. Energy* **2017**, *8*, 395–403. [CrossRef]
232. Nasiri, M.; Milimonfared, J.; Fathi, S.H. A review of low-voltage ride-through enhancement methods for permanent magnet synchronous generator based wind turbines. *Renew. Sustain. Energy Rev.* **2015**, *47*, 399–415. [CrossRef]
233. Xie, Z.; Shi, Q.; Song, H.; Zhang, X.; Yang, S. High voltage ride through control strategy of doubly fed induction wind generators based on active resistance. In Proceedings of the 7th International Power Electronics and Motion Control Conference, Harbin, China, 10–12 May 2012; pp. 2193–2196.
234. Bollen, M.H.J. *Understanding Power Quality Problems: Voltage Sags and Interruptions*; Series on Power Engineering; IEEE Press: New York, NY, USA, 2000.
235. Wessels, C.; Fuchs, F.W. High voltage ride through with FACTS for DFIG based wind turbines. In Proceedings of the 2009 13th European Conference on Power Electronics and Applications, Barcelona, Spain, 8–10 September 2009; pp. 1–10.
236. Mehmood, K.K.; Kim, C.-H.; Khan, S.U.; Haider, Z.M. Unified planning of wind generators and switched capacitor banks: A multiagent clustering-based distributed approach. *IEEE Trans. Power Syst.* **2018**, *33*, 6978–6988. [CrossRef]
237. Justo, J.J.; Mwasilu, F.; Jung, J.-W. Doubly-fed induction generator based wind turbines: A comprehensive review of fault ride-through strategies. *Renew. Sustain. Energy Rev.* **2015**, *45*, 447–467. [CrossRef]

238. Fang, Y.; Sun, D.; Xiong, P. A coordinated control strategy of DFIG-based WECS for high voltage ride-through enhancement. In Proceedings of the 2014 17th International Conference on Electrical Machines and Systems (ICEMS), Hangzhou, China, 22–25 October 2014; pp. 2808–2814.
239. Mohseni, M.; Masoum, M.A.S.; Islam, S.M. Low and high voltage ride-through of DFIG wind turbines using hybrid current controlled converters. *Electr. Power Syst. Res.* **2011**, *81*, 1456–1465. [CrossRef]
240. Alharbi, Y.M.; Yunus, A.M.S.; Abu-Siada, A. Application of STATCOM to improve the high-voltage-ride-through capability of wind turbine generator. In Proceedings of the 2011 IEEE PES Innovative Smart Grid Technologies, Perth, WA, Australia, 13–16 November 2011; pp. 1–5.
241. Ahmad, A.; Loganathan, R. Development of LVRT and HVRT control strategy for DFIG based wind turbine system. In Proceedings of the 2010 IEEE International Energy Conference, Manama, Bahrain, 18–22 December 2010; pp. 316–321.
242. Zheng, Z.; Yang, G.; Geng, H. High voltage ride-through control strategy of grid-side converter for DFIG-based WECS. In Proceedings of the IECON 2013—39th Annual Conference of the IEEE Industrial Electronics Society, Vienna, Austria, 10–13 November 2013; pp. 5282–5287.
243. Liu, C.; He, J.; Xie, Z. High voltage ride-through of grid-side converter for PMSG based directly driven wind turbines. In Proceedings of the 2016 35th Chinese Control Conference (CCC), Chengdu, China, 27–29 July 2016; pp. 8528–8532.
244. Mehmood, K.K.; Khan, S.U.; Lee, S.-J.; Haider, Z.M.; Rafique, M.K.; Kim, C.-H. A real-time optimal coordination scheme for the voltage regulation of a distribution network including an OLTC, capacitor banks, and multiple distributed energy resources. *Int. J. Electr. Power Energy Syst.* **2018**, *94*, 1–14. [CrossRef]
245. Coughlan, P.; O'sullivan, J.; Kamaluddin, N. High Wind Speed Shutdown analysis. In Proceedings of the 2012 47th International Universities Power Engineering Conference (UPEC), London, UK, 4–7 September 2012; pp. 1–5.
246. Folly, K.A.; Sheetekela, S.P.N. Impact of fixed and variable speed wind generators on the transient stability of a power system network. In Proceedings of the 2009 IEEE/PES Power Systems Conference and Exposition, Seattle, WA, USA, 15–18 March 2009; pp. 1–7.
247. Shehata, E.G. Active and Reactive Power Control of Doubly Fed Induction Generators for Wind Energy Generation under Unbalanced Grid Voltage Conditions. *Electr. Power Compon. Syst.* **2013**, *41*, 619–640. [CrossRef]
248. Ibrahim, A.O.; Nguyen, T.H.; Lee, D.C.; Kim, S.C. A Fault Ride-Through Technique of DFIG Wind Turbine Systems Using Dynamic Voltage Restorers. *IEEE Trans. Energy Convers.* **2011**, *26*, 871–882. [CrossRef]
249. Meegahapola, L.; Littler, T.; Fox, B.; Kennedy, J.; Flynn, D. Voltage and power quality improvement strategy for a DFIG wind farm during variable wind conditions. In Proceedings of the 2010 Modern Electric Power Systems, Wroclaw, Poland, 20–22 September 2010; pp. 1–6.
250. Yang, L.; Xu, Z.; Ostergaard, J.; Dong, Z.Y.; Wong, K.P. Advanced Control Strategy of DFIG Wind Turbines for Power System Fault Ride Through. *IEEE Trans. Power Systems* **2012**, *27*, 713–722. [CrossRef]
251. Cheng, P.; Nian, H. An improved control strategy for DFIG system and dynamic voltage restorer under grid voltage dip. In Proceedings of the 2012 IEEE International Symposium on Industrial Electronics, Hangzhou, China, 28–31 May 2012; pp. 1868–1873.
252. Dong, B.; Asgarpoor, S.; Qiao, W. Voltage analysis of distribution systems with DFIG wind turbines. In Proceedings of the 2009 IEEE Power Electronics and Machines in Wind Applications, Lincoln, NE, USA, 24–26 June 2009; pp. 1–5.
253. Wang, D.; Hou, Y.; Hu, J. Stability of DC-link voltage control for paralleled DFIG-based wind turbines connected to Weak AC grids. In Proceedings of the 2016 IEEE Power and Energy Society General Meeting (PESGM), Boston, MA, USA, 17–21 July 2016; pp. 1–5.
254. Banawair, K.S.; Pasupuleti, J. DFIG wind-turbine modeling with reactive power control integrated to large distribution network. In Proceedings of the 2014 IEEE International Conference on Power and Energy (PECon), Kuching, Malaysia, 1–3 December 2014; pp. 298–303.
255. Abdou, A.F.; Abu-Siada, A.; Pota, H.R. Impact of VSC faults on dynamic performance and low voltage ride through of DFIG. *Int. J. Electr. Power Energy Syst.* **2015**, *65*, 334–347. [CrossRef]
256. Ananth, D.V.N.; Kumar, G.V.N. Fault ride-through enhancement using an enhanced field oriented control technique for converters of grid-connected DFIG and STATCOM for different types of faults. *ISA Trans.* **2016**, *62*, 2–18. [CrossRef]
257. Xiao, S.; Yang, G.; Geng, H. Analysis of the control limit of crowbar-less LVRT methods for DFIG-based wind power systems under asymmetrical voltage dips. In Proceedings of the 2011 IEEE PES Innovative Smart Grid Technologies, Perth, WA, Australia, 13–16 November 2011; pp. 1–7.
258. Rodriquez, M.; Abad, G.; Sarasola, I.; Gilabert, A. Crowbar control algorithms for doubly fed induction generator during voltage dips. In Proceedings of the 2005 European Conference on Power Electronics and Applications, Dresden, Germany, 1–14 September 2005; p. 10.
259. Londero, R.R.; Affonso, C.d.M.; Vieira, J.P.A. Long-Term Voltage Stability Analysis of Variable Speed Wind Generators. *IEEE Trans. Power Syst.* **2015**, *30*, 439–447. [CrossRef]
260. Londero, R.R.; Affonso, C.d.M.; Vieira, J.P.A. Effects of operational limits of DFIG wind turbines on long-term voltage stability studies. *Electr. Power Syst. Res.* **2017**, *142*, 134–140. [CrossRef]
261. Kim, J.; Park, G.; Seok, J.K.; Lee, B.; Kang, Y.C. Hierarchical voltage control of a wind power plant using the adaptive IQ-V characteristic of a doubly-fed induction generator. *J. Electr. Eng. Technol.* **2015**, *10*, 504–510. [CrossRef]

262. Hansen, A.D.; Michalke, G.; Sørensen, P.E.; Lund, T.; Iov, F. Co-ordinated voltage control of DFIG wind turbines in uninterrupted operation during grid faults. *Wind Energy* **2007**, *10*, 51–68. [CrossRef]
263. Gkavanoudis, S.I.; Demoulias, C.S. Fault ride-through capability of a DFIG in isolated grids employing DVR and supercapacitor energy storage. *Int. J. Electr. Power Energy Syst.* **2015**, *68*, 356–363. [CrossRef]
264. Liao, Y.; Li, H.; Yao, J.; Zhuang, K. Operation and control of a grid-connected DFIG-based wind turbine with series grid-side converter during network unbalance. *Electr. Power Syst. Res.* **2011**, *81*, 228–236. [CrossRef]
265. Masood, N.A.; Yan, R.; Saha, T.K. Estimation of maximum wind power penetration level to maintain an adequate frequency response in a power system. In Proceedings of the 8th International Conference on Electrical and Computer Engineering, Dhaka, Bangladesh, 20–22 December 2014; pp. 587–590.
266. Le, H.T.; Santoso, S. Analysis of Voltage Stability and Optimal Wind Power Penetration Limits for a Non-radial Network with an Energy Storage System. In Proceedings of the 2007 IEEE Power Engineering Society General Meeting, Tampa, FL, USA, 24–28 June 2007; pp. 1–8.
267. Nguyen, N.; Mitra, J. An Analysis of the Effects and Dependency of Wind Power Penetration on System Frequency Regulation. *IEEE Trans. Sustain. Energy* **2016**, *7*, 354–363. [CrossRef]
268. Toma, R.; Gavrilas, M. The impact on voltage stability of the integration of renewable energy sources into the electricity grids. In Proceedings of the 2014 International Conference and Exposition on Electrical and Power Engineering (EPE), Iasi, Romania, 16–18 October 2014; pp. 1051–1054.
269. Jiapaer, A.; Zeng, G.; Ma, S.; Xie, H.; Tong, Y.; Huang, M. Study on capacity of wind power integrated into power grid based on static voltage stability. In Proceedings of the 2012 IEEE 15th International Conference on Harmonics and Quality of Power, Hong Kong, China, 17–20 June 2012; pp. 855–859.
270. Cocina, V.; Di Leo, P.; Pastorelli, M.; Spertino, F. Choice of the most suitable wind turbine in the installation site: A case study. In Proceedings of the 2015 International Conference on Renewable Energy Research and Applications (ICRERA), Palermo, Italy, 22–25 November 2015; pp. 1631–1634.
271. Multazam, T.; Putri, R.I.; Pujiantara, M.; Priyadi, A.; Hery, P.M. Wind farm site selection base on fuzzy analytic hierarchy process method; Case study area Nganjuk. In Proceedings of the 2016 International Seminar on Intelligent Technology and Its Applications (ISITIA), Lombok, Indonesia, 28–30 July 2016; pp. 545–550.
272. Zhu, J.; Cheung, K. Selection of wind farm location based on fuzzy set theory. In Proceedings of the IEEE PES General Meeting, Minneapolis, MN, USA, 25–29 July 2010; pp. 1–6.
273. Seguro, J.V.; Lambert, T.W. Modern estimation of the parameters of the Weibull wind speed distribution for wind energy analysis. *J. Wind Eng. Ind. Aerodyn.* **2000**, *85*, 75–84. [CrossRef]
274. Cetinay, H.; Kuipers, F.A.; Guven, A.N. Optimal siting and sizing of wind farms. *Renew. Energy* **2017**, *101*, 51–58. [CrossRef]
275. Gigoviç, L.; Pamuìar, D.; Božaniç, D.; Ljubojeviç, S. Application of the GIS-DANP-MABAC multi-criteria model for selecting the location of wind farms: A case study of Vojvodina, Serbia. *Renew. Energy* **2017**, *103*, 501–521. [CrossRef]
276. Bezbradica, M.; Kerkvliet, H.; Borbolla, I.M.; Lehtimäki, P. Introducing multi-criteria decision analysis for wind farm repowering: A case study on Gotland. In Proceedings of the 2016 International Conference Multidisciplinary Engineering Design Optimization (MEDO), Belgrade, Serbia, 14–16 September 2016; pp. 1–8.
277. Hofer, T.; Sunak, Y.; Siddique, H.; Madlener, R. Wind farm siting using a spatial Analytic Hierarchy Process approach: A case study of the Städteregion Aachen. *Appl. Energy* **2016**, *163*, 222–243. [CrossRef]
278. Putranto, L.M.; Hara, R.; Kita, H.; Tanaka, E. WAMS hybrid configuration for real-time voltage stability monitoring application. In Proceedings of the 2016 17th International Scientific Conference on Electric Power Engineering (EPE), Prague, Czech Republic, 16–18 May 2016; pp. 1–6.
279. Zhang, K.; Zhang, S.; Liu, Y. Structure and Applications of a Novel WAMS. *Procedia Eng.* **2011**, *15*, 4492–4498. [CrossRef]
280. Nohac, K.; Tesarova, M.; Nohacova, L.; Veleba, J.; Majer, V. Utilization of Events Measured by WAMS-BIOZE-Detector for System Voltage Stability Evaluation. *IFAC-PapersOnLine* **2016**, *49*, 364–369. [CrossRef]
281. Saha, K.; Meera, K.S.; Shivaprasad, V. Smart grid and WAMS in Indian context—A review. In Proceedings of the 2015 Clemson University Power Systems Conference (PSC), Clemson, SC, USA, 10–13 March 2015; pp. 1–6.
282. Zhou, Y.; Wang, K.; Xiong, Y.; Zhang, B. Study of Power System Online Dynamic Equivalent Based on Wide Area Measurement System. *Energy Procedia* **2012**, *16*, 1768–1775.
283. Putranto, L.M.; Hara, R.; Kita, H.; Tanaka, E. Risk-based voltage stability monitoring and preventive control using wide area monitoring system. In Proceedings of the 2015 IEEE Eindhoven PowerTech, Eindhoven, The Netherlands, 29 June–2 July 2015; pp. 1–6.
284. Popelka, A.; Jurik, D.; Marvan, P.; Povolny, V. Advanced applications of WAMS. In Proceedings of the 22nd International Conference and Exhibition on Electricity Distribution (CIRED 2013), Stockholm, Sweden, 10–13 June 2013; pp. 1–4.
285. Putranto, L.M.; Hoonchareon, N. Wide Area Monitoring System implementation in securing voltage stability based on phasor measurement unit data. In Proceedings of the 2013 10th International Conference on Electrical Engineering/Electronics, Computer, Telecommunications and Information Technology, Krabi, Thailand, 15–17 May 2013; pp. 1–6.
286. Ahmadi, A.; Beromi, Y.A. Neuro-fuzzy based algorithm for online dynamic voltage stability status prediction using wide-area phasor measurements. In Proceedings of the 2015 30th International Power System Conference (PSC), Tehran, Iran, 23–25 November 2015; pp. 14–20.

287. Mousavi-Seyedi, S.S.; Aminifar, F.; Afsharnia, S. Application of WAMS and SCADA Data to Online Modeling of Series-Compensated Transmission Lines. *IEEE Trans. Smart Grid* **2017**, *8*, 1968–1976. [CrossRef]

288. Rahmatian, M.; Dunford, W.G.; Palizban, A.; Moshref, A. Transient Stability Assessment of power systems through Wide-Area Monitoring System. In Proceedings of the 2015 IEEE Power & Energy Society General Meeting, Denver, CO, USA, 26–30 July 2015; pp. 1–5.

289. Liu, Y.; Fan, R.; Terzija, V. Power system restoration: A literature review from 2006 to 2016. *J. Mod. Power Syst. Clean Energy* **2016**, *4*, 332–341. [CrossRef]

290. Lei, J.; Li, Y.; Zhang, B.; Liu, W. A WAMS based adaptive load shedding control strategy using a novel index of transient voltage stability. In Proceedings of the 33rd Chinese Control Conference, Nanjing, China, 28–30 July 2014; pp. 8164–8169.

291. Chen, H.; Jiang, T.; Yuan, H.; Jia, H.; Bai, L.; Li, F. Wide-area measurement-based voltage stability sensitivity and its application in voltage control. *Int. J. Electr. Power Energy Syst.* **2017**, *88*, 87–98. [CrossRef]

292. Hering, P.; Juřík, D.; Janeček, P.; Popelka, A. Applications of Czech PMU/WAMS Systems in Distribution Systems. *IFAC-PapersOnLine* **2016**, *49*, 358–363. [CrossRef]

293. Arias, D.; Vargas, L.; Rahmann, C. WAMS-Based Voltage Stability Indicator Considering Real-Time Operation. *IEEE Lat. Am. Trans.* **2015**, *13*, 1421–1428. [CrossRef]

294. Zhao, J.; Zeng, Y.; Wei, W.; Jia, H. Fast assessment of regional voltage stability based on WAMS. In Proceedings of the APCCAS 2008—2008 IEEE Asia Pacific Conference on Circuits and Systems, Macao, China, 30 November–3 December 2008; pp. 635–638.

*Article*

# Controller Hardware-in-the-Loop Testbed of a Distributed Consensus Multi-Agent System Control under Deception and Disruption Cyber-Attacks

Ibtissam Kharchouf * and Osama A. Mohammed *

Energy Systems Research Laboratory, Department of Electrical and Computer Engineering, Florida International University, Miami, FL 33174, USA
* Correspondence: ikhar002@fiu.edu (I.K.); mohammed@fiu.edu (O.A.M.); Tel.: +1-305-348-3040 (O.A.M.)

**Abstract:** The impact of communication disturbances on microgrids (MGs) needs robust and scalable Information Communication Technology (ICT) infrastructure for efficient MG control. This work builds on advances in the Internet of Things (IoT) to provide a practical platform for testing the impact of various cyber-attacks on a distributed control scheme for a Multi-Agent System (MAS). This paper presents a Controller Hardware-in-the-Loop (CHIL) testbed to investigate the impact of various cyber-attacks and communication disruptions on MGs. A distributed consensus secondary control scheme for a MAS within an MG cyber-physical system (CPS) is proposed. The proposed cyber-physical testbed integrates a real-time islanded AC microgrid on RT-Lab, secondary controllers implemented on single-board computers, and an attacker agent on another single-board computer. Communication occurs via a UDP/IP network between OPAL-RT and controller agents, as well as between the agents. Through meticulous experimentation, the efficacy of the proposed control strategy using the developed platform is validated. Various attacks were modeled and launched including deception attacks on sensors, actuators, and their combinations, as well as disruption attacks. The ramifications of both deception and disruption cyber-attacks on system performance are analyzed.

**Keywords:** microgrids; consensus algorithm; distributed secondary control; real-time simulation; deception attack; denial-of-service attack (DoS); cyber-physical system (CPS); multi-agent system (MAS); OPAL-RT; controller hardware-in-the-loop (CHIL)

**Citation:** Kharchouf, I.; Mohammed, O.A. Controller Hardware-in-the-Loop Testbed of a Distributed Consensus Multi-Agent System Control under Deception and Disruption Cyber-Attacks. *Energies* **2024**, *17*, 1669. https://doi.org/10.3390/en17071669

Academic Editors: Quynh Thi Tu Tran and Saeed Sepasi

Received: 8 February 2024
Revised: 24 March 2024
Accepted: 28 March 2024
Published: 31 March 2024

## 1. Introduction

In the face of increasing concerns about the environmental impact of fossil fuel-based power plants and the commitment of many countries to achieving net zero carbon emissions by 2050, microgrids have emerged as a practical solution to integrate renewable energy and ensure energy security. The term microgrids refers to a group of distributed generators (DGs), loads, and energy storage systems capable of seamlessly transitioning between islanded and grid-connected modes. The control architecture of microgrids is hierarchically structured, involving primary, secondary, and tertiary control levels [1]. The primary control (PC) often has a droop-based design to stabilize the frequency and voltage and to achieve active/reactive power-sharing using local measurements. The secondary control (SC) is implemented as centralized or distributed [2]. It addresses deviations caused by primary control by restoring frequency and voltage. At the top level, tertiary control (TC) is used to manage the power flow and for optimal dispatch operation. Communication networks become vital as secondary controllers exchange critical information such as voltage, frequency, and active and reactive power among DGs [3]. Some standard communication protocols used in microgrids are DNP3, Modbus, TCP/IP, XML, CAN Bus, IEC 61850 [4], etc. The complex interdependency between cyber and physical systems makes them vulnerable to cyber threats. Therefore, a Controller

Hardware-In-the-Loop (CHIL) testbed is developed in the real-time environment using OPAL-RT to analyze the impact of different cyber-attacks on the AC microgrid [5–7].

Cyber-physical systems simulation techniques can be classified into three types, i.e., co-simulation, semi-physical simulation, and embedded simulation. The embedded simulation technique aims to design the communication modules in the power system simulation software. However, the difficulty of this technique lies in the communication module design. In the semi-physical simulation method, one of the systems is simulated using simulation software while a real hardware object replaces the other system. This technique presents an increased simulation authenticity and high simulation accuracy. However, real physical devices make it expensive. Finally, the co-simulation technique aims to build a joint simulation platform by using power system and communication network simulation software, and to realize information exchange between both systems. It is divided into real-time co-simulation and non-real-time co-simulation. The physical and communication simulators have different time management mechanisms in the non-real-time co-simulation. Therefore, a time synchronization method should be designed to achieve collaborative simulation between the two simulators. Even though a collaborative study of a CPS can be achieved through non-real-time co-simulation with a good time synchronization method, real-time co-simulation is given more attention. Real-time co-simulation means that the simulation software runs in real time and that the system can be divided into various sub-models for parallel computation purposes. Hardware-based real-time power system simulation platforms are required for real-time co-simulation, such as OPAL-RT and RTDS [8]. CHIL is a specific co-simulation type involving a hardware controller interacting with a simulation [9]. In our study, we built a CHIL testbed using OPAL-RT and implemented multiple secondary controllers using single-board computers. The developed platform is suitable for studying the impact of communication network disturbances, such as time delays, packet loss, limited bandwidth, and cyber-attacks on cyber-physical systems.

Cyber threats present a serious risk to CPS-based microgrids [10]. For instance, a breach in the communication link may result in miscommunication among DERs, deteriorating the power-sharing objectives and MG stability. Compared to conventional cyber security; attacks on CPSs manipulate data transmission and the physical entities within the system. These attacks can be categorized into replay attacks, DoS, and deception attacks. Replay attacks disrupt authentication by intercepting and retransmitting valid messages [11,12], while DoS attackers jam the communication links among agents to prevent data from reaching their destination. DoS attacks are widely discussed in the literature. Deception attacks compromise sensors' and actuators' data integrity through injection or modification. These attacks do not necessarily require deep knowledge of system dynamics and may result in catastrophic consequences [13]. Recent cyber-attacks on industrial infrastructure, such as the Stuxnet worm's attack on Iran's nuclear power plant and coordinated attacks on power grids in Ukraine and Venezuela, underline the urgent need for robust defensive strategies in CPS security. For instance, the cyber-attack on Ukraine's power grid in 2015 began with an initial compromise as early as eight months before. Initially, the attacker managed to compromise the information technology (IT) network through spear fishing emails. Once inside the network, variants of BlackEnergy 3 malware were remotely controlled to penetrate the industrial control systems (ICSs) and supervisory control and data acquisition (SCADA) systems responsible for managing the power grid. With control over these systems, the hacker remotely took control and executed commands to manipulate the operation of electrical substations and power distribution equipment. At the same time, the operator was prevented from regaining control of the network using a modified KillDisk firmware attack and customers were prevented from reporting the outages by launching a distributed denial of service (DDoS) attack on call centers. The prolonged impact of the attack underscored the vulnerability of critical infrastructure to cyber-physical threats and highlighted the need for robust defense mechanisms to mitigate such risks effectively [14]. Similarly, Venezuela experienced blackouts caused by cyber-attacks.

Hence, the motivation for this paper arises from the critical need to analyze the impact of different cyber-attacks on microgrids and further develop robust defensive strategies to protect microgrids' cyber-physical systems.

The contributions of this research paper can be summarized as follows:

- Development of a novel cyber-physical platform integrating a real-time islanded AC microgrid model running on OPAL-RT, distributed consensus secondary control on Raspberry Pis, and an attacker agent for disruption and deception attacks;
- Development of a distributed consensus secondary controller for frequency and voltage restoration and accurate power sharing;
- Implementation of a communication network using graph theory and the Laplacian matrix to enable information exchange among agents and assess network vulnerabilities;
- Modeling and implementation of disruption and deception attacks on the microgrid communication network using an attacker agent deployed on a Raspberry Pi;
- Assessment of multi-agent system operation under various scenarios of disruption and deception cyber-attacks.

Figure 1 shows a multilayered MAS framework, where each agent receives and shares information with neighboring agents through a communication network. The hardware setup includes an OPAL-RT real-time simulator and independent hardware agents running on Raspberry Pis. The communication between the OPAL-RT and the agents, as well as the communication between the agents, is all through the UDP/IP protocol.

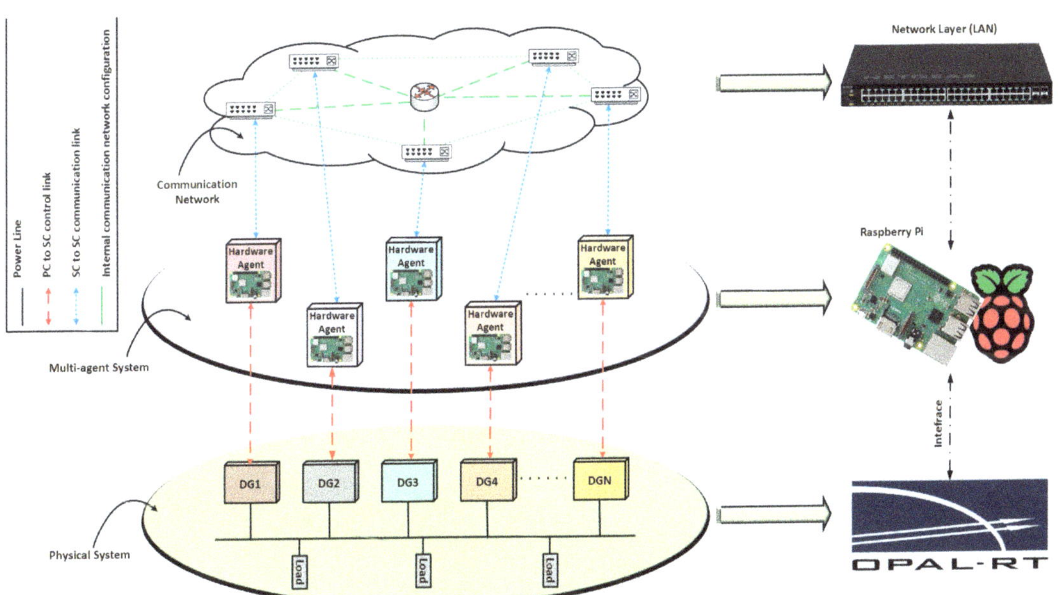



**Figure 1.** Multilayered framework of a multi-agent system.

The rest of the paper is organized as follows. Section 2 introduces preliminaries on the primary control, secondary control, and communication network. The cyber-attacks model is explained in Section 3. The testbed setup is presented in Section 4. Section 5 presents experimental results and discussion to validate the testbed setup and show the impact of deception and disruption attacks on the proposed consensus control strategy. Section 6 concludes the research paper.

## 2. Preliminaries

Voltage Source Inverters (VSIs) are usually used to connect DGs to the network. Figure 2 depicts the block diagram of a VSI, its components, the primary and secondary control loops, and the communication network. The primary and secondary control loops are presented in the following.

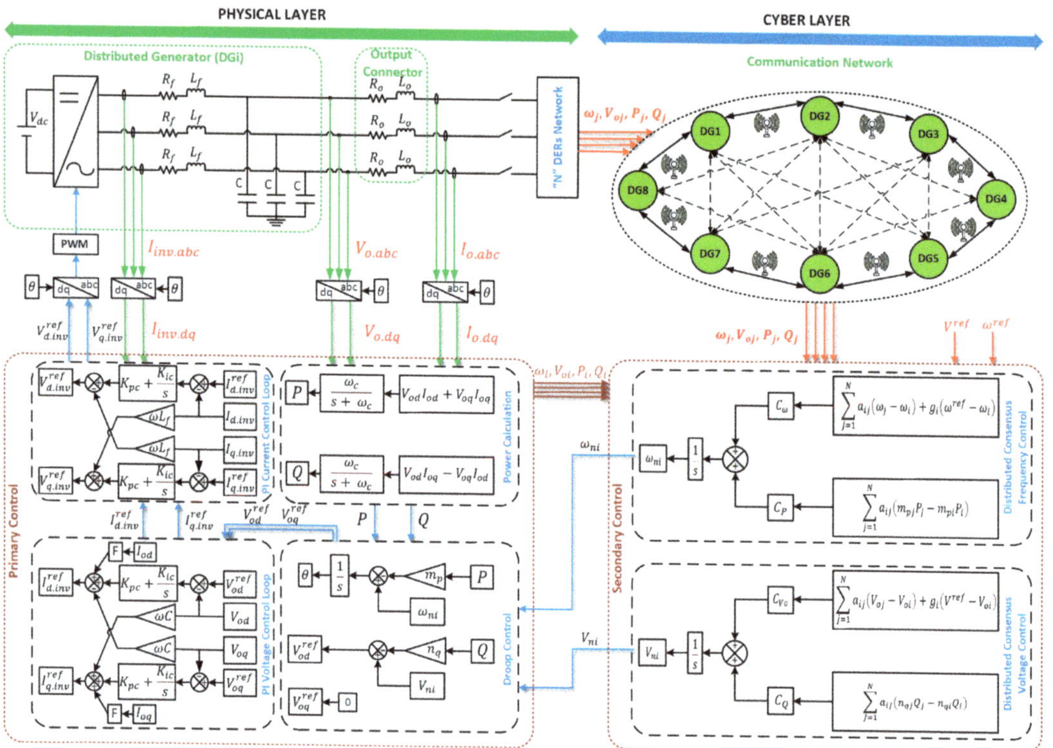



**Figure 2.** Diagram of cyber-physical AC microgrid controlled by a distributed consensus secondary control scheme.

### 2.1. Primary Control

The primary control aims to stabilize the microgrid and ensure power sharing. The proportional control loops are employed locally at each inverter to enable plug-and-play functionality and enhance redundancy. While this decentralized control strategy enables power sharing, it also impacts voltage and frequency regulation.

It consists of three control loops: power control loop, voltage control loop, and current control loop.

#### 2.1.1. Power Control Loop

It is widely employed to adjust the frequency and voltage magnitude in the case of inverter-based DGs in islanded MGs. This adjustment is based on droop characteristics associated with both real and reactive power. The concept of droop control is derived from emulating the behavior of synchronous generators in conventional power systems. Rotating machines respond to an increase in demand by decreasing the system frequency, governed by their droop characteristics. Similarly, inverters implement this principle by reducing the reference frequency as the load increases. The reactive power sharing is managed through the implementation of a droop characteristic in the voltage magnitude [15–17]. The block diagram of the power control loop is shown in Figure 2.

The active ($P$) and reactive ($Q$) power can be calculated from the measured output voltage and current and then passed through low-pass filters as given in Equation (1), where $\omega_c$ is the cut-off frequency and $s$ is the Laplace variable.

$$\begin{cases} P = \frac{\omega_c}{s+\omega_c}\left(V_{od}I_{od} + V_{oq}I_{oq}\right) \\ Q = \frac{\omega_c}{s+\omega_c}\left(V_{od}I_{oq} - V_{oq}I_{od}\right) \end{cases} \tag{1}$$

The active and reactive power sharing between VSIs is achieved by using an artificial droop, introduced, respectively, in the frequency and the voltage magnitude as given in Equation (2), where $\omega_n$ and $V_n$ are the nominal frequency and voltage amplitude, respectively. $\omega$ and $V_o$ are the reference frequency and voltage, respectively [18,19].

$$\begin{cases} \omega = \omega_n - m_p P \\ V_o = V_n - n_q Q \end{cases} \tag{2}$$

The droop coefficients ($m_p$ and $n_q$) are calculated in Equation (3) based on the output power rating. The power control loop provides the voltage reference for the voltage control loop ($V_o^{ref}$). Note that the output voltage reference is chosen to be aligned to the direct axis of the inverter reference frame (d-axis), and the quadrature axis (q-axis) reference is set to zero.

$$m_p = \frac{\Delta\omega}{P_{max}} \ , \ n_q = \frac{\Delta V_o}{Q_{max}} \tag{3}$$

### 2.1.2. Voltage and Current Control Loops

Voltage and current control loops provide the output current and input voltage references ($I_{inv}^{ref}$ and $V_{inv}^{ref}$). The block diagram of the internal voltage and current control loops is shown in Figure 2.

### 2.2. Preliminaries and Communication Network

This section briefly describes graph theory properties. The microgrid is essentially envisioned as a MAS, where the DGs take on the roles of communicating agents or nodes, while the communication links are seen as edges forming a sparse communication network. Each DG can exchange information with its neighboring DGs through this sparse communication network. In our case, the studied system is an islanded microgrid consisting of N DGs, where the communication among them is visually represented by a directed (one-way) or undirected (two-way) communication graph [20,21]. This graph is mathematically represented as $\mathcal{G} = (\mathcal{V}, E, A)$ where $\mathcal{V} = \{v_1, v_2, \ldots, v_N\}$ is a set of $\mathcal{N}$ nodes, $E \subseteq V \times V$ is a set of edges, $A \triangleq [a_{ij}] \in \mathbb{R}^{N \times N}$ is the Adjacency matrix, and it is defined as follows:

$$A \triangleq [a_{ij}] \quad where \quad a_{ij} \triangleq \begin{cases} 0 & \forall \ i = j \\ > 0 & \forall \ i \neq j \end{cases} \tag{4}$$

The edge $(v_j, v_i)$ means that node $j$ transmits information to node $i$. The weight of edge $a_{ij} > 0$ if $(v_j, v_i) \in E$, otherwise $a_{ij} = 0$. $N_i = \{j | (v_j, v_i) \in E\}$ is the set of neighbors of the $i$th node where $j$ is called the neighbor of $i$ if $(v_j, v_i) \in E$. Every node in a graph has an in-degree matrix $D \triangleq diag\{d_i\}$, defined as follows:

$$D \triangleq diag\{d_i\} \quad where \quad d_i \triangleq \sum_{j \in \mathcal{N}_i} a_{ij} \ \forall \ i = j \tag{5}$$

where the Laplacian matrix $L \triangleq D - A$ is defined as follows:

$$L \triangleq [\ell_{ij}] \quad where \ \ell_{ij} \triangleq \begin{cases} d_i = \sum_{j \in \mathcal{N}_i} a_{ij} & \forall \ i = j \\ -a_{ij} & \forall \ i \neq j \end{cases} \tag{6}$$

A weighted graph is called balanced if and only if all the included nodes are balanced such that $\sum_{j=1}^{\mathcal{N}_i} a_{ij} = \sum_{j=1}^{\mathcal{N}_i} a_{ji}$. A graph $\mathcal{G}$ is said to be strongly connected (SC) if there is a connection path of edges between each two separate nodes with accurate direction. The graph is said to have a spanning tree if there is a directed path from a root node $i_r$ to every other node in the graph [22].

The adjacency, in-degree and, Laplacian matrices, as well as other parameters, can effectively improve the control algorithm [23]. Equation (7) is always used in control algorithms based on graph theory where any scalar $x_i$ satisfies the consensus principle in continuous time.

$$\dot{x}_i = u_i = \sum_{j \in \mathcal{N}_i} a_{ij}(x_j - x_i) \tag{7}$$

The microgrid model is shown in Figure 3a, its equivalent weighted graph is shown in Figure 3b, and its corresponding adjacency matrix is shown in Figure 3c.

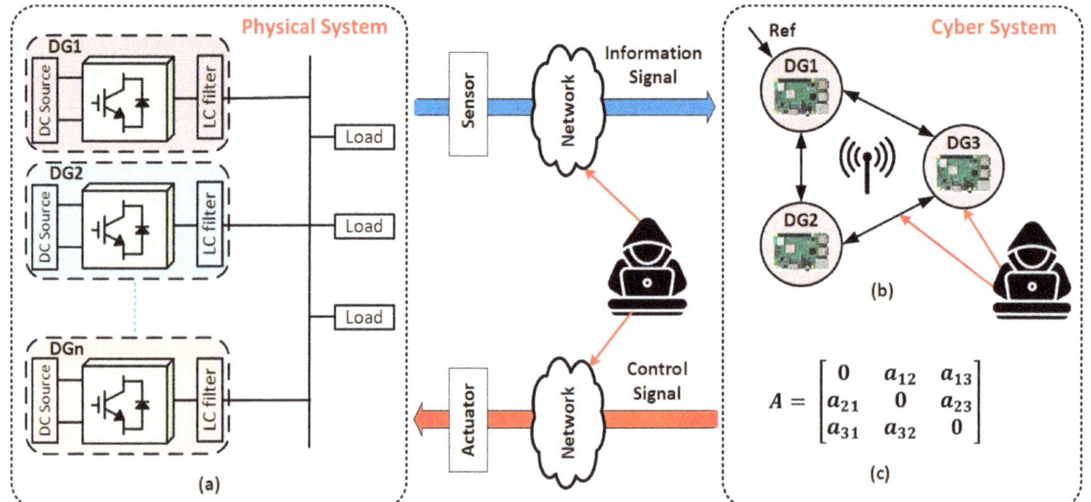


**Figure 3.** (**a**) MG model, (**b**) Weighted graph, (**c**) Adjacency matrix.

*2.3. Distributed Secondary Control*

The secondary control objectives include frequency and voltage restoration and contributing to the power sharing. Each DG communicates with its neighboring DGs to exchange information. Differentiating both terms in Equation (2) gives:

$$\begin{cases} \dot{\omega}_i = \dot{\omega}_{ni} - m_{pi}\dot{P}_i \equiv u_{\omega i} \\ \dot{V}_{oi} = \dot{V}_{ni} - n_{qi}\dot{Q}_i \equiv u_{Voi} \end{cases} \tag{8}$$

The SC sets the nominal set-points $\omega_{ni}$ and $V_{ni}$ as follows:

$$\omega_{ni} = \int \left(\dot{\omega}_i + m_{pi}\dot{P}_i\right)dt = \int (u_{\omega i} + u_{Pi})dt \tag{9}$$

$$V_{ni} = \int \left(\dot{V}_{oi} + n_{qi}\dot{Q}_i\right)dt = \int (u_{Voi} + u_{Qi})dt \tag{10}$$

The accurate power sharing problem can be expressed as follows: $u_{Pi} = m_{pi}\dot{P}_i$ and $u_{Qi} = n_{qi}\dot{Q}_i$.

$\omega_{ni}$ has $u_{\omega i}$ and $u_{Pi}$ as secondary control inputs and $V_{ni}$ has $u_{Voi}$ and $u_{Qi}$ as secondary control inputs, where $u_{\omega i}$ and $u_{Voi}$ are the auxiliary controls.

The proposed distributed SC control objectives are as follows:

1.  Frequency and voltage restoration:

$$\lim_{t \to \infty} \left| \omega_i(t) - \omega_{ref} \right| = 0 \qquad \forall i = 1, 2, \ldots, N. \tag{11}$$

$$\lim_{t \to \infty} \left| V_{oi}(t) - V_{ref} \right| = 0 \qquad \forall i = 1, 2, \ldots, N. \tag{12}$$

2.  Accurate Power Sharing:

$$\lim_{t \to \infty} \left| m_{pi} P_i(t) - m_{pj} P_j(t) \right| = 0 \qquad \forall i \neq j. \tag{13}$$

$$\lim_{t \to \infty} \left| n_{qi} Q_i(t) - n_{qj} Q_j(t) \right| = 0 \qquad \forall i \neq j. \tag{14}$$

Achieving these control objectives involves adjusting the control inputs for each agent: $u_{\omega i}$, $u_{Voi}$, $u_{Pi}$, and $u_{Qi}$.

### 2.3.1. Frequency and Voltage Control

For a microgrid composed of N DGs, the secondary voltage and frequency control for a first order and linear MAS are transformed into the tracking synchronization problem.

$$\begin{cases} \dot{V}_{o1} = u_{Vo1} \\ \dot{V}_{o2} = u_{Vo2} \\ \quad \cdot \\ \quad \cdot \\ \quad \cdot \\ \dot{V}_{oN} = u_{VoN} \end{cases} \tag{15}$$

$$\begin{cases} \dot{\omega}_1 = u_{\omega 1} \\ \dot{\omega}_2 = u_{\omega 2} \\ \quad \cdot \\ \quad \cdot \\ \quad \cdot \\ \dot{\omega}_N = u_{\omega N} \end{cases} \tag{16}$$

As mentioned earlier, DGs communication is achieved through the designed communication graph shown in Figure 3b. The control signals $u_{\omega i}$ and $u_{Voi}$ are calculated using the DGs' own information and the neighbors' information as follows:

$$u_{\omega i} = C_\omega \left[ \sum_{j=1}^{N} a_{ij} (\omega_j - \omega_i) + g_i \left( \omega^{ref} - \omega_i \right) \right] \tag{17}$$

$$u_{Voi} = C_{Vo} \left[ \sum_{j=1}^{N} a_{ij} (V_{oj} - V_{oi}) + g_i \left( V^{ref} - V_{oi} \right) \right] \tag{18}$$

$C_\omega$ and $C_{Vo}$ represent the control gains; both are greater than zero. The pinning gain $g_i$ is set to 1 if a DG can directly receive set points, otherwise $g_i$ is set to 0.

In a global form, Equations (17) and (18) can be written as:

$$u_\omega = C_\omega \left[ -L\omega + G \left( \omega^{ref} 1_{n \times 1} - \omega \right) \right] \tag{19}$$

$$u_{Vo} = C_{Vo} \left[ -LV_o + G \left( V^{ref} 1_{n \times 1} - V_o \right) \right] \tag{20}$$

where $u_{Vo} = [u_{Vo1}, \ldots, u_{VoN}]^T$, $u_\omega = [u_{\omega 1}, \ldots, u_{\omega N}]^T$, $V_o = [V_{o1}, \ldots, V_{oN}]^T$, $\omega = [\omega_1, \ldots, \omega_N]^T$, $C_\omega = diag(C_{\omega 1}, \ldots, C_{\omega N})$, $C_{Vo} = diag(C_{Vo1}, \ldots, C_{VoN})$, and $1_N$ is an all-ones vector of length $N$.

### 2.3.2. Active and Reactive Power Sharing

According to the power sharing objective in Equations (13) and (14), the power ratio among DGs will be equalized in steady state.

$$m_{pi}P_i(t) = m_{pj}P_j(t) \tag{21}$$

$$n_{qi}Q_i(t) = n_{qj}Q_j(t) \tag{22}$$

Substituting the droop coefficient equations in Equation (3), the real and reactive power among DGs is shared as in Equations (23) and (24).

$$\frac{P_i}{P_j} = \frac{m_{pj}}{m_{pi}} = \frac{\frac{\Delta\omega}{Pj_{max}}}{\frac{\Delta\omega}{Pi_{max}}} = \frac{Pi_{max}}{Pj_{max}} \tag{23}$$

$$\frac{Q_i}{Q_j} = \frac{n_{qj}}{n_{qi}} = \frac{\frac{\Delta V_o}{Qj_{max}}}{\frac{\Delta V_o}{Qi_{max}}} = \frac{Qi_{max}}{Qj_{max}} \tag{24}$$

The auxiliary controls $u_{Pi}$ and $u_{Qi}$ are chosen based on the DGs' own information and their neighbors' information as follows:

$$u_{Pi} = C_P \left[ \sum_{j=1}^{N} a_{ij} \left( m_{pj}P_j - m_{pi}P_i \right) \right] \tag{25}$$

$$u_{Qi} = C_Q \left[ \sum_{j=1}^{N} a_{ij} \left( n_{qj}Q_j - n_{qi}Q_i \right) \right] \tag{26}$$

In a global form, Equations (25) and (26) can be written as:

$$u_P = -C_P L\left(m_p P\right) \tag{27}$$

$$u_Q = -C_Q L\left(n_q Q\right) \tag{28}$$

where $u_P = [u_{P1},\ldots,u_{PN}]^T$, $u_Q = [u_{Q1},\ldots,u_{QN}]^T$, $m_p P = [m_{p1}P_1,\ldots,m_{pN}P_N]^T$, $n_q Q = [n_{q1}Q,\ldots,n_{qN}Q_N]^T$, $C_P = diag(C_{P1},\ldots,C_{PN})$, and $C_Q = diag(C_{Q1},\ldots,C_{QN})$. Note that $L$ and $a_{ij}$, the Laplacian matrix and the elements of the adjacent matrix A, are defined in the Preliminaries section.

## 3. Cyber-Attacks Model

### 3.1. Deception and Disruption Cyber-Attacks

Deception and disruption attacks have been widely discussed in the networked control literature. Disruption attacks, known as DoS or jamming attacks, primarily target data availability, whereas deception attacks target the integrity of packets [24,25]. In deception attacks, intruders may manipulate sensor measurements and control commands within networked agents by accessing physical agents, sensors, controllers, actuators, and/or communication channels. This latter proves challenging to detect and handle especially if the attack sequences are strategically launched. Deception attacks can be classified based on attack types and attack points [26].

### 3.1.1. Attack Types

- Linear additive deception attack:
  - Characteristics: The attacker injects false data $\mathfrak{A}_{ij}(t)$ into the normal data $\mathfrak{d}_{ij}(t)$ sent by agent $i$;

- Effect: The corrupted data $\widetilde{\mathfrak{d}}_{ij}(t)$ received by agent $j$ are the addition of the original data and the injected false data.

$$\widetilde{\mathfrak{d}}_{ij}(t) = \mathfrak{d}_{ij}(t) + \mathfrak{A}_{ij}(t) \tag{29}$$

- Multiplicative deception attack:
  - Characteristics: This attack involves scaling up or down the original data $\mathfrak{d}_{ij}(t)$ by a scaling factor $\mathfrak{s}_{ij}(t)$;
  - Effect: The received data $\widetilde{\mathfrak{d}}_{ij}(t)$ are a scaled version of the original data, potentially altered in magnitude.

$$\widetilde{\mathfrak{d}}_{ij}(t) = \mathfrak{d}_{ij}(t)\mathfrak{s}_{ij}(t) \tag{30}$$

- Combined additive and multiplicative deception attack:
  - Characteristics: This type of attack combines both additive and multiplicative deception. It scales the original data and adds injected false data;
  - Effect: The received data $\widetilde{\mathfrak{d}}_{ij}(t)$ are a combination of the scaled original data and the injected false data.

$$\widetilde{\mathfrak{d}}_{ij}(t) = \mathfrak{d}_{ij}(t)\mathfrak{s}_{ij}(t) + \mathfrak{A}_{ij}(t) \tag{31}$$

- Replacement attack:
  - Characteristics: In a replacement attack, the attacker completely replaces the normal data $\mathfrak{d}_{ij}(t)$ with an arbitrary signal $\mathfrak{r}_{ij}(t)$;
  - Effect: The received data $\widetilde{\mathfrak{d}}_{ij}(t)$ are entirely replaced by the arbitrary signal, disregarding the original data.

$$\widetilde{\mathfrak{d}}_{ij}(t) = \mathfrak{r}_{ij}(t) \tag{32}$$

- Impulsive false data attack:
  - Characteristics: This attack involves injecting impulsive false data using Dirac impulses $\delta(.)$ at designated time instances $\{t_k\}_{k=1}^{\infty}$, with destabilizing impulse parameters $\mathfrak{T}_k$;
  - Effect: The received data $\widetilde{\mathfrak{d}}_{ij}(t)$ include the normal data $\mathfrak{d}_{ij}(t)$ along with impulsive false data at designated time instances.

$$\widetilde{\mathfrak{d}}_{ij}(t) = \mathfrak{d}_{ij}(t) + \sum_{k=1}^{\infty} \mathfrak{T}_k \mathfrak{d}_{ij}(t)\delta(t - t_k) \tag{33}$$

### 3.1.2. Attack Points

Deception attacks can be classified based on the attack points. Intruders may launch attacks on sensors, actuators, or both. Equations (34) and (35) model attacks on actuators and sensors, respectively [27].

$$\widetilde{x}_i(t) = x_i(t) + \alpha_i(t)x_i^a(t) \tag{34}$$

$$\widetilde{u}_i(t) = u_i(t) + \beta_i(t)u_i^a(t) \tag{35}$$

where, $x_i^a(t)$ and $u_i^a(t)$ denote the attack signals injected into the sensor and the actuator of agent $i$, respectively. $\widetilde{x}_i(t)$ and $\widetilde{u}_i(t)$ are the corrupted state and control protocol of agent $i$. $\alpha_i(t) = 1$ when agent $i$ is under sensor attack, otherwise $\alpha_i(t) = 0$. Similarly, $\beta_i(t) = 1$ when agent $i$ is under actuator attack, otherwise $\beta_i(t) = 0$.

### 3.2. Attacks Model

#### 3.2.1. Disruption Attack Model

Various strategies can be employed to conduct a DoS attack such as data packet loss, network flooding, zero input, etc. Let us consider the communication channel between $DG_i$ and $DG_j$ during $[t_1, t_2] \subset [0, \infty]$. The states communicated from the OPAL-RT simulation to the agents as well as between the agents are $x_i \in \left[\omega_i, V_{oi}, m_{pi}P_i, n_{qi}Q_i\right]$ and from the agents to the OPAL-RT simulation are $y_i \in [u_{\omega i}, u_{Voi}, u_{Pi}, u_{Qi}]$. Let $m_\mu$ be the number of DoS attacks that might occur during $[t_1, t_2] \subset [0, \infty]$ and $I_k = [t_a, t_a + \tau_a]$ is the kth interval in which a DoS attack take place, where $t_a$, $t_a + \tau_a$, and $\tau_a$ are, respectively, the start, end, and length of DoS attack [7]. The total DoS time intervals of DoS between two DGs can be given as:

$$\Gamma_{DoS}^{(i,j)} = (t_1, t_2) \bigcap (\bigcup_{a=1}^{m_\mu} I_k^{(i,j)}) \tag{36}$$

In order to achieve stealthiness, which is a property under which attacks are not detected, intruders may impose some additional constraints, for instance on $\tau_a$.

#### 3.2.2. Deception Attack Model

Let $\delta_i^S$ denotes the potential attacks applied by the adversary on the *ith* agent sensors. The consensus control in Equations (17) and (18) in the presence of such an attack can be written as follows:

$$\tilde{u}_{\omega i} = C_\omega \left[\sum_{j=1}^{N} a_{ij}\left(\left(\omega_j + \delta_{\omega i}^S\right) - \omega_i\right) + g_i\left(\omega^{ref} - \omega_i\right)\right] \tag{37}$$

$$\tilde{u}_{Voi} = C_{Vo} \left[\sum_{j=1}^{N} a_{ij}\left(\left(V_{oj} + \delta_{Vi}^S\right) - V_{oi}\right) + g_i\left(V^{ref} - V_{oi}\right)\right] \tag{38}$$

Similarly let $\delta_i^a$ denote the potential attacks applied by the adversary on the ith agent actuators. The consensus control in Equations (17) and (18) in the presence of such attack can be written as follows:

$$\tilde{u}_{\omega i} = C_\omega \left[\sum_{j=1}^{N} a_{ij}(\omega_j - \omega_i) + g_i\left(\omega^{ref} - \omega_i\right)\right] + \delta_{\omega i}^a \tag{39}$$

$$\tilde{u}_{Voi} = C_{Vo} \left[\sum_{j=1}^{N} a_{ij}(V_{oj} - V_{oi}) + g_i\left(V^{ref} - V_{oi}\right)\right] + \delta_{Vi}^a \tag{40}$$

The attack signals $\delta_\omega^a$ and $\delta_V^a$ can be designed to cause the microgrid instability while remaining stealthy to adversary-detection systems.

### 4. Testbed Setup

This section introduces an evaluation framework for the implemented agents featuring the proposed distributed secondary control. It outlines the components of the examined three-bus islanded Microgrid (MG), providing insights into various aspects such as real-time simulation, Multi-Agent System (MAS) control platform, communication network design, and protocols.

As depicted in Figure 4, the Controller Hardware-in-the-Loop (CHIL) experimental testbed comprises two main interconnected parts: (i) the proposed physical system, encompassing AC microgrid elements and local controllers, implemented in the OPAL-RT real-time simulator; (ii) the distributed consensus secondary controller, where hardware agents operate independently on Raspberry Pi devices. Initially, real-time measurements are transmitted from the OPAL-RT simulation to the corresponding external control hardware (Raspberry Pi agents) through the UDP/IP protocol and transmitted to neighboring agents via UDP/IP.

### 4.1. Real-Time Simulation

The simulated AC microgrid is modeled using MATLAB/Simulink and implemented in OPAL-RT. The developed model for this experiment contains the power system model along with all primary controllers and is implemented in RT-Lab version 2023.1. The primary controllers consist of inner and outer loops for current and voltage control. During each control iteration, the local data packets $[\omega_i, V_{oi}, m_{pi}P_i, n_{qi}Q_i]$ from the primary controllers are transmitted to the corresponding Raspberry Pis via UDP. The control input packets $[u_{\omega i}, u_{Voi}, u_{Pi}, u_{Qi}]$ from the secondary controllers are sent back to OPAL-RT via UDP to calculate $\omega_{ni}$ and $V_{ni}$ setpoints. The exchange of data between external secondary controllers, the real-time simulator, and neighboring agents occurs through the LAN network. This interconnected system enables the essential communication and coordination for the functioning of the AC microgrid model.

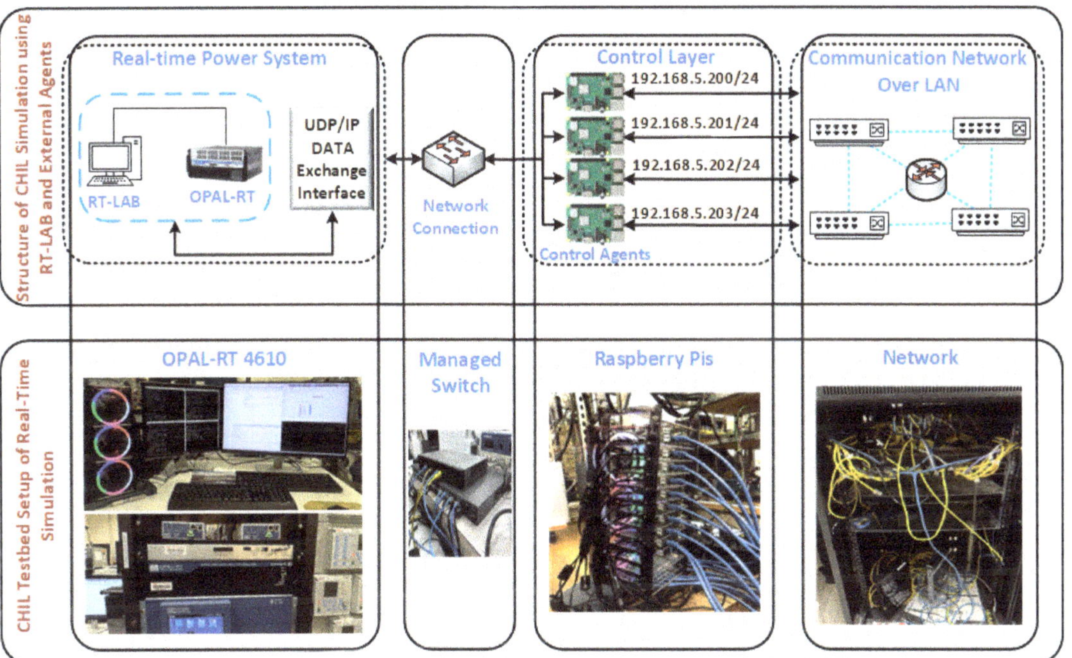


**Figure 4.** CHIL testbed setup.

### 4.2. Cyber Layer-Based Raspberry Pi Agents

The agents designed with the proposed distributed consensus secondary control can update the power network state, perform calculations, and provide control decisions. Each DG is represented by an agent responsible for managing the distributed secondary control algorithm based on predefined control objectives. The secondary agent collects voltage, frequency, active, and reactive power data from local measurements and sends control signals to the primary controller (PC). The communication topology among the Raspberry Pi agents is depicted in Figure 2.

For this implementation, the Raspberry Pi 3 Model B+ is utilized. The control action is programmed within each agent (Rpi) using a Python script, with all agents assigned static IP addresses. Communication ports are established when running the Python script. Each Raspberry Pi opens a communication channel for each device to facilitate data exchange with neighboring agents and establishes a client socket with OPAL-RT. Once the connection is established, the client sends commands to the runtime. The flow of the consensus algorithm implemented in every agent is shown in Figure 5.

### 4.3. Attack Agent

The attack agents were programmed using Python scripts and implemented in a separate agent. The objective of DoS depends on the targeted agent's IP address and port. An agent can be flooded with many packets to consume its resources and, therefore, make it out of service. Also, the communication links between DGs or between a DG and the real-time simulator can be attacked, which results in modifying the communication topology.

The result of deception and disruption attacks and their impact on the physical system are demonstrated using the CHIL testbed setup. The network traffic during a sequence of DoS attacks with varying packet rates and attack lengths targeting one of the DGs can be shown in Figure 6.

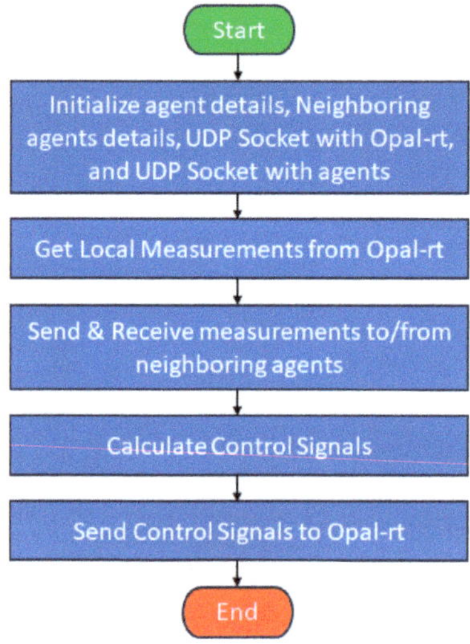


**Figure 5.** The flow of the consensus algorithm.

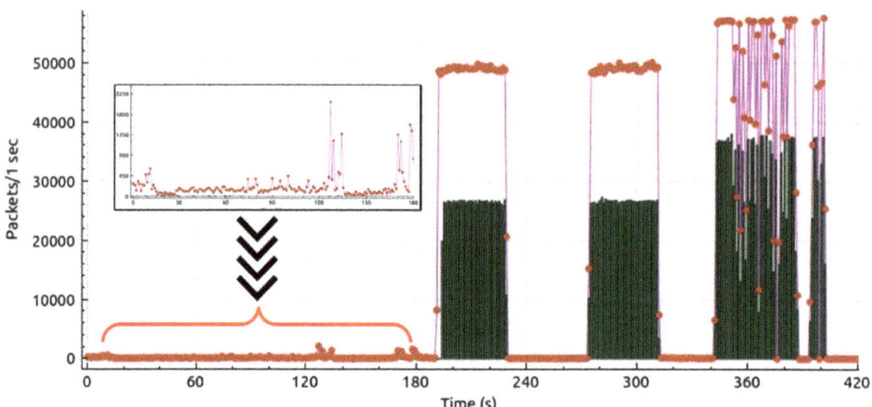


**Figure 6.** Network traffic during a sequence of DoS attacks (Purple line and red dots: All Packets, Green Impulse: UDP Packets).

## 5. Results and Discussion

Extensive real-time digital simulations on OPAL-RT are performed to evaluate the effect of various cyber-attacks on the proposed distributed consensus secondary control of islanded AC MG. Using MATLAB/Simulink, an AC microgrid, structured by three parallel inverters with power ratings of 500 KW, 300 KW, and 200 KW connected to the PCC bus, is modeled. The parameters of distributed secondary controllers are all set to $C_\omega = 0.2$, $C_{VO} = 0.1$, $C_P = 4$, and $C_Q = 100$. The droop coefficients are all set to $m_p = 0.01$ and $n_q = 0.04$.

In this section there are three study cases conducted, including:

- Performance Under Normal Operation;
- Performance Under Linear Additive Deception Attacks on Sensor, Actuator, and Combined;
- Performance Under Disruption Attack.

### 5.1. Performance under Normal Operation

Under normal operation, the performance of the proposed distributed consensus secondary control strategy in case of load variations is presented in Figure 7. At t = 20 s, droop activation initiated proportional power sharing among the three DGs, maintaining the frequency and voltage around 60.15 Hz and 598 V, respectively. Following a load increase at t = 40 s, the DGs adjusted their power outputs, resulting in a frequency and voltage drop to 59.98 Hz and 694 V, respectively. The activation of the distributed consensus secondary controller at t = 60 s restored frequency and voltage to reference values (60 Hz and 600 V) without disrupting proportional power sharing, indicating stable microgrid operation. Further load variations at t = 80 s, t = 100 s, and t = 120 s prompted additional power output adjustments by the DGs, ensuring continued stability with restored frequency and voltage levels.

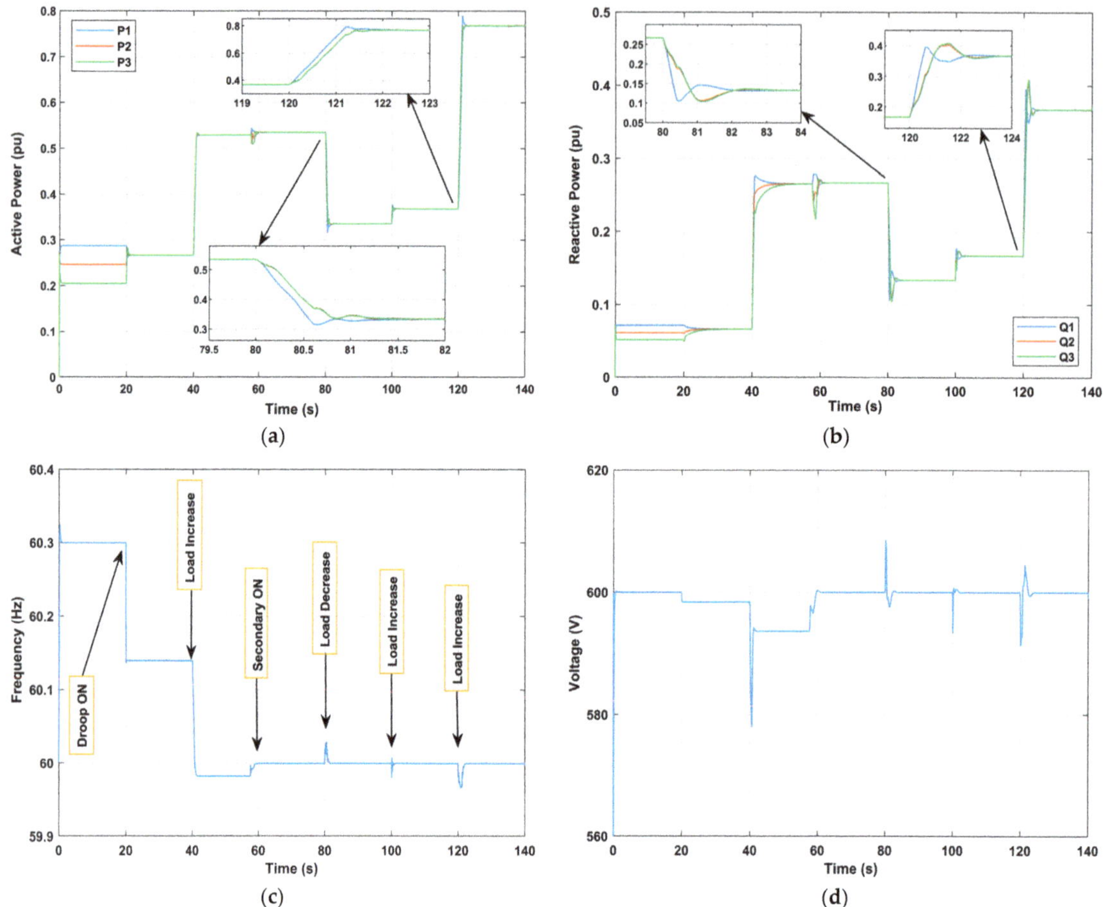

**Figure 7.** Performance of the proposed distributed consensus secondary control under normal operation: (**a**) Active power, (**b**) Reactive power, (**c**) Frequency, (**d**) Voltage.

*5.2. Performance under Deception Attacks*

This section evaluates the effect of linear additive deception attacks on sensors, actuators, and both.

5.2.1. Deception Attack on Actuators' Frequency

In this study case, $\delta_1^{a,\omega}$ denotes the actuator attack injections to the frequency control loop of DG #1. The attack signal is $\delta_1^{a,\omega} = \delta_1^{a,\omega 1}$ for $59.8 < t(s) < 109.8$ and $\delta_1^{a,\omega} = \delta_1^{a,\omega 2}$ for $109.8 < t(s) < 160$ where $\delta_1^{a,\omega 1} > \delta_1^{a,\omega 2}$.

The simulation scenario for this case is as follows:

- At t = 20 s, the Droop control is activated;
- At t = 40 s, the distributed consensus secondary control is activated;
- At t = 59.8 s, the first actuator attack is launched;
- At t = 90 s, the load is increased;
- At t = 109.8 s, the second actuator attack is launched;
- At t = 140 s, the load is decreased.

The results are shown in Figure 8. At t = 40 s, when there is no cyber-attack, the frequency and voltage of the islanded microgrid restore to their reference values while the active and reactive power of DGs are accurately sharing. The cyber-attack is initiated at approximately t = 59.8 s, resulting in a notable transient impact on the system. However, both voltage and frequency stabilize, maintaining alignment with the consensus power sharing objectives. Remarkably, despite the ongoing attack, the system recovered. By t = 90 s, even with an increase in load, the microgrid operates as if it were under normal conditions. At t = 109.8 s, the attack signal is slightly reduced. As a result, the system shows behavior similar to that observed during the initial attack. The performed real-time tests demonstrated that the control objectives were achieved successfully with bounded attack signals. However, with unbounded attack signals, the system became unstable. This underscores the vital role of cybersecurity measures in keeping the microgrid stable and dependable.

### 5.2.2. Deception Attack on Actuators' Voltage

In this study case, $\delta_1^{a,V}$ denotes the actuator attack injections to the voltage control loop of DG #1. The attack signal is $\delta_1^{a,V} = \delta_1^{a,V1}$ for $59.8 < t(s) < 109.8$ and $\delta_1^{a,V} = \delta_1^{a,V2}$ for $109.8 < t(s) < 160$ where $\delta_1^{a,V1} > \delta_1^{a,V2}$. The simulation scenario is as follows:

- At t = 20 s, the droop control is activated;
- At t = 40 s, the distributed consensus secondary control is activated;
- At t = 59.2 s, the first actuator attack is launched;
- At t = 90 s, the load is increased;
- At t = 109.2 s, the second actuator attack is launched;
- At t = 140 s, the load is decreased.

In this experiment, we focused on the vulnerability of the actuator voltage control loop to cyber-attacks within the islanded AC microgrid system. The setup mirrored the first experiment with adjustments made to the voltage control loop instead of the frequency control loop. As depicted in Figure 9, the results showed a similar pattern to the previous experiment.

### 5.2.3. Deception Attack on Sensors' Frequency

In this study case, $\delta_1^{S,\omega}$ denotes the sensor attack injections to the frequency measurements of DG #1. The attack signal is $\delta_1^{S,\omega} = \delta_1^{S,\omega1}$ between $60 < t(s) < 80$ and $\delta_1^{S,\omega} = \delta_1^{S,\omega2}$ between $120 < t(s) < 140$ where $\delta_1^{S,\omega1} > \delta_1^{S,\omega2}$. The simulation scenario for this case is as follows:

- At t = 20 s, the droop control is activated;
- At t = 40 s, the distributed consensus secondary control is activated;
- At t = 60 s, the first frequency sensor attack is launched;
- At t = 80 s, the first attack is removed;
- At t = 100 s, the total load is increased;
- At t = 120 s, the second frequency sensor attack is launched;
- At t = 140 s, the second attack is removed;
- At t = 160 s, the load is decreased.

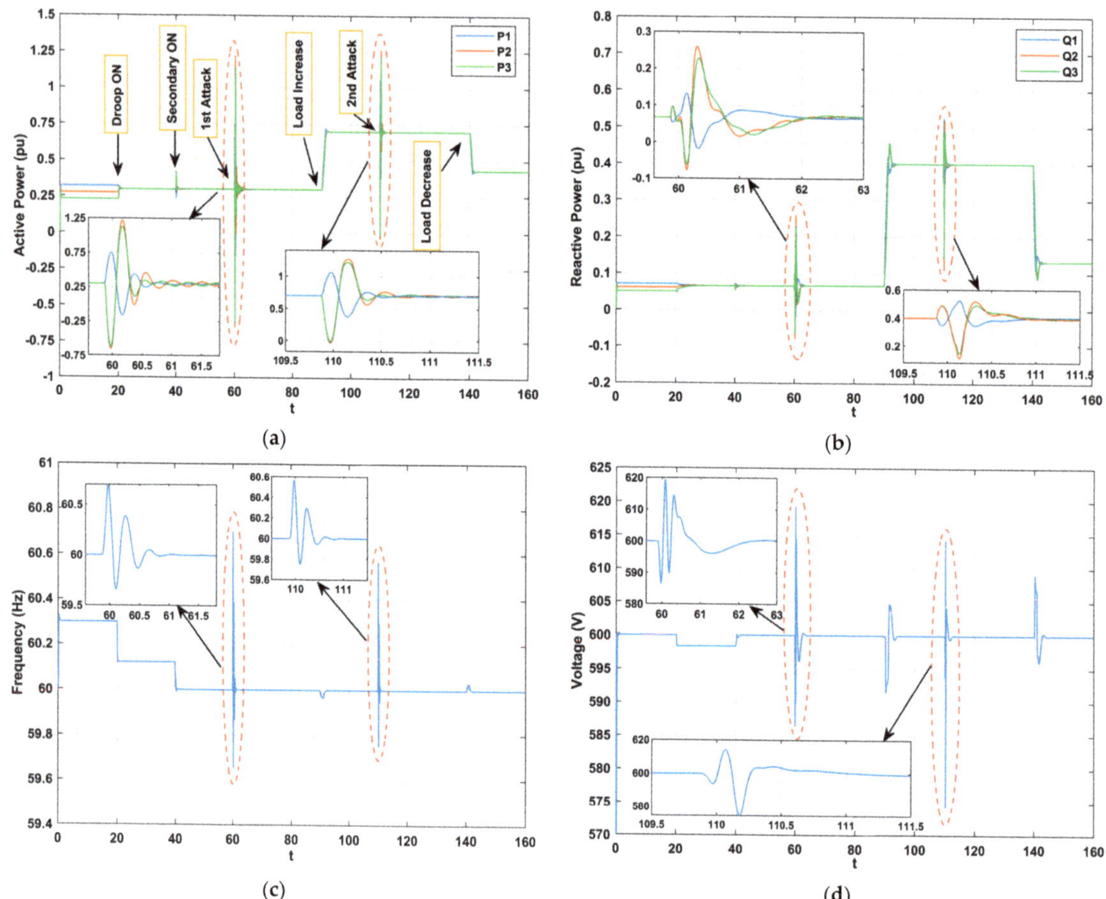

**Figure 8.** Performance of the proposed distributed consensus secondary control under deception attack on actuators' frequency: (**a**) Active power, (**b**) Reactive power, (**c**) Frequency, (**d**) Voltage.

The experiment conducted using the test setup demonstrates results when subject to deception attacks on sensor frequency, as shown in Figure 10. Following the activation of droop and secondary controllers, the first attack is launched at t = 60 s. The attack induced a notable impact on the system dynamics; it caused a frequency drop from its rated value. Upon initiation of the attack, the frequency dropped, and the equitable distribution of active power among DGs was impacted. Notably, the attacked DG1's active power output was minimal while DG2 and DG3 maintained their active power sharing and even increased compared to pre-attack levels. Simultaneously, transient voltage and reactive power disturbances were observed, though quickly mitigated. Upon the removal of the attack at t = 80 s, an instantaneous restoration of the frequency to its nominal value was observed alongside the restoration of active power sharing objectives. Afterwards, a similar behavior was noticed during the second attack at t = 120 s, though with a smaller amplitude, yet still impacting the frequency and active power sharing. These results underscore the importance of understanding the effects of attacks on MG systems and highlight the need for robust defense mechanisms to protect against potential disruptions.

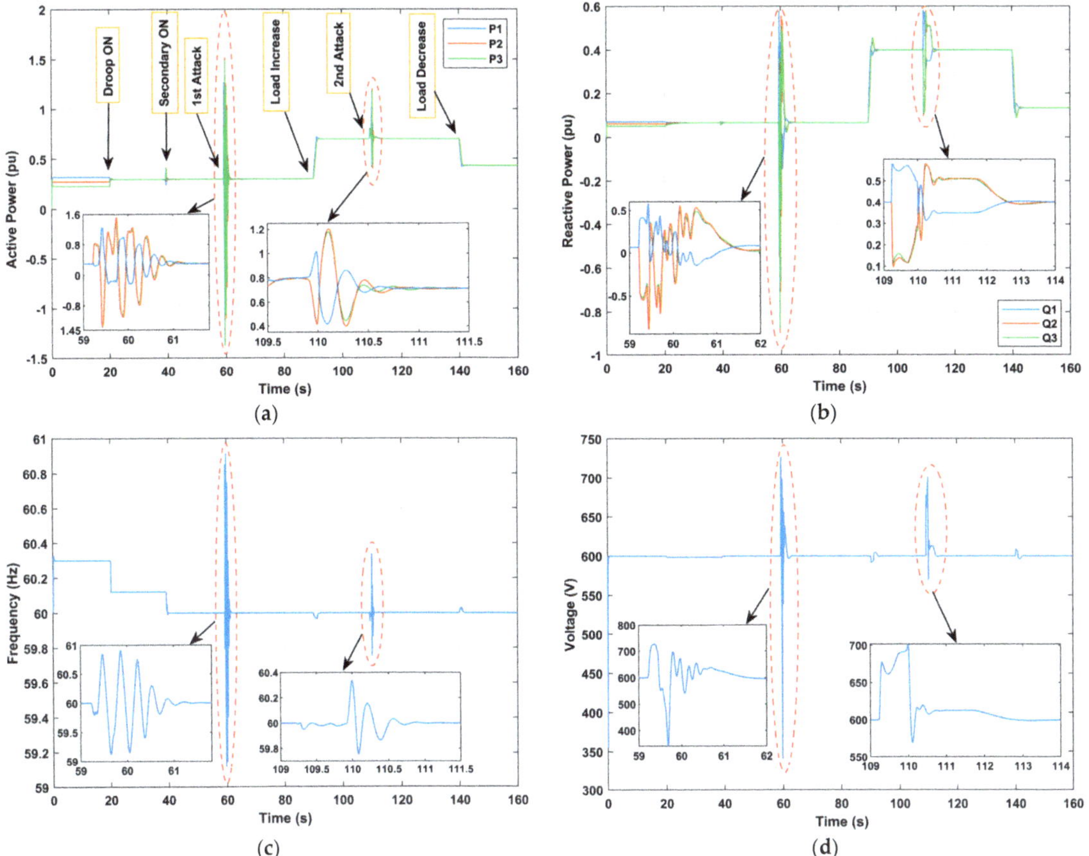

**Figure 9.** Performance of the proposed distributed consensus secondary control under deception attack on actuators' voltage: (**a**) Active power, (**b**) Reactive power, (**c**) Frequency, (**d**) Voltage.

### 5.2.4. Deception Attack on Sensors' Voltage

In this study case, $\delta_1^{S,V}$ denotes the sensor attack injections to the voltage measurements of DG #1. The attack signal is $\delta_1^{S,V} = \delta_1^{S,V1}$ between $60 < t(s) < 80$ and $\delta_1^{S,V} = \delta_1^{S,V2}$ between $120 < t(s) < 140$ where $\delta_1^{S,V1} > \delta_1^{S,V2}$.

The simulation scenario for this case is as follows:

-   At t = 20 s, the droop control is activated;
-   At t = 40 s, the distributed consensus secondary control is activated;
-   At t = 59 s, the first voltage sensor attack is launched;
-   At t = 79 s, the first attack is removed;
-   At t = 100 s, the total load is increased;
-   At t = 119 s, the second voltage sensor attack is launched;
-   At t = 139 s, the second attack is removed;
-   At t = 160 s, the load is decreased.

The conducted experiment shows the effects of deception attacks on sensor's voltage measurement within the distributed control framework. Notable consequences were observed upon launching the attack at t = 59 s, where the voltage dropped significantly from its rated value of 600 V to 530 V as shown in Figure 11. The consensus reactive power sharing deteriorated, and there was a decrease in active power sharing compared to pre-attack conditions. However, an unexpected behavior was observed upon ceasing the attack at t = 79 s. Instead of a smooth recovery of the control objectives as seen in the previous experiment, the voltage surged beyond its rated value, accompanied by an increase in both active and reactive power as shown in Figure 11. While the control objectives were eventually restored after approximately 14 s, this deviation highlights the sensitivity of the system to this kind of attack, even with small injected values. After a load increase and the launch of a second attack at t = 119 s, a similar situation was observed, yet with a crucial difference. Despite the attack being stopped at t = 139 s, the system took a longer time to respond. It was not until t = 180 s that the control objectives were eventually restored. This suggests that the severity and duration of the attack directly influence the system's ability to recover and maintain stability.

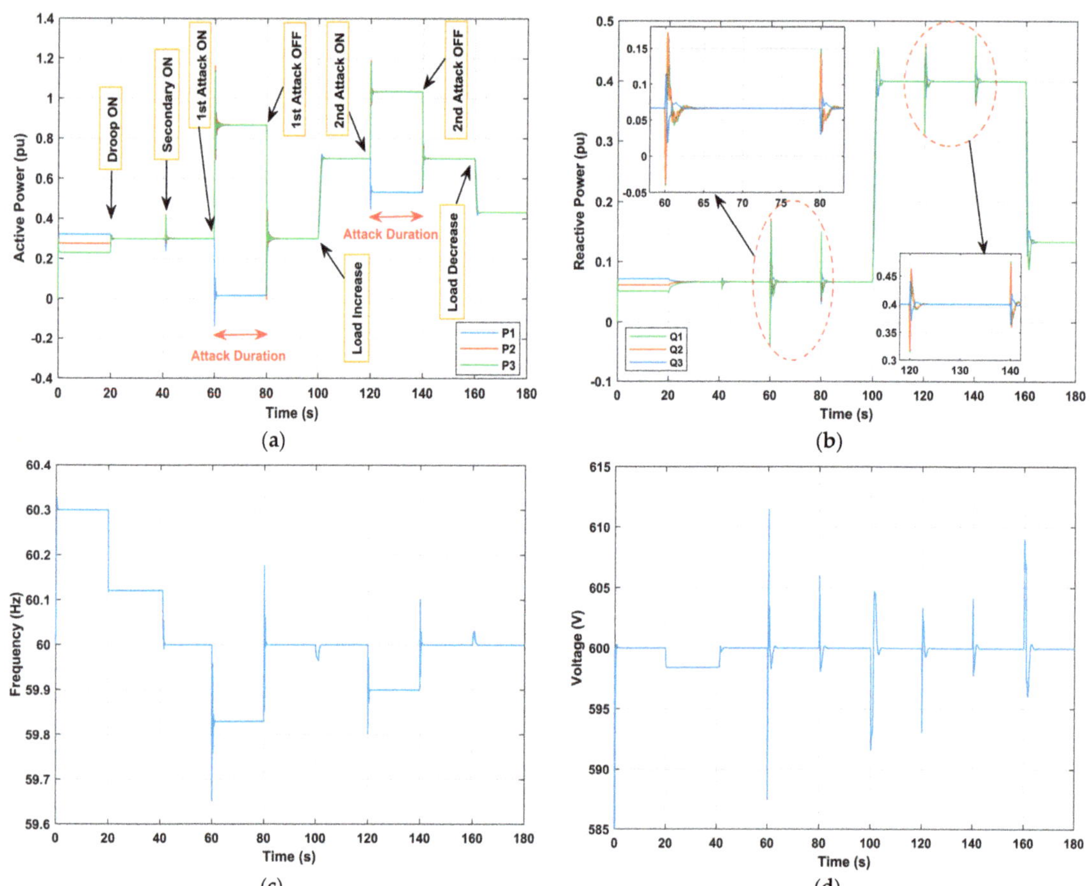


**Figure 10.** Performance of the proposed distributed consensus secondary control under deception attack on sensor's frequency: (**a**) Active power, (**b**) Reactive power, (**c**) Frequency, (**d**) Voltage.

### 5.2.5. Combined Deception Attack on Frequency

In this study case, $\delta_1^{a,\omega}$ denotes the actuator attack injections to the frequency control loop of DG #1. The attack signal is $\delta_1^{a,\omega} = \delta_1^{a,\omega 1}$ between $51.6 < t(s) < 71.6$ and $\delta_1^{a,\omega} = \delta_1^{a,\omega 2}$ between $111.6 < t(s) < 131.6$ where $\delta_1^{a,\omega 1} > \delta_1^{a,\omega 2}$, and $\delta_1^{S,\omega}$ denotes the sensor attack injections to the frequency measurements of DG #1. The attack signal is $\delta_1^{S,\omega} = \delta_1^{S,\omega 1}$ between $57.6 < t(s) < 77.6$ and $\delta_1^{S,\omega} = \delta_1^{S,\omega 2}$ between $117.6 < t(s) < 137.6$ where $\delta_1^{S,\omega 1} > \delta_1^{S,\omega 2}$.

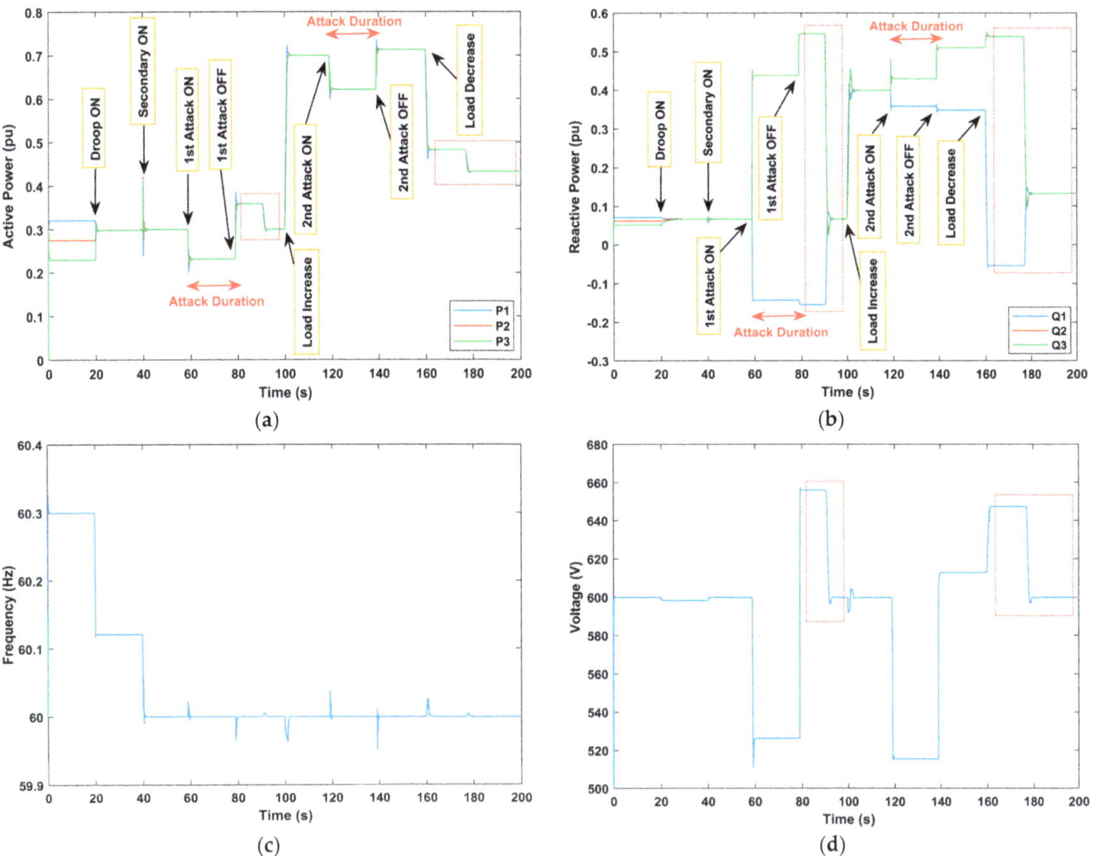

**Figure 11.** Performance of the proposed distributed consensus secondary control under deception attack on voltage: (**a**) Active power, (**b**) Reactive power, (**c**) Frequency, (**d**) Voltage.

The simulation scenario for this case is as follows:

- At t = 20 s, the droop control is activated;
- At t = 30.6 s, the distributed consensus secondary control is activated;
- At t = 51.6 s, the first frequency actuator attack is launched;
- At t = 57.6 s, the first frequency sensor attack is launched;
- At t = 71.6 s, the first frequency actuator attack is removed;
- At t = 77.6 s, the first frequency sensor attack is removed;
- At t = 100 s, the total load is increased;
- At t = 111.6 s, the second frequency actuator attack is launched;
- At t = 117.6 s, the second frequency sensor attack is launched;
- At t = 131.6 s, the second frequency actuator attack is removed;
- At t = 137.6 s, the second frequency sensor attack is removed.
- At t = 160 s, the load is decreased.

This experiment involves both actuator and sensor deception attacks on the frequency control loop of DG #1. It demonstrates a sequence of events where actuator attacks are initiated first, followed by sensor attacks as depicted in Figure 12. The attack signals for both actuator and sensor attacks vary in time intervals and magnitudes. During the attack periods, the system exhibited similar behavior to that observed in experiments 5.2.1 and 5.2.3. Upon removal of the attacks, the system restores to its nominal operating conditions.

However, the restoration may not be instantaneous and may depend on the attack's severity and duration as shown in the next experiment.

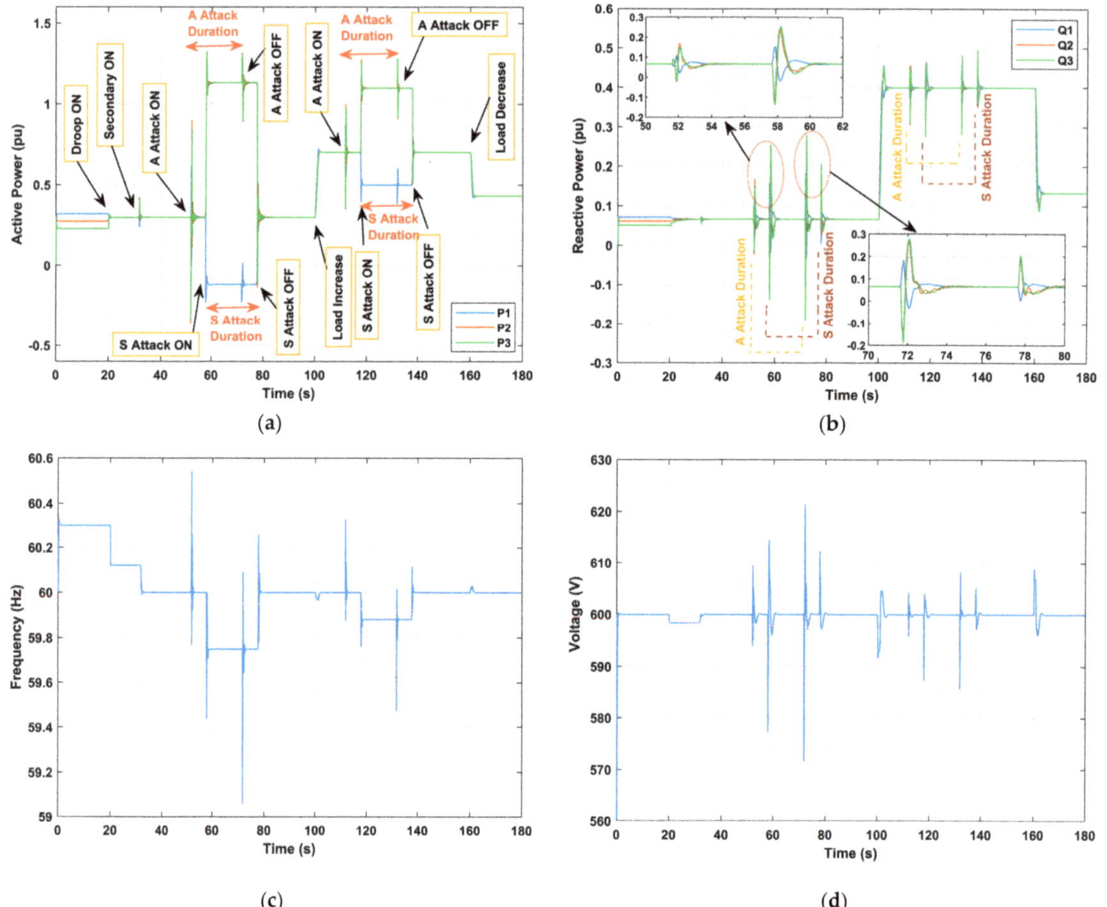


**Figure 12.** Performance of the proposed distributed consensus secondary control under combined deception attack on frequency: (**a**) Active power, (**b**) Reactive power, (**c**) Frequency, (**d**) Voltage.

## 5.2.6. Combined Deception Attack on Voltage

In this study case, $\delta_1^{a,V}$ denotes the actuator attack injections to the voltage control loop of DG #1. The attack signal is $\delta_1^{a,V} = \delta_1^{a,V1}$ between $62.7 < t(s) < 82.7$ and $\delta_1^{a,V} = \delta_1^{a,V2}$ between $122.7 < t(s) < 142.7$ where $\delta_1^{a,V1} > \delta_1^{a,V2}$, and $\delta_1^{S,V}$ denotes the sensor attack injections to the voltage measurements of DG #1. The attack signal is $\delta_1^{S,V} = \delta_1^{S,V1}$ between $68.7 < t(s) < 88.7$ and $\delta_1^{S,V} = \delta_1^{S,V2}$ between $128.7 < t(s) < 148.7$ where $\delta_1^{S,V1} > \delta_1^{S,V2}$.

The simulation scenario for this case is as follows:

- At t = 20 s, the droop control is activated;
- At t = 41.7 s, the distributed consensus secondary control is activated;
- At t = 62.7 s, the first voltage actuator attack is launched;
- At t = 68.7 s, the first voltage sensor attack is launched;
- At t = 82.7 s, the first voltage actuator attack is removed;
- At t = 88.7 s, the first voltage sensor attack is removed;
- At t = 100 s, the total load is increased;

- At t = 122.7 s, the second voltage actuator attack is launched;
- At t = 128.7 s, the second voltage sensor attack is launched;
- At t = 142.7 s, the second voltage actuator attack is removed;
- At t = 148.7 s, the second voltage sensor attack is removed;
- At t = 160 s, the load is decreased.

This experiment combines scenarios of both experiments 5.2.2 and 5.2.4. Actuator attacks on the voltage control loop are initiated first, followed by sensor attacks. During the attack periods and similar to the individual experiments, deviations from nominal voltage levels occur, impacting the consensus power sharing among DGs. As shown in Figure 13, the recovery of the system after the removal of the attacks at t = 148.7 s depends on severity of the attack.

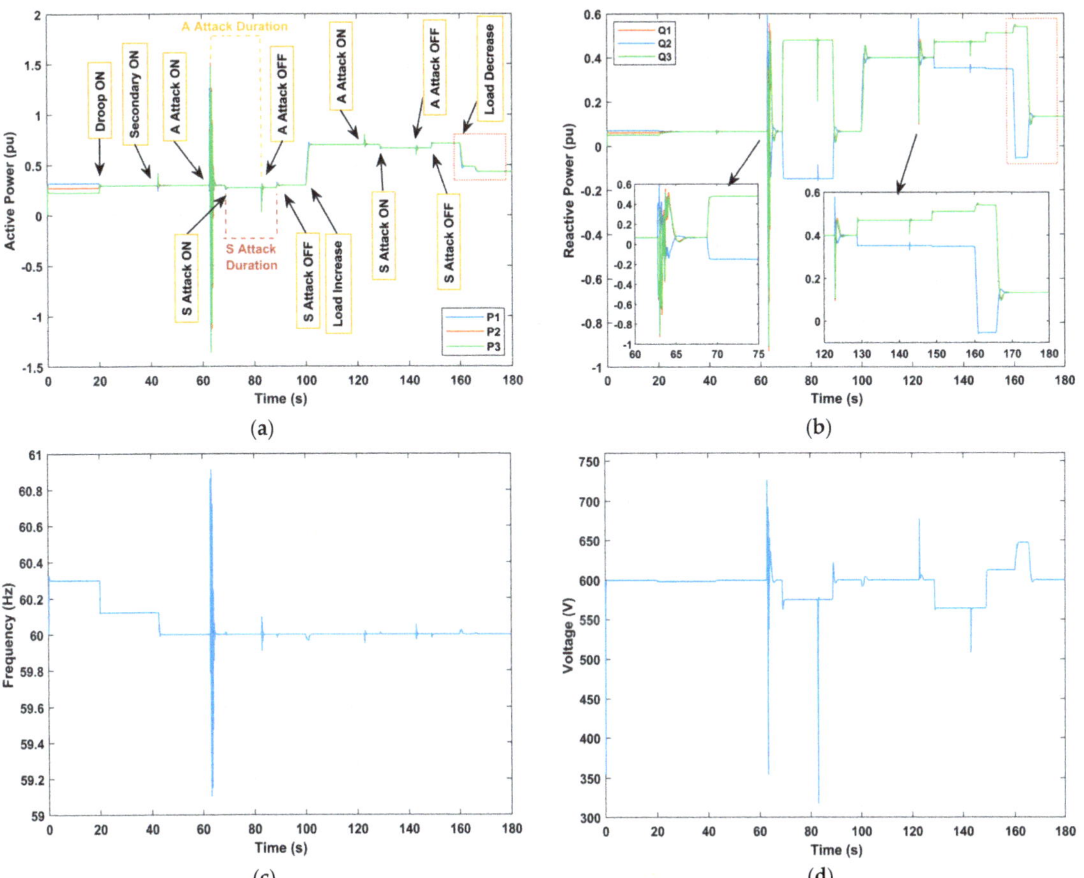


**Figure 13.** Performance of the proposed distributed consensus secondary control under combined deception attack on frequency: (**a**) Active power, (**b**) Reactive power, (**c**) Frequency, (**d**) Voltage.

### 5.3. Performance under Disruption Attacks

The simulation scenario for this case is as follows:

At t = 20 s, the droop control is activated;

At t = 40 s, the total load is increased;

At t = 60 s, the distributed consensus secondary control is activated;

At t = 80 s, the total load is decreased;

At t = 100 s and t = 120 s, the total load is increased.

The DoS attack occurs at t = 20 s, t = 65 s, t = 80 s, and t = 100 s. The attack targeted agent #2 with short lengths (around $\tau_a = 5$ s) except the last one ($\tau_a > 10$ s). Due to local droop control and DG1 reference signal, system frequency and voltage can still be restored to 60 Hz and 600 V as shown in Figure 14. However, it can be noticed that when $\tau_a > 5$ s the DoS attack has a direct impact on the MG stability.

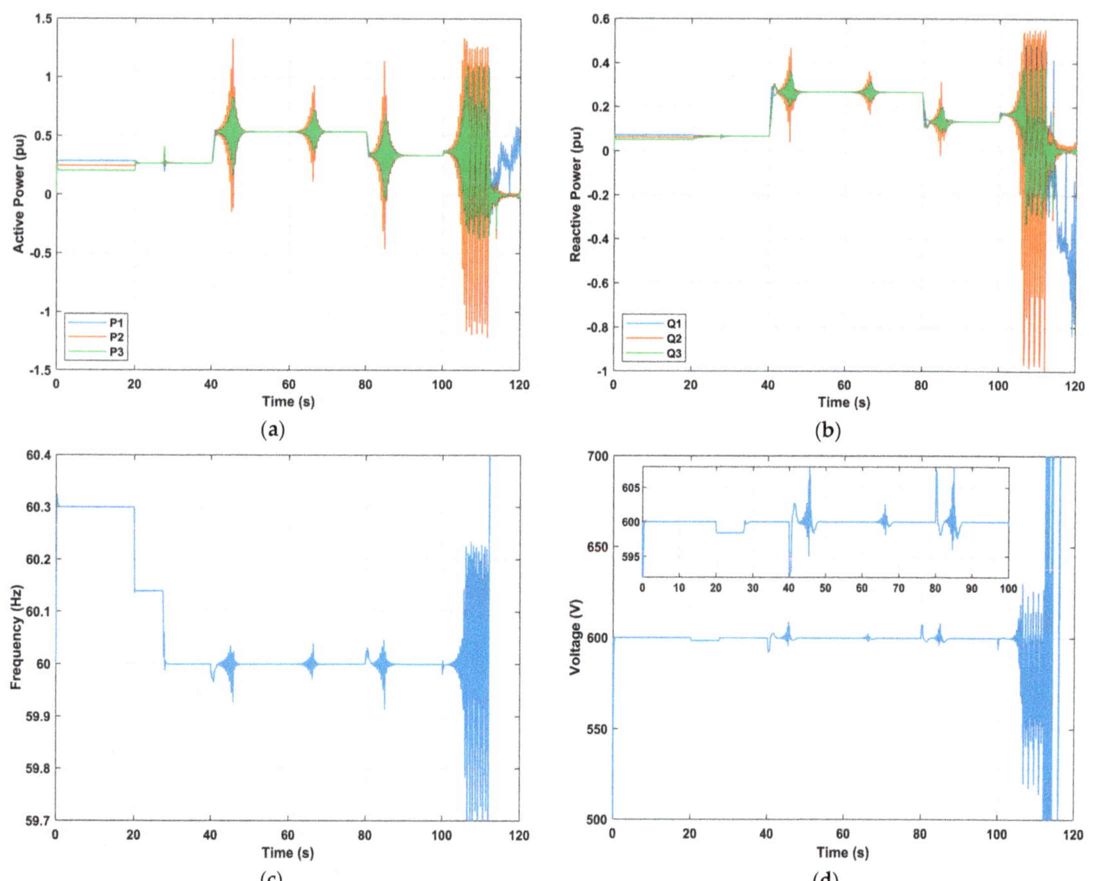

**Figure 14.** Performance of the proposed distributed consensus secondary control under combined deception attack on frequency: (**a**) Active power, (**b**) Reactive power, (**c**) Frequency, (**d**) Voltage.

It is worth mentioning that the effectiveness of DoS attacks often depends on the volume of traffic generated and the capacity of the target system to handle it. Attackers may adjust the packet rate based on their own resources and the target capabilities.

## 6. Conclusions

Considering the growing reliance on interconnected systems, the vulnerabilities of microgrid cyber-physical systems need careful attention. Through the development of a CHIL testbed and the implementation of distributed consensus secondary control strategy, this research not only highlights the potential risks posed by cyber threats but also is a building block for developing concrete solutions to bolster the resilience of microgrids. The proposed platform includes, a real-time islanded AC MG implemented on OPAL-RT, controllers implemented on Raspberry Pis, and a separate Raspberry Pi to launch the attacks. Extensive real-time digital simulations on OPAL-RT are performed to evaluate the effect of various cyber-attacks on the proposed distributed consensus secondary control of the islanded AC MG. The experiments include modeling and launching linear additive deception attacks on sensors, actuators, and their combinations, as well as disruption attacks. The outcomes of our real-time tests vividly illustrated the effects of both bounded and unbounded cyber-attacks on control objectives and system stability. These findings underscore the critical importance of implementing robust cybersecurity measures to uphold the stability and reliability of microgrid cyber-physical systems.

**Author Contributions:** Conceptualization, I.K. and O.A.M.; methodology, I.K.; software, I.K.; validation, I.K. and O.A.M.; formal analysis, I.K.; investigation, O.A.M.; resources, O.A.M.; writing—original draft preparation, I.K.; writing—review and editing, O.A.M.; visualization, I.K.; supervision, O.A.M. All authors have read and agreed to the published version of the manuscript.

**Funding:** This work was partially supported by grants from the Office of Naval Research, National Science Foundation, and the US Department of Energy. The authors are with the Energy Systems Research Laboratory, Department of Electrical and Computer Engineering, Florida International University, Miami, FL, USA.

**Data Availability Statement:** Data are contained within the article.

**Conflicts of Interest:** The authors declare no conflicts of interest.

# References

1. Abianeh, A.J.; Mardani, M.M.; Ferdowsi, F.; Gottumukkala, R.; Dragicevic, T. Cyber-Resilient Sliding-Mode Consensus Secondary Control Scheme for Islanded AC Microgrids. *IEEE Trans. Power Electron.* **2022**, *37*, 6074–6089. [CrossRef]
2. Yang, T.; He, Y.; Liu, G.-P. Distributed Voltage Restoration of AC Microgrids Under Communication Delays: A Predictive Control Perspective. *IEEE Trans. Circuits Syst. Regul. Pap.* **2022**, *69*, 2614–2624. [CrossRef]
3. Liu, Y.; Li, Y.; Wang, Y.; Zhang, X.; Gooi, H.B.; Xin, H. Robust and Resilient Distributed Optimal Frequency Control for Microgrids Against Cyber Attacks. *IEEE Trans. Ind. Inform.* **2022**, *18*, 375–386. [CrossRef]
4. Chlela, M.; Joos, G.; Kassouf, M.; Brissette, Y. Real-time testing platform for microgrid controllers against false data injection cybersecurity attacks. In Proceedings of the 2016 IEEE Power and Energy Society General Meeting (PESGM), Boston, MA, USA, 17–21 July 2016; IEEE: Piscataway, NJ, USA, 2016; pp. 1–5. [CrossRef]
5. Rath, S.; Pal, D.; Sharma, P.S.; Panigrahi, B.K. A Cyber-Secure Distributed Control Architecture for Autonomous AC Microgrid. *IEEE Syst. J.* **2021**, *15*, 3324–3335. [CrossRef]
6. Yao, W.; Wang, Y.; Xu, Y.; Deng, C. Cyber-Resilient Control of an Islanded Microgrid Under Latency Attacks and Random DoS Attacks. *IEEE Trans. Ind. Inform.* **2023**, *19*, 5858–5869. [CrossRef]
7. Tadepalli, P.S.; Pullaguram, D. Distributed Control Microgrids: Cyber-Attack Models, Impacts and Remedial Strategies. *IEEE Trans. Signal Inf. Process. Netw.* **2022**, *8*, 1008–1023. [CrossRef]
8. Fan, H.; Wang, H.; Xia, S.; Li, X.; Xu, P.; Gao, Y. Review of Modeling and Simulation Methods for Cyber Physical Power System. *Front. Energy Res.* **2021**, *9*, 642997. [CrossRef]
9. Li, Y.; Fan, L.; Dao, L.; Miao, Z. CHIL Testbed of Consensus Control-Based Battery Energy Storage Systems. In Proceedings of the 2020 52nd North American Power Symposium (NAPS), Tempe, AZ, USA, 11–13 April 2021; IEEE: Piscataway, NJ, USA, 2021; pp. 1–6. [CrossRef]
10. Naderi, E.; Asrari, A. Hardware-in-the-Loop Experimental Validation for a Lab-Scale Microgrid Targeted by Cyberattacks. In Proceedings of the 2021 9th International Conference on Smart Grid (icSmartGrid), Setubal, Portugal, 29 June–1 July 2021; IEEE: Piscataway, NJ, USA, 2021; pp. 57–62. [CrossRef]
11. Ren, X.; Liang, J.; Liu, Q. A Defensive Strategy for Integrity Detection in Cyber-Physical Systems Subject to Deception Attacks. In Proceedings of the 2020 10th International Conference on Information Science and Technology (ICIST), Bath, London, and Plymouth, UK, 9–15 September 2020; IEEE: Piscataway, NJ, USA, 2020; pp. 241–246. [CrossRef]
12. Shabad, P.K.R.; Alrashide, A.; Mohammed, O. Anomaly Detection in Smart Grids using Machine Learning. In Proceedings of the IECON 2021—47th Annual Conference of the IEEE Industrial Electronics Society, Toronto, ON, Canada, 13–16 October 2021; IEEE: Piscataway, NJ, USA, 2021; pp. 1–8. [CrossRef]
13. Afshari, A.; Karrari, M.; Baghaee, H.R.; Gharehpetian, G.B. Resilient Synchronization of Voltage/Frequency in AC Microgrids Under Deception Attacks. *IEEE Syst. J.* **2021**, *15*, 2125–2136. [CrossRef]
14. Liang, G.; Weller, S.R.; Zhao, J.; Luo, F.; Dong, Z.Y. The 2015 Ukraine Blackout: Implications for False Data Injection Attacks. *IEEE Trans. Power Syst.* **2017**, *32*, 3317–3318. [CrossRef]
15. Lai, J.; Zhou, H.; Lu, X.; Yu, X.; Hu, W. Droop-Based Distributed Cooperative Control for Microgrids with Time-Varying Delays. *IEEE Trans. Smart Grid* **2016**, *7*, 1775–1789. [CrossRef]
16. Pogaku, N.; Prodanovic, M.; Green, T.C. Modeling, Analysis and Testing of Autonomous Operation of an Inverter-Based Microgrid. *IEEE Trans. Power Electron.* **2007**, *22*, 613–625. [CrossRef]

17. Mohammadi, F.D.; Vanashi, H.K.; Feliachi, A. State-Space Modeling, Analysis, and Distributed Secondary Frequency Control of Isolated Microgrids. *IEEE Trans. Energy Convers.* **2018**, *33*, 155–165. [CrossRef]
18. Jamali, M.; Sadabadi, M.S.; Davari, M.; Sahoo, S.; Blaabjerg, F. Resilient Cooperative Secondary Control of Islanded AC Microgrids Utilizing Inverter-Based Resources Against State-Dependent False Data Injection Attacks. *IEEE Trans. Ind. Electron.* **2024**, *71*, 4719–4730. [CrossRef]
19. Kharchouf, I.; Abdelrahman, M.S.; Mohammed, O.A. ANN-Based Secure Control of Islanded Microgrid Under False Data Injection Cyber-Attack. In Proceedings of the 2023 IEEE Industry Applications Society Annual Meeting (IAS), Nashville, TN, USA, 29 October–2 November 2023; IEEE: Piscataway, NJ, USA, 2023; pp. 1–6. [CrossRef]
20. Raman, G.; Liao, K.; Peng, J.C.-H. Improving AC Microgrid Stability Under Cyberattacks Through Timescale Separation. *IEEE Trans. Circuits Syst. II Express Briefs* **2023**, *70*, 2191–2195. [CrossRef]
21. Mi, Y.; Deng, J.; Wang, X.; Lin, S.; Su, X.; Fu, Y. Multiagent Distributed Secondary Control for Energy Storage Systems with Lossy Communication Networks in DC Microgrid. *IEEE Trans. Smart Grid* **2013**, *14*, 1736–1749. [CrossRef]
22. Bidram, A.; Davoudi, A.; Lewis, F.L. A Multiobjective Distributed Control Framework for Islanded AC Microgrids. *IEEE Trans. Ind. Inform.* **2014**, *10*, 1785–1798. [CrossRef]
23. Han, Y.; Li, H.; Shen, P.; Coelho, E.A.A.; Guerrero, J.M. Review of Active and Reactive Power Sharing Strategies in Hierarchical Controlled Microgrids. *IEEE Trans. Power Electron.* **2017**, *32*, 2427–2451. [CrossRef]
24. Zhuang, J.; Peng, S.; Wang, Y. Secure Consensus of Stochastic Multi-agent Systems Subject to Deception Attacks via Impulsive Control. In Proceedings of the 2022 4th International Conference on Control and Robotics (ICCR), Guangzhou, China, 2–4 December 2022; IEEE: Piscataway, NJ, USA, 2022; pp. 1–6. [CrossRef]
25. Lu, L.-Y.; Liu, J.-H.; Lin, S.-W.; Chu, C.-C. Concurrent Cyber Deception Attack Detection of Consensus Control in Isolated AC Microgrids. *IEEE Trans. Ind. Appl.* **2023**, *59*, 7584–7596. [CrossRef]
26. He, W.; Xu, W.; Ge, X.; Han, Q.-L.; Du, W.; Qian, F. Secure Control of Multiagent Systems Against Malicious Attacks: A Brief Survey. *IEEE Trans. Ind. Inform.* **2022**, *18*, 3595–3608. [CrossRef]
27. Mustafa, A.; Moghadam, R.; Modares, H. Resilient Synchronization of Distributed Multi-agent Systems under Attacks. *arXiv* **2019**, arXiv:1807.02856. [CrossRef]

MDPI

St. Alban-Anlage 66

4052 Basel

Switzerland

www.mdpi.com

*Energies* Editorial Office

E-mail: energies@mdpi.com

www.mdpi.com/journal/energies